Bioinspired Catechol-Based Systems: Chemistry and Applications

Special Issue Editors

Marco d'Ischia
Daniel Ruiz-Molina

MDPI • Basel • Beijing • Wuhan • Barcelona • Belgrade

MDPI

Special Issue Editors
Marco d'Ischia
University of Naples Federico II
Italy

Daniel Ruiz-Molina
Catalan Institute of Nanoscience and Nanotechnology (ICN2)
Spain

Editorial Office
MDPI
St. Alban-Anlage 66
Basel, Switzerland

This edition is a reprint of the Special Issue published online in the open access journal *Biomimetics* (ISSN 2313-7673) in 2017 (available at: http://www.mdpi.com/journal/biomimetics/special_issues/catechol).

For citation purposes, cite each article independently as indicated on the article page online and as indicated below:

Lastname, F.M.; Lastname, F.M. Article title. *Journal Name* **Year**, *Volume*, Article number.

ISBN 978-3-03842-732-2 (Pbk)
ISBN 978-3-03842-731-5 (PDF)

Table of Contents

About the Special Issue Editors v

Preface to "Bioinspired Catechol-Based Systems: Chemistry and Applications" vii

Chapter 1: Chemistry of Catechol-Based Systems

Vincenzo Barone, Ivo Cacelli, Alessandro Ferretti and Giacomo Prampolini
Noncovalent Interactions in the Catechol Dimer
Reprinted from: *Biomimetics* **2017**, 2(3), 18; doi: 10.3390/biomimetics2030018 1

Orlando Crescenzi, Marco d'Ischia and Alessandra Napolitano
Kaxiras's Porphyrin: DFT Modeling of Redox-Tuned Optical and Electronic Properties in a
Theoretically Designed Catechol-Based Bioinspired Platform
Reprinted from: *Biomimetics* **2017**, 2(4), 21; doi: 10.3390/biomimetics2040021 14

Riccardo Amorati, Andrea Baschieri, Adam Cowden and Luca Valgimigli
The Antioxidant Activity of Quercetin in Water Solution
Reprinted from: *Biomimetics* **2017**, 2(3), 9; doi: 10.3390/biomimetics2030009 42

Chapter 2: Catechol-Based Biomechanisms and Bioactivity

Natalie A. Hamada, Victor A. Roman, Steven M. Howell and Jonathan J. Wilker
Examining Potential Active Tempering of Adhesive Curing by Marine Mussels
Reprinted from: *Biomimetics* **2017**, 2(3), 16; doi: 10.3390/biomimetics2030016 57

Raffaella Micillo, Valeria Pistorio, Elio Pizzo, Lucia Panzella, Alessandra Napolitano and
Marco d'Ischia
2-*S*-Lipoylcaffeic Acid, a Natural Product-Based Entry to Tyrosinase Inhibition via Catechol
Manipulation
Reprinted from: *Biomimetics* **2017**, 2(3), 15; doi: 10.3390/biomimetics2030015 68

Matteo Ramazzotti, Paolo Paoli, Bruno Tiribilli, Caterina Viglianisi, Stefano Menichetti and
Donatella Degl'Innocenti
Catechol-Containing Hydroxylated Biomimetic 4-Thiaflavanes as Inhibitors of
Amyloid Aggregation
Reprinted from: *Biomimetics* **2017**, 2(2), 6; doi: 10.3390/biomimetics2020006 79

Chapter 3: Catechol Applications in Materials Science

Vincent Ball
Composite Materials and Films Based on Melanins, Polydopamine, and Other
Catecholamine-Based Materials
Reprinted from: *Biomimetics* **2017**, 2(3), 12; doi: 10.3390/biomimetics2030012 93

Salvio Suárez-García, Josep Sedó, Javier Saiz-Poseu and Daniel Ruiz-Molina
Copolymerization of a Catechol and a Diamine as a Versatile Polydopamine-Like Platform for
Surface Functionalization: The Case of a Hydrophobic Coating
Reprinted from: *Biomimetics* **2017**, 2(4), 22; doi: 10.3390/biomimetics2040022 109

Jun Feng, Xuan-Anh Ton, Shifang Zhao, Julieta I. Paez and Aránzazu del Campo
Mechanically Reinforced Catechol-Containing Hydrogels with Improved Tissue
Gluing Performance
Reprinted from: *Biomimetics* **2017**, 2(4), 23; doi: 10.3390/biomimetics2040023 126

Maria P. Sousa and João F. Mano
Cell-Adhesive Bioinspired and Catechol-Based Multilayer Freestanding Membranes for Bone
Tissue Engineering
Reprinted from: *Biomimetics* **2017**, 2(4), 19; doi: 10.3390/biomimetics2040019 142

Devang R. Amin, Caroline Sugnaux, King Hang Aaron Lau and Phillip B. Messersmith
Size Control and Fluorescence Labeling of Polydopamine Melanin-Mimetic Nanoparticles for
Intracellular Imaging
Reprinted from: *Biomimetics* **2017**, 2(3), 17; doi: 10.3390/biomimetics2030017 162

Eunkyoung Kim, Zhengchun Liu, Yi Liu, William E. Bentley and Gregory F. Payne
Catechol-Based Hydrogel for Chemical Information Processing
Reprinted from: *Biomimetics* **2017**, 2(3), 11; doi: 10.3390/biomimetics2030011 181

About the Special Issue Editors

Marco d'Ischia, Professor, received the Laurea degree at the University of Naples. Since 2001, he has been Professor of Organic Chemistry at the Department of Chemical Sciences of Naples University Federico II where he leads the group of Bioinspired Product Chemistry. He is the author or co-author of more than 260 publications in the fields of organic and bioorganic chemistry. His research interests include structure, synthesis, physicochemical properties and reactivity of melanins, polydopamine and related bioinspired functional materials for surface functionalization and hybrid nanostructures for bioelectronics and biomedical applications; design, antioxidant properties and reactivity of natural phenolic and quinone compounds; free radical oxidations and nature-inspired redox-active systems for biomedical and technological applications; chemistry and physicochemical properties of natural or bioinspired heterocyclic compounds; bioorganic chemistry of sulphur and selenium compounds; model reactions and transformation pathways of polycyclic aromatic hydrocarbons and their derivatives of astrochemical relevance. In 2011, he was awarded the Raper Medal by the International Federation of Pigment Cell Societies and the European Society for Pigment Cell Research.

Daniel Ruiz-Molina, Ph.D., obtained his PhD on polyradical dendrimers at the Institute of Materials Science of Barcelona (ICMAB). Afterwards, he took a postdoctoral position at the University of California, San Diego (UCSD) working on single-molecule magnets and molecular switches for three years. Since 2001, he has obtained a permanent position at the Spanish National Research Council (CSIC). More recently, he moved to the new Catalan Institute of Nanoscience and Nanotechnology (ICN2) where he is leading the Nanostructured Functional Materials group. His main research areas are fabrication of hybrid colloids and surfaces, biomimetic functional nanostructures, coordination polymers and micro-/nanoparticles for smart applications and encapsulation/delivery systems.

Preface to "Bioinspired Catechol-Based Systems: Chemistry and Applications"

Catechols are widely found in nature taking part in a variety of biological functions, ranging from the aqueous adhesion of marine organisms to the storage of transition metal ions. This has been achieved thanks to their (i) rich redox chemistry and ability to cross-link through complex and irreversible oxidation mechanisms, (ii) excellent chelating properties, and (iii) the diverse modes of interaction of the vicinal hydroxyl groups with all kinds of surfaces of remarkably different chemical and physical nature [1]. Therefore, guided by such interest, catechol-based systems have been subject in recent years to intense research (at the laboratory scale) aimed at mimicking and translating these natural concepts into new functional adhesives and coatings with enhanced properties.

This Special Issue collects contributions from different laboratories working on both basic research and applications of bioinspired catechol systems presented by cutting edge specialists in this growing field. Taking advantage of its open access publication, this collection of papers, influenced by biomimetic approaches, will bring about new avenues for new research and innovative solutions in biomedicine and technology. Main topics addressed in the field of basic catechol chemistry include (i) a computational investigation by Barone et al. [2] of noncovalent interactions in catechol dimers, which are of central importance in determining the overall properties of catechol base systems, (ii) a theoretical analysis of indole-based porphyrin structures proposed as a model for eumelanin biopolymers (Crescenzi et al. [3]), and (iii) a detailed insight into the mechanism of the antioxidant activity of quercetin in water by Amorati et al. [4]. Of both basic and applicative interest for adhesion is the study by Hamada et al. [5] addressing the issue of whether mussels manage byssus mechanical properties via control of catechol chemistry. The design of films for surface functionalization and energy applications based on polydopamine-inorganic and polydopamine-organic composites is reviewed by Ball [6]. The potential of a cross-linking reaction between catechol and hexamethylenediamine for surface functionalization and coating under oxidative conditions is demonstrated by Suarez-Garcia et al. [7]. Catechol-containing hydrogels with enhanced gluing properties for tissue engineering are reported by Feng et al. [8]. Sousa and Mano [9] synthesized cell-adhesive membranes for bone tissue engineering via a mussel-inspired conjugated polymer obtained by covalent modification of hyaluronic acid with dopamine. Amin et al. [10] studied melanin-mimetic nanoparticles based on polydopamine for multimodal cell imaging, opening interesting perspectives for drug delivery applications and surface chemistry-dependent cellular interactions. The scope of catechol–chitosan redox-capacitors and other systems for chemical information and signal processing is illustrated by Kim et al. [11]. In the field of bioinspired bioactive compounds, Micillo et al. [12] designed a lipoyl–caffeic acid conjugate as a new type of tyrosinase inhibitor for the control of melanogenesis. Ramazzotti et al. [13] report the anti-aggregating properties of five biomimetic 4-thiaflavanes on an amyloid model, suggesting further studies of this class of compounds as anti-amyloid agents.

Integrating more than replacing the many excellent reviews, the present collection will provide the reader with a concise panorama of the status quo and perspectives in the increasingly expanding field of basic and applied research on bioinspired catechol systems. It is clear that the interest for catechol-based materials is experiencing a steady burst, perfectly represented by polydopamine (two contributions in this special issue deal with this research area). Several patents based on bioinspired catechol systems and different products are already commercialized and available the market. We believe that this special issue may fulfill an important function in promoting biomimetic catechol chemistry for an increasing range of applications.

Conflicts of Interest: The authors declare no conflict of interest.

References

1. Sedó, J.; Saiz-Poseu, J.; Busqué, F.; Ruiz-Molina, D. Catechol-based biomimetic functional materials. *Adv. Mater.* **2013**, *25*, 653–701.
2. Barone, V.; Cacelli, I.; Ferretti, A.; Prampolini, G. Noncovalent interactions in the catechol dimer. Biomimetics **2017**, 2, 18.
3. Crescenzi, O.; d'Ischia, M.; Napolitano, A. Kaxiras's porphyrin: DFT modeling of redox-tuned optical and electronic properties in a theoretically designed catechol-based bioinspired platform. Biomimetics **2017**, 2, 21.
4. Amorati, R.; Baschieri, A.; Cowden, A.; Valgimigli, L. The antioxidant activity of quercetin in water solution. Biomimetics **2017**, 2, 9.
5. Hamada, N.A.; Roman, V.A.; Howell, S.M.; Wilker, J.J. Examining potential active tempering of adhesive curing by marine mussels. Biomimetics **2017**, 2, 16.
6. Ball, V. Composite materials and films based on melanins, polydopamine, and other catecholamine-based materials. Biomimetics **2017**, 2, 12.
7. Suárez-García, S.; Sedó, J.; Saiz-Poseu, J.; Ruiz-Molina, D. Copolymerization of a catechol and a diamine as a versatile polydopamine-like platform for surface functionalization: The case of a hydrophobic coating. Biomimetics **2017**, 2, 22.
8. Feng, J.; Ton, X.-A.; Zhao, S.; Paez, J.I.; del Campo, A. Mechanically reinforced catechol-containing hydrogels with improved tissue gluing performance. Biomimetics **2017**, 2, 23.
9. Sousa, M.P.; Mano, J.F. Cell-adhesive bioinspired and catechol-based multilayer freestanding membranes for bone tissue engineering. Biomimetics **2017**, 2, 19.
10. Amin, D.R.; Sugnaux, C.; Lau, K.H.A.; Messersmith, P.B. Size control and fluorescence labeling of polydopamine melanin-mimetic nanoparticles for intracellular imaging. Biomimetics **2017**, 2, 17.
11. Kim, E.; Liu, Z.; Liu, Y.; Bentley, W.E.; Payne, G.F. Catechol-based hydrogel for chemical information processing. Biomimetics **2017**, 2, 11.
12. Micillo, R.; Pistorio, V.; Pizzo, E.; Panzella, L.; Napolitano, A.; d'Ischia, M. 2-S-Lipoylcaffeic acid, a natural product-based entry to tyrosinase inhibition via catechol manipulation. Biomimetics **2017**, 2, 15.
13. Ramazzotti, M.; Paoli, P.; Tiribilli, B.; Viglianisi, C.; Menichetti, S.; Degl'Innocenti, D. Catechol-containing hydroxylated biomimetic 4-thiaflavanes as inhibitors of amyloid aggregation. Biomimetics **2017**, 2, 6.

Marco d'Ischia and Daniel Ruiz-Molina

Special Issue Editors

Chapter 1:
Chemistry of Catechol-Based Systems

biomimetics

MDPI

Article

Noncovalent Interactions in the Catechol Dimer

Vincenzo Barone [1,*], Ivo Cacelli [2,3], Alessandro Ferretti [3] and Giacomo Prampolini [3]

[1] Scuola Normale Superiore di Pisa, Piazza dei Cavalieri, I-56126 Pisa, Italy
[2] Dipartimento di Chimica e Chimica Industriale, Università di Pisa, Via G. Moruzzi 13, I-56124 Pisa, Italy; ivo.cacelli@unipi.it
[3] Istituto di Chimica dei Composti OrganoMetallici (ICCOM-CNR), Area della Ricerca, Via G. Moruzzi 1, I-56124 Pisa, Italy; ferretti@iccom.cnr.it (A.F.); giacomo.prampolini@pi.iccom.cnr.it (G.P.)
* Correspondence: vincenzo.barone@sns.it

Academic Editors: Marco d'Ischia and Daniel Ruiz-Molina
Received: 21 June 2017; Accepted: 5 September 2017; Published: 13 September 2017

Abstract: Noncovalent interactions play a significant role in a wide variety of biological processes and bio-inspired species. It is, therefore, important to have at hand suitable computational methods for their investigation. In this paper, we report on the contribution of dispersion and hydrogen bonds in both stacked and T-shaped catechol dimers, with the aim of delineating the respective role of these classes of interactions in determining the most stable structure. By using second-order Møller–Plesset (MP2) calculations with a small basis set, specifically optimized for these species, we have explored a number of significant sections of the interaction potential energy surface and found the most stable structures for the dimer, in good agreement with the highly accurate, but computationally more expensive coupled cluster single and double excitation and the perturbative triples (CCSD(T))/CBS) method.

Keywords: noncovalent interactions; catechol; aromatic dimers; computation; electronic correlation; dispersion

1. Introduction

Nowadays, there is a general consensus about the primary role played by noncovalent interactions, in particular those involving aromatic rings, in molecular, life, and materials sciences. In addition to being responsible for key biological processes that range from base stacking in deoxyribonucleic acid (DNA) [1], to the color of red wine [2] and, more generally, food quality [3], it is of the foremost importance to understand, rationalize and, hence, exploit their features in cutting-edge applications as advanced catalysis [4,5], biomedical materials [6,7] and novel drugs design [8], advanced organic photovoltaics [9–13], complex self-assembled structures [14], or bio-nano-materials [15,16]. Such ubiquity of the aromatic interactions has often inspired multidisciplinary research [17], aimed to exploit their peculiar features in the design and construction of biomimetic materials. From a physical point of view, noncovalent interactions among molecules bearing aromatic moieties originate from a variety of different forces, including π-stacking, XH–π or charge-transfer (CT), besides the ubiquitous dispersion. Furthermore, the presence of additional functional groups can introduce other kinds of interactions (like e.g., hydrogen (HB) or halogen bonds), leading to nontrivial interference effects, which tune both the structure and the properties of the resulting material. In this framework, computational methods can play a crucial role for rational design and interpretation, provided that they are able to couple reliability, feasibility, and ability to unravel the different contributions [18,19]. It should be also mentioned that, although the embedding environment is often neglected, or only roughly approximated in most computational studies, its effect can be significant or even decisive in biomimetic processes. However, comprehensive studies of pairs of interacting species in the gas phase are a mandatory starting point for unraveling the weight of the different effects.

In the past few years, catechol has attracted increasing attention as a precursor of bio-inspired materials [20–26]. From a theoretical point of view, catechol is an ideal candidate to test the capability of new computational approaches to accurately represent the delicate balance among the different kinds of noncovalent interactions, occurring in the presence of catechol units. In fact, apart from the π-stacking and XH–π interactions due to the aromatic core, interactions between these species are also characterized by the insurgence of strong (OH–H) and weak (OH–π) HB patterns, which may play an important role in the supramolecular assembling. The main problem is that aromatic interactions are dominated by dispersion forces that standard electronic calculations have difficulty to reproduce. Indeed, in the past ten years, much effort has been devoted to the development of approaches that overcome the problem [27–43]. Within the framework of density functional theory (DFT), attempts have been made to set appropriate functionals which incorporate the effects of dispersion, such as that of Truhlar et al. [43] or to introduce semi-empirical atomistic corrections, as suggested by Grimme and coworkers [30,32,33]. Among wave function (WF)-based approaches, the most accurate but also computationally most expensive method is the coupled cluster approach including a full account of single and double excitations together with perturbative inclusion of connected triple excitations, and extrapolation to the complete basis set limit (CCSD(T)/CBS) [4,19,34,37,38,41,44–50]. Still within a WF framework, perturbative second-order Møller–Plesset (MP2) calculations could be carried out at a much lower computational cost, yet it is well known [40] that they tend to overestimate aromatic binding energies, especially when employed with large basis sets. These inaccuracies can be overcome by resorting to an idea proposed almost forty years ago by Kroon-Batenburg and Van Duijneveldt [51] and successively refined by Hobza and Zahradnik [52], based on the use MP2 calculations with the small 6-31G* basis set, modified by reducing to 0.25 the exponent of the d polarization function placed on each carbon atom of the benzene dimer. Such an approach, often referred to as MP2/6-31G*(0.25), was then fully validated with reference to interaction energies of benzene and a few other aromatic dimers computed at the CCSD(T)/CBS level [53–61]. More recently, the method has been generalized to different basis sets, and applied to several molecular prototypes, including liquid crystals [62,63], pyridine [64], quinhydrone [27], dihydroxyindole derivatives relevant in eumelanin formation [65], and, very recently, to small aromatic heterocycles [66], where the procedure to find the suitable modified basis sets, labeled MP2^{mod}, has been automated and extended to the optimization of the orbital exponents of d functions on heteroatoms and p functions on hydrogen, within the 6-31G** basis set.

Here, the MP2^{mod} method is applied to the catechol dimer in the gas phase. First, MP2^{mod} accuracy is validated against high-quality CCSD(T)/CBS predictions, purposely carried out for a number of selected geometries of catechol dimers. Next, MP2^{mod} is employed in the exploration of the catechol's interaction potential energy surface (IPES), with the aim of finding the optimal structure of the dimer by a comparison of different possible arrangements. This allows us to investigate the different roles played by HB and π-stacking interactions in the dimer formation. Incidentally, it might also be of interest, following Wheeler group's suggestions [4,44,45], to verify if noncovalent interactions in catechol can be correlated to the simple direct interaction between the (hydroxyl) substituents, or if, on the contrary, a rationalization of the resulting interaction patterns requires a more complex analysis, taking into account the specific role of each contribution.

The catechol dimer has also been studied at the DFT level by Estévez et al. [67], who considered structures determined either by X-ray measurements or by geometry optimizations at the MPW1B95/6-311++G(2d,2p) level. In the following these results will also be discussed in comparison with our findings.

2. Computational Details

The full geometry optimization of the catechol monomer has been performed by DFT, at the B3LYP/*aug*-cc-pvDZ level, by minimizing the energy with respect to all internal coordinates. Unless otherwise stated, the internal monomer's geometry was kept unaltered in all subsequent calculations.

As far as the intermolecular energy is concerned, reference CCSD(T)/CBS calculations have been carried out on catechol dimers following the protocol adopted in previous works [27,36,66], which can be summarized as follows:

1. The difference $\Delta_{CC\text{-}MP2}$ between CCSD(T) and MP2 interaction energy is evaluated using for both calculations the Dunning's correlated *aug*-cc-pvDz basis sets:

$$\Delta_{CC-MP2} = \left| \Delta E^{CCSD(T)} \right|_{aug-cc-pvDz} - \left| \Delta E^{MP2} \right|_{aug-cc-pvDz} \tag{1}$$

2. The MP2 energy in the CBS limit, ΔE_{CBS}^{MP2}, is computed through the extrapolation scheme proposed by Halkier et al. [68], making use of the *aug*-cc-pvDz and *aug*-cc-pvTz basis sets. Despite the state-of-the-art extrapolation procedure [37,41,50] is often carried out with the larger *aug*-cc-pvTz and *aug*-cc-pvQz basis sets, it has been recently shown that, for similar aromatic dimers, the use of the smaller *aug*-cc-pvDz and *aug*-cc-pvTz affects the computed interaction energies by few hundredths of kcal/mol [66]. In consideration of the fairly large number of dimers investigated and the computational cost of a CCSD(T) calculation at the *aug*-cc-pvQz level, the smaller sets (Dz and Tz) were chosen as the best compromise between accuracy and feasibility.

3. Finally, the CCSD(T)/CBS interaction energy, $\Delta E_{CBS}^{CCSD(T)}$, is recovered as:

$$\Delta E_{CBS}^{CCSD(T)} = \Delta E_{CBS}^{MP2} + \Delta_{CC-MP2} \tag{2}$$

4. All energies were corrected for the basis set superposition error (BSSE) with the standard counterpoise (CP) correction [69].

The MP2^{mod} exponent optimization was performed by means of the EXOPT code [27,36,66], by minimizing the objective function I:

$$I(\overline{P}) = \frac{1}{N_{geom}} \sum_{k=1}^{N_{geom}} \left[\Delta E_{CBS}^{CCSD(T)} - \Delta E^{MP2^{mod}}(\overline{P}) \right]^2 \tag{3}$$

where N_{geom} is the number of considered dimer geometries and $\overline{P}$ the vector containing the basis sets exponents to be optimized. All the MP2^{mod} calculations were carried out with the 6-31G** basis set, and the exponents of the *d* functions on heavy atoms and the *p* functions on H were optimized. Further details on the optimization protocol can be found in [66] and are also briefly commented in the next section. In all MP2^{mod} calculations, the CP correction was applied to take care of the basis set superposition error.

Finally, to better compare with the results reported by Estévez et al. [67], the interaction energy of selected dimer arrangements was also computed at the DFT level, using the same procedure employed in [67]: the MPW1B95 functional was employed, together with the 6-311++G(2*d*,2*p*), while no correction was applied to take care of the BSSE.

All CCSD(T), MP2, MP2^{mod} and DFT calculations were carried out with the Gaussian09 software package [70].

3. Results and Discussion

3.1. MP2^{mod} Tuning and Validation

After geometry optimization, the catechol monomer is planar with the two hydroxyl hydrogens pointing in the same direction (see Figure 1a).

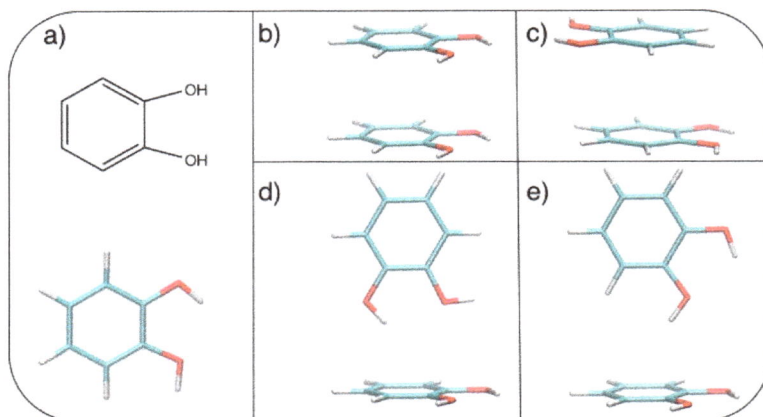

Figure 1. (**a**) Catechol structural formula (**top**) and graphical representation (**bottom**). Stacked dimers: (**b**) face-to-face (FF) and (**c**) antiparallel face-to-face (AFF); T-shaped (TS) dimers: (**d**) TS_1 and (**e**) TS_2. C: Cyan; H: White; O: Red.

Based on the results recently achieved for several heteroaromatic dimers, where stacked and T-shaped (TS) conformers where found to be the most stable, four starting arrangements have been set up by placing the two monomers at different distances and relative orientation. Namely, the face-to-face (FF, Figure 1b), the antiparallel face-to-face (AFF, Figure 1c), and two TS conformations, one with both hydroxyls (TS_1, Figure 1d) and one with only one hydroxyl (TS_2, Figure 1e) pointing towards the other ring. Following the protocol recently developed in our group [66], the $MP2^{mod}$ best exponents were determined as follows: starting from each of the four selected conformations, a set of dimer arrangements was created by displacing one monomer along a selected coordinate R, defined as the line connecting the centers of the two rings, as shown in the insets of Figure 2. Next, an estimate (data not shown) of the interaction energy (ΔE) of the resulting dimer geometries was obtained at the $MP2^{mod}$ level, employing the basis set recently optimized by us for quinhydrone [27], thus obtaining preliminary interaction energy profiles. Three points (displayed as blue squares in Figure 2) were selected for each profile (namely one in the minimum, one in the short distance range and one in the attractive branch of the curve) and the corresponding CCSD(T)/CBS interaction energies were computed and used to build a reference database containing 12 elements. This database was then used for the optimization of the exponents of the polarization functions of the 6-31G** basis sets suitable for $MP2^{mod}$ calculations. The starting exponents of the standard 6-31G** basis set are 0.80 for *d* functions on carbon and oxygen and 1.1 for *p* functions on hydrogen. After optimization, the best exponents were found to be 0.27 and 0.34 for the *d* functions on carbon and oxygen, respectively, and 0.36 for *p* functions on hydrogen. The final standard deviation, $\sqrt{I}$, see Equation (3), resulted to be less than 0.3 kcal/mol with respect to the CCSD(T)/CBS energies.

The resulting $MP2^{mod}$ curves are displayed in Figure 2, together with the reference values. The excellent agreement between the two methods, in line with the results previously obtained for similar molecules, allows us to apply rather confidently the $MP2^{mod}$ method to the study of the catechol dimer. According to both CCSD(T)/CBS and $MP2^{mod}$ results, the most stable structure is the TS_2 one (around -5.0 kcal/mol), with the minimum at a slightly smaller value of R (5.4 Å), with respect to the similar TS_1 conformer (5.6 Å), which is in turn almost as stable (≈ -4.0 kcal/mol) as the antiparallel stacked conformer (AFF, -3.8 kcal/mol). Among the two stacked conformations, FF and AFF, the second one is more stable, in agreement with the repulsive interaction between the OH dipoles in the FF form.

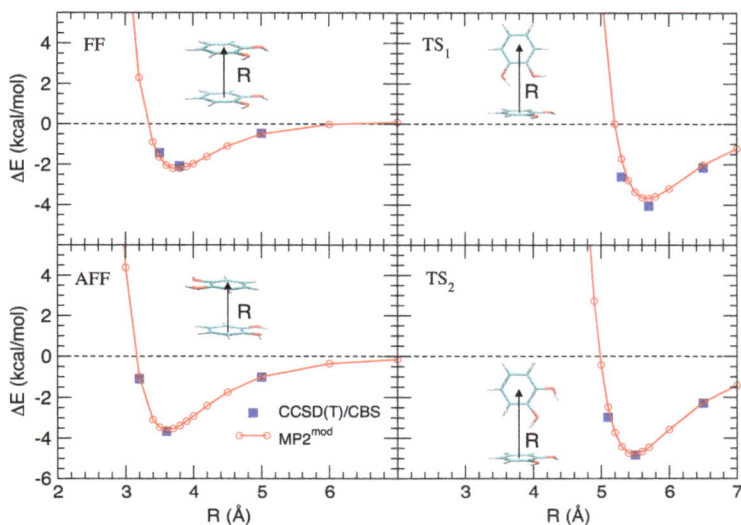

Figure 2. Comparison between the 'best exponent' and CCSD(T)/CBS for the interaction energy profiles obtained by displacement of the four structures shown in Figure 1.

3.2. Stacked Cathecol Dimers

Due to its importance, the stacked arrangement has been studied with some care as a function of the ring–ring distance R and of the angle β, which expresses, as shown in Figure 3b, the relative rotation of the two rings with respect to the line connecting their centers. The relevant results are reported in Figure 3. In the left panel, the interaction energy, reported vs. R for assigned rotation angles, shows minima at similar R values for all angles, and a marked dependence on β at low vales (from 0 to 60°), whereas for $\beta > 90°$ the curves are close to each other: at the minimum the interaction energy changes by only ≈ 0.25 kcal/mol in the range 90–180°. Although this behavior seems roughly consistent with a dipole–dipole interaction, the resemblance of the 90, 120, 150, and 180° curves is an indication that higher multipoles, or, equivalently, local dipoles, should play a role in an electrostatic rationalization of the observed energy curves. This is in agreement with the idea of Wheeler and coworkers [45,46] that stacking interaction in substituted aromatic species is strongly influenced by the local interaction of the substituents, rather than to changes induced in the π electronic density upon substitution, as suggested by older models.

Figure 3b shows the energy variation as a function of β and connects the FF ($\beta = 0°$) to the AFF ($\beta = 180°$) arrangement at a fixed ring-ring distance ($R = 3.5$ Å). The curve shows a not monotonic behavior, probably due to the presence of two functional groups, with an absolute minimum near 110°, rather than at 180°, as could be expected for single substituted benzene rings. However, despite the perturbations triggered by the specific interaction among the two strong local dipoles of the monomers, the transition from FF to AFF arrangements along β is rather marked and clearly indicates a preference for antiparallel stacked arrangements, as already put in evidence in Figure 2.

In order to gain a deeper insight into the orientation dependence of the stacking forces in the catechol dimer, taking advantage from the low computational cost of the MP2^{mod} method, we can explore different sections of the catechol IPES. For instance, in Figure 4 a two-dimensional contour plot of the interaction energy (ΔE) is reported as a function of the horizontal displacement of the two rings (R) and of the rotation angle (β) of one of the two rings around the perpendicular axis, at the inter-ring distance of 3.5 Å (i.e., the position of the minimum for the stacked energy curves reported in Figure 3).

a)   b)

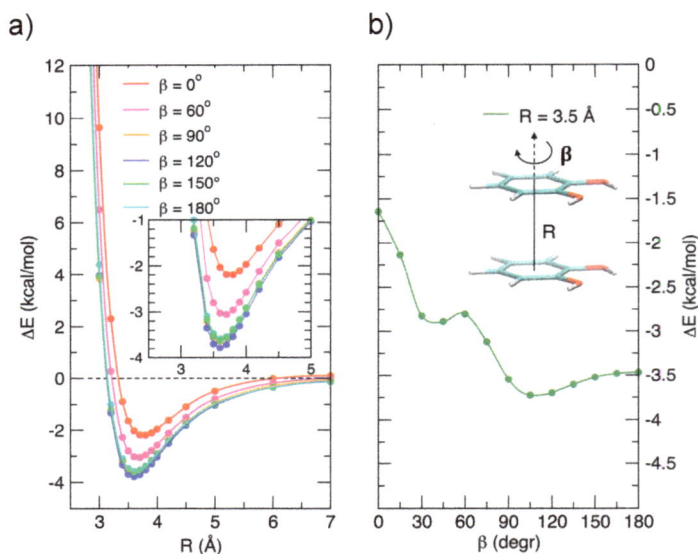

Figure 3. MP2^{mod} results for the stacked configurations. (**a**) Interaction energy as a function of the inter-ring separation R for different β angles. (**b**) Interaction energy as a function of the β angle at the ring–ring separation ($R = 3.5$ Å) corresponding to the minimum energy.

a)   b)

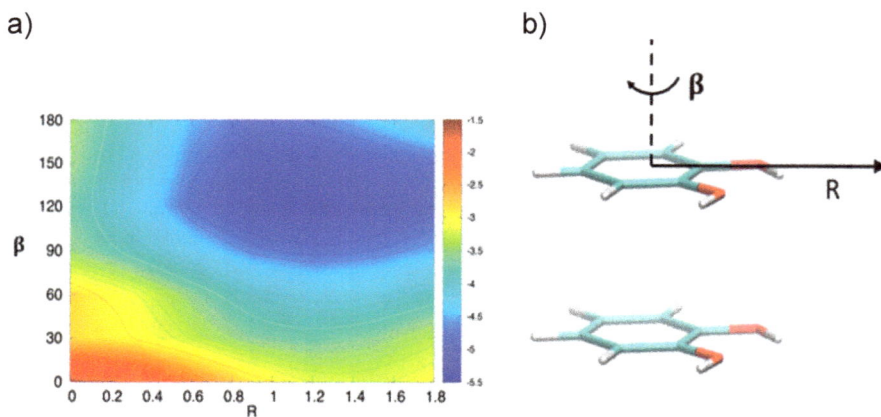

Figure 4. (**a**) Two-dimensional scan of the catechol interaction potential energy surface (IPES) in stacked conformations, performed at the MP2^{mod} level. (**b**) The IPES section was sampled by varying the angle β; the displacement R is also shown.

Figure 4 clearly shows that the dimer is much more stable when displaced and rotated with respect to the FF arrangement, with a minimum at $R \approx 1.2$ Å and $\beta \approx 130°$. It is noteworthy that the effects of horizontal displacement (i.e., varying R) and β rotation can be ascribed to different origins, closely related to the catechol molecular structure. In fact, the increase of the binding energy upon displacement closely resembles the well-known behavior of the benzene dimer [47,49,50] originated from a "pure" aromatic interaction: shifting one monomer along the R coordinate diminishes the quadrupolar repulsion between the two rings [49], whereas the attractive dispersion interaction decreases to a lesser extent, hence resulting in a global increase of the binding energy [47,49]. As discussed above, the energy profile vs. β rotation is strictly connected with the presence of OH substituents, as suggested by the net increase of the interaction energy in going from a parallel ($\beta = 0°$) to an antiparallel ($\beta = 180°$) arrangement.

This simple picture is consistent with the minimum of -5.2 kcal/mol ($R = 1.2$ Å, $\beta = 120°$) in a displaced near antiparallel configuration, not coincident with the perfect antiparallel arrangement ($\beta = 180°$) where the MP2^{mod} interaction energy is -4.7 kcal/mol. This subtle difference can find a rationale at a closer look of the molecular structure, embracing Wheeler's idea that unexpected substituents effects can be explained by considering their direct interaction with the neighboring cloud of the other ring [44–46]. The $\beta = 120°$ and $\beta = 180°$ conformers are displayed in Figure 5. In Figure 5b,d, where a top view of both dimers is shown, the positions of the oxygen atoms are marked with colored circles, to put in evidence the differences between the two arrangements. It appears as in the $\beta = 180°$ geometry all oxygen atoms lie approximately above a C=C bond of the other ring, resulting in an unfavorable electrostatic interaction with the carbon π orbitals, while at $\beta = 120°$ only three oxygen atoms contribute to such repulsive term. Consistently, the Hartree–Fock contribution to the total MP2^{mod} energy, which is repulsive in both cases, increases by 1 kcal/mol, in going from $\beta = 120°$ (3.3 kcal/mol) to $\beta = 180°$ (4.3 kcal/mol). Finally, another possible source of attractive interaction comes from the HB interaction between the hydrogen atom of one hydroxyl group and the closest oxygen of the other ring, as evidenced in Figure 5a,c, where it appears as in the $\beta = 120°$ conformer the hydrogen atoms lie at much closer distances (3.7 Å).

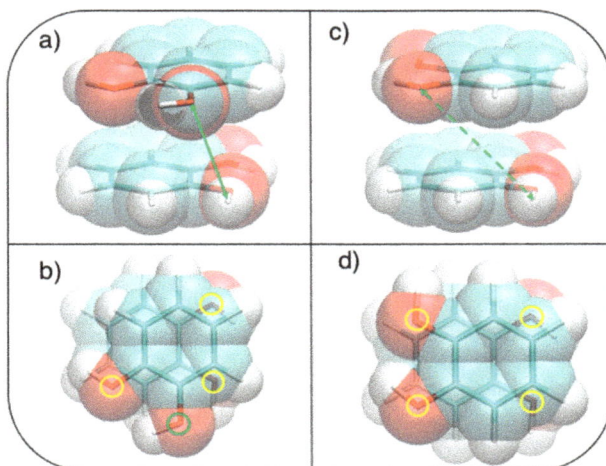

Figure 5. Stacked displaced ($R = 1.2$ Å) geometries at (**a,b**) $\beta = 120°$ and (**c,d**) $\beta = 180°$. (**a,c**): Side view, H–O distances of (**a**) 3.7 and (**c**) 5.0 Å are indicated with a green arrow ; (**b,d**): Top view, the position of oxygen atoms is shown with colored circles, distinguishing more (yellow) or less (green) interacting ones.

3.3. T-Shaped Cathecol Dimers

As shown in Figure 2, another kind of arrangement which can compete with the stacked geometries discussed above is the TS configuration. In this case, most of the interaction energy is expected to come from XH–π forces, in particular when two or one hydroxyl groups point towards the other ring's plane, as in the TS$_1$ and TS$_2$ geometries.

In order to verify this assumption, the MP2^{mod} computational feasibility has been exploited once again to explore an additional IPES section, related to the TS conformers and shown in Figure 6.

At small inter-ring distances, the dependence on β-rotation is striking and the most favorite conformer at R = 4.9 Å is found at β = 0° (i.e., the TS arrangement shown in Figure 6b), with the interaction energy (−2.4 kcal/mol) very similar to the value reported for the benzene dimer in the same configuration [49,50,53,57,71]. Conversely, due to the small distance between the H hydroxyl atom and the other catechol ring (see for instance TS$_1$ in Figure 2), the interaction energy in the 180–300° range is repulsive, with a maximum of almost 25 kcal/mol at β = 270°. The situation changes dramatically by increasing R, as in the 180–300° range the interaction energy shows a much steeper gradient. In fact, the IPES section minimum is found in a TS conformation at β = 270° and R = 5.5 Å, where the hydroxyl group points towards the other ring plane similarly to the TS$_2$ arrangement shown in the right bottom panel of Figure 2, resulting in a total interaction energy of −5.1 kcal/mol.

a)   b)

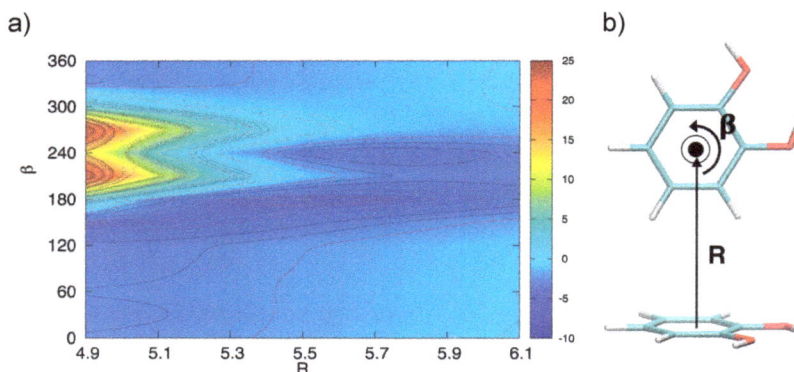

Figure 6. (**a**) Two-dimensional scan of the catechol IPES in TS conformation, performed at MP2^{mod} level. The IPES section was sampled by varying the angle β and the displacement R shown in (**b**), where the TS arrangement at β = 0° is displayed. The β rotation is performed as indicated by the black arrow (e.g., for β = 240° the dimer is found in the TS$_1$ geometry shown in the right top panel of Figure 2).

3.4. Effect of the Hydrogen Bond

The above described competition between stacked and TS geometries misses although another player, which could significantly alter the delicate balance between them. In fact, apart for a small contribution to the stability of the β = 120° conformer in the stacked conformations, the HB contribution was never decisive to the total interaction, due to too large distances between the involved hydrogen and oxygen atoms, which could be reduced by allowing internal rotation around C–O bonds. In order to find even more stable structures, we have released such constraint and performed a full optimization at MP2^{mod} level, starting from four different conformations (see Figure 7, top panels). The first starting geometry is a displaced AFF (AFFD). Next, two TS structures were prepared, with one or both hydroxyl groups pointing down towards the other ring (TS$_d$ and TS$_u$, respectively). Notice that the latter is very similar to that taken from crystallographic data and investigated by Estévez et al. [67]. Finally, a fourth arrangement was built from scratch, where the two rings are placed in side-by-side (SS) conformation, with both hydroxyl groups resulting at close distance, thus maximizing the effect of HBs. All optimizations ended up successfully in four different local minima, as confirmed by a frequency calculation purposely carried out for each of the resulting structures. The corresponding optimized structures are shown in the bottom panel of Figure 7 as I, II, III, and IV, respectively.

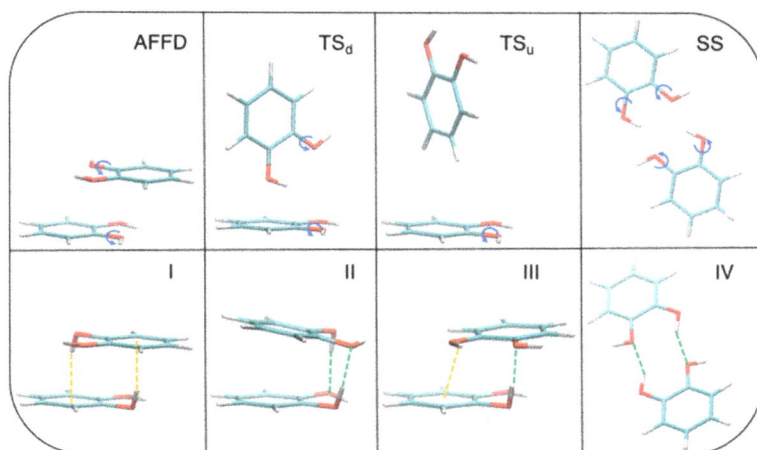

Figure 7. MP2^{mod} geometry optimization starting from the displaced AFF (AFFD), TS$_d$, and TS$_u$ conformations (top panels). The corresponding optimized structures, I, II, III, and IV, are displayed in the bottom row. The rotated hydroxyl groups are evidenced in the top panel with a blue arrow, while the atoms involved in OH–O and OH–π interactions are connected in the bottom panels by green and orange dashed lines, respectively.

Dimer formation does not result in large changes in the internal geometry of each catechol monomer. Bond lengths within each monomer change by less than 0.03 Å and the backbone remains planar. For each ring, only one hydroxyl hydrogen moves out of plane, establishing OH–O or OH–π interactions, while the other O–H bond remains nearly coplanar with the ring, due to the formation of an intramolecular OH–O HB with the closest oxygen atom (in Figure 7, geometry II it is out of plane by only 13°). The dihedral angle which drives the position of the out of plane hydrogen is 66° for I, 71.5° for II, and 68.7° for III, whereas the other ring hydrogen does not rotate. The final conformations reveal that the internal rotation has a significant effect on the interplay of the different interaction terms. In fact, as evident from Figure 7, both TS conformations are not stable upon a full optimization, and eventually end up in a stacked arrangement, whereas the AFFD conformer undergoes to the expected rotation from $\beta = 180°$ to $\beta \approx 120°$, but maintains the stacking arrangements. The OH–O interaction plays the major role in TS$_d$, which becomes II, while OH–π weak HBs guide the hydroxyl rotation and are prevalent in AFFD, which becomes I. Although less stable, the last optimized conformer III, is characterized by a single hydroxyl rotation, which allows the insurgence of a HB (green dashed line in Figure 7), while the other hydrogen remains coplanar to the ring, yet interacting with the other monomer establishing a OH–π noncovalent bond. For the conformation IV, geometry optimization results in a structure which is again less stable than I and II and somewhat more stable than III (see Table 1). In this case, the hydroxyl hydrogens undergo a small rotation with respect to the initial conformation and the two rings are slightly displaced out of the plane that initially (see SS) contained both monomers.

The interaction energies for the four final structures are reported in Table 1, along with the value computed at the same geometries with the MPW1B95/6-311++G(2*d*,2*p*) in [67], as well as with the "gold standard" CCSD(T)/CBS. From these data, it is clear that the most stable structure is II, which differs by only 1.6 kcal/mol from I, whereas III and IV are far higher in energy. The agreement between the MP2^{mod} values and their CCSD(T)/CBS counterparts is very good, especially considering that these latter geometries are outside the MP2^{mod} training set, while the computational advantage of using MP2^{mod} with small basis sets is apparent from the last three columns. Surprisingly, the MPW1B95 functional severely underestimates the reference CCSD(T)/CBS interaction energies, yielding, in the present case, only a qualitative correct description, at least according to the protocol provided in [67].

Table 1. Interaction energies, in kcal/mol, for the four optimized conformations shown in Figure 7, computed with MP2^{mod}, CCSD(T)/CBS and MPW1B95/6-311++G(2d,2p). Central processing unit (CPU) times on a single 2.60 GHz Intel® Xeon CPU are also given for an evaluation of the computational cost of the different methods.

Geometry	Energies (kcal/mol)			CPU Time (min)		
	MP2^{mod}	CCSD(T)/CBS	MPW1B95	MP2^{mod}	CCSD(T)/CBS	MPW1B95
I	−10.7	−11.1	−8.1	27	50,640	145
II	−12.4	−12.6	−8.3	25	49,740	180
III	−5.3	−5.7	−2.2	27	50,820	79
IV	−6.1	−7.3	−5.7	18	51,720	142

Finally, it is interesting to investigate the different HB contributions in the two most stable conformations I and II. This can be done by performing a rigid scan of the rotation angle δ of the two hydrogen atoms with respect to the C–O bond in both conformations (δ). The results are shown in Figure 8. For $\delta = 0°$ (i.e., when each hydrogen is coplanar to the aromatic ring), dispersion interactions are the main source of attraction, although perturbed by the electrostatic interaction between the dipoles, which favors dimer I (in an antiparallel alignment) by ≈1 kcal/mol with respect to dimer II. As δ increases, both the hydrogen atoms involved in the rotation come to closer distances from the other monomer, and may establish HBs. These noncovalent interactions remarkably stabilize both complexes, by almost 7 kcal/mol in I and more than 10 kcal/mol in II. These differences can find a rationale by looking at the insets of Figure 8. In dimer I, since each hydrogen points approximately towards the center of the neighboring ring, two weak HBs of the OH–π type are settled whereas, in dimer II, both hydrogen atoms are involved in a stronger OH–O HB. As a consequence, the minimum of the latter conformer is stabilized by ≈2 kcal/mol with respect to I.

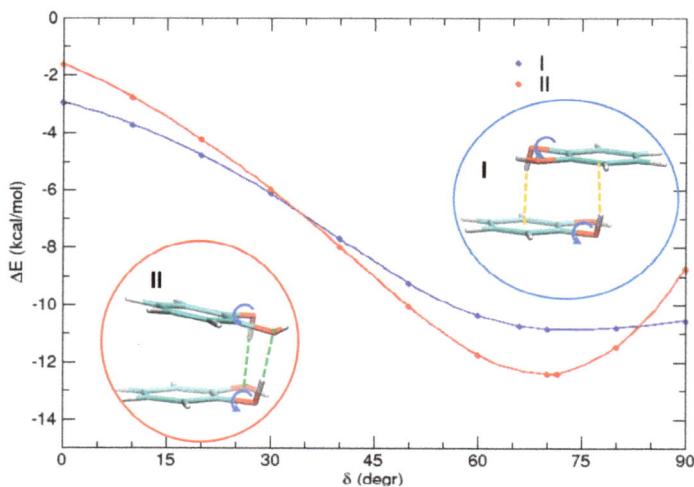

Figure 8. MP2^{mod} scans of the HOCC dihedral (δ) for conformers I (blue) and II (red), highlighting the role of OH–O and OH–π interactions, indicated with orange and green dashed lines, respectively.

4. Conclusions

In this paper, we have reported our study of the intermolecular landscape of a catechol dimer with a two-fold interest. On the one hand, noncovalent interactions, and especially those involving aromatic rings, govern many biological processes and it is, therefore, of basic importance to reach a good comprehension of the different role that the various forces play in specific systems. On the other hand, noncovalent interactions are still a challenging benchmark for standard computational methods, hence, it can be significant to exploit dedicated approaches.

Catechol is well known to be a precursor of many bioinspired materials and it is, therefore, a good candidate to investigate on the interplay between dispersion interactions, essentially due to aromaticity, and strong (OH–O) or weak (OH–π) HBs, settled by the hydroxyl substituents. The employed MP2^{mod} computational route consists in MP2 calculations with a small 6-31G** basis set, in which the exponents of the polarization functions are suitably modified. This has been done through a validation procedure based on the comparison with the highly accurate CCSD(T)/CBS calculations, resulting in new exponents for polarization functions on carbon (0.27), hydrogen (0.36), and oxygen (0.34).

Within the IPES sections explored, two minima were identified, held together by a network of stacking, OH–O, and OH–π interactions, whose relative weight has been analyzed in some detail. The two catechol units tend to aggregate in stacked conformation, which eventually result more stable than the TS ones, thanks to their ability to establish strong and weak HBs.

A final remark should be made concerning the effects that solvation can have in these systems. Despite most computational approaches designed for noncovalent interactions only focus on two isolated molecules, we are aware that water might affect the results and change the picture that we report here (see, for instance, [27,68]). It is, however, important to have a preliminary reference to guide the more complex study in solution, which is a natural continuation of the one presented here.

Author Contributions: The present article has been conceived by the four authors during a nice dinner. Computations have been shared to be more efficient as it was with the writing. All the authors have therefore equally contributed to the final work.

Conflicts of Interest: The authors declare no conflict of interest.

References

1. Mak, C.H. Unraveling base stacking driving forces in DNA. *J. Phys. Chem. B* **2016**, *120*, 6010–6020. [CrossRef] [PubMed]
2. Trouillas, P.; Sancho-García, J.C.; De Freitas, V.; Gierschner, J.; Otyepka, M.; Dangles, O. Stabilizing and modulating color by copigmentation: Insights from theory and experiment. *Chem. Rev.* **2016**, *116*, 4937–4982. [CrossRef] [PubMed]
3. Lv, C.; Zhao, G.; Ning, Y. The interactions between plant proteins/enzymes and other food components and their effects on food quality. *Crit. Rev. Food Sci. Nutr.* **2015**, *57*, 1718–1728. [CrossRef] [PubMed]
4. Wheeler, S.E.; Seguin, T.J.; Guan, Y.; Doney, A.C. Noncovalent interactions in organocatalysis and the prospect of computational catalyst design. *Acc. Chem. Res.* **2016**, *49*, 1061–1069. [CrossRef] [PubMed]
5. Neel, A.J.; Hilton, M.J.; Sigman, M.S.; Toste, F.D. Exploiting non-covalent π interactions for catalyst design. *Nature* **2017**, *543*, 637–646. [CrossRef] [PubMed]
6. Ding, X.; Wang, Y. Weak bond-based injectable and stimuli responsive hydrogels for biomedical applications. *J. Mater. Chem. B* **2017**, *5*, 887–906. [CrossRef]
7. Yilmazer, N.; Korth, M. Prospects of applying enhanced semi-empirical QM methods for 2101 virtual drug design. *Curr. Med. Chem.* **2016**, *23*, 2101–2111. [CrossRef] [PubMed]
8. Meanwell, N.A. A synopsis of the properties and applications of heteroaromatic rings in medicinal chemistry. *Adv. Heterocycl. Chem.* **2017**, *123*, 245–361.
9. Ghosh, T.; Panicker, J.; Nair, V. Self-assembled organic materials for photovoltaic application. *Polymers* **2017**, *9*, 112. [CrossRef]
10. Würthner, F. Dipole–dipole interaction driven self-assembly of merocyanine dyes: From dimers to nanoscale objects and supramolecular materials. *Acc. Chem. Res.* **2016**, *49*, 868–876. [CrossRef] [PubMed]

11. Chandra, B.K.C.; D'Souza, F. Design and photochemical study of supramolecular donor–acceptor systems assembled via metal–ligand axial coordination. *Coord. Chem. Rev.* **2016**, *322*, 104–141.

12. Bredas, J.-L.; Marder, S.R.; Reichmanis, E. Preface to the chemistry of materials special issue on π-functional materials. *Chem. Mater.* **2011**, *23*, 309. [CrossRef]

13. Shewmon, N.T.; Watkins, D.L.; Galindo, J.F.; Zerdan, R.B.; Chen, J.; Keum, J.; Roitberg, A.E.; Xue, J.; Castellano, R.K. Enhancement in organic photovoltaic efficiency through the synergistic interplay of molecular donor hydrogen bonding and π-stacking. *Adv. Funct. Mater.* **2015**, *25*, 5166–5177. [CrossRef]

14. Rest, C.; Kandanelli, R.; Fernández, G. Strategies to create hierarchical self-assembled structures via cooperative non-covalent interactions. *Chem. Soc. Rev.* **2015**, *44*, 2543–2572. [CrossRef] [PubMed]

15. Hammann, F.; Schmid, M. Determination and quantification of molecular interactions in protein films: A review. *Materials* **2014**, *7*, 7975–7996. [CrossRef] [PubMed]

16. Makwana, K.M.; Mahalakshmi, R. Implications of aromatic-aromatic interactions: From protein structures to peptide models. *Protein Sci.* **2015**, *24*, 1920–1933. [CrossRef] [PubMed]

17. Cragg, P.J. *Supramolecular Chemistry—From Biological Inspiration to Biomedical Applications*; Springer: Amsterdam, The Netherlands, 2010.

18. Wun Hwang, J.; Li, P.; Shimizu, K.D. Synergy between experimental and computational studies of aromatic stacking interactions. *Org. Biomol. Chem.* **2017**, *15*, 1554–1564. [CrossRef] [PubMed]

19. Sherrill, C.D. Energy component analysis of π interactions. *Acc. Chem. Res.* **2013**, *46*, 1020–1028. [CrossRef] [PubMed]

20. Heinzmann, C.; Weder, C.; de Espinosa, L.M. Supramolecular polymer adhesives: Advanced materials inspired by nature. *Chem. Soc. Rev.* **2015**, *342*, 342–358. [CrossRef] [PubMed]

21. Jenkins, C.L.; Siebert, H.M.; Wilker, J.J. Integrating mussel chemistry into a bio-based polymer to create degradable adhesives. *Macromolecules* **2017**, *50*, 561–568. [CrossRef]

22. Kord Forooshani, P.; Lee, B.P. Recent approaches in designing bioadhesive materials inspired by mussel adhesive protein. *J. Polym. Sci. Part A: Polym. Chem.* **2017**, *55*, 9–33. [CrossRef] [PubMed]

23. North, M.A.; Del Grosso, C.A.; Wilker, J.J. High strength underwater bonding with polymer mimics of mussel adhesive proteins. *ACS Appl. Mater. Interfaces* **2017**, *9*, 7866–7872. [CrossRef] [PubMed]

24. Saiz-Poseu, J.; Sedó, J.; García, B.; Benaiges, C.; Parella, T.; Alibés, R.; Hernando, J.; Busqué, F.; Ruiz-Molina, D. Versatile nanostructured materials via direct reaction of functionalized catechols. *Adv. Mater.* **2013**, *25*, 2066–2070. [CrossRef] [PubMed]

25. Sedó, J.; Saiz-Poseu, J.; Busqué, F.; Ruiz-Molina, D. Catechol-based biomimetic functional materials. *Adv. Mater.* **2013**, *25*, 653–701. [CrossRef] [PubMed]

26. Wang, X.; Jing, S.; Liu, Y.; Liu, S.; Tan, Y. Diblock copolymer containing bioinspired borneol and dopamine moieties: Synthesis and antibacterial coating applications. *Polymer* **2017**, *116*, 314–323. [CrossRef]

27. Barone, V.; Cacelli, I.; Crescenzi, O.; D'Ischia, M.; Ferretti, A.; Prampolini, G.; Villani, G. Unraveling the interplay of different contributions to the stability of the quinhydrone dimer. *RSC Adv.* **2014**, *4*, 876–885. [CrossRef]

28. Burns, L.A.; Vázquez-Mayagoitia, A.; Sumpter, B.G.; Sherrill, C.D. Density-functional approaches to noncovalent interactions: A comparison of dispersion corrections (DFT-D), exchange-hole dipole moment (XDM) theory, and specialized functionals. *J. Chem. Phys.* **2011**, *134*, 084107. [CrossRef] [PubMed]

29. Corminboeuf, C. Minimizing density functional failures for non-covalent interactions beyond van der Waals complexes. *Acc. Chem. Res.* **2014**, *47*, 3217–3224. [CrossRef] [PubMed]

30. Goerigk, L.; Grimme, S. Efficient and accurate double-hybrid-meta-GGA density functionals-evaluation with the extended GMTKN30 database for general main group thermochemistry, kinetics, and noncovalent interactions. *J. Chem. Theory Comput.* **2011**, *7*, 291–309. [CrossRef] [PubMed]

31. Goldey, M.B.; Belzunces, B.; Head-Gordon, M. Attenuated MP2 with a long-range dispersion correction for treating nonbonded interactions. *J. Chem. Theory Comput.* **2015**, *11*, 4159–4168. [CrossRef] [PubMed]

32. Grimme, S. Density functional theory with London dispersion corrections. *WIREsComput. Mol. Sci.* **2011**, *1*, 211–228. [CrossRef]

33. Grimme, S.; Hansen, A.; Brandenburg, J.G.; Bannwarth, C. Dispersion-corrected mean-field electronic structure methods. *Chem. Rev.* **2016**, *116*, 5105–5154. [CrossRef] [PubMed]

34. Hobza, P.; Zahradník, R.; Müller-Dethlefs, K. The world of non-covalent interactions: 2006. *Collect. Czechoslov. Chem. Commun.* **2006**, *71*, 443–531. [CrossRef]

35. Piton, M. Accurate intermolecular interaction energies from a combination of MP2 and TDDFT response theory. *J. Chem. Theory Comput.* **2010**, *6*, 168–178. [CrossRef] [PubMed]
36. Prampolini, G.; Cacelli, I.; Ferretti, A. Intermolecular interactions in eumelanins: A computational bottom-up approach. I. Small building blocks. *RSC Adv.* **2015**, *5*, 38513–38526. [CrossRef]
37. Řezáč, J.; Hobza, P. Describing noncovalent interactions beyond the common approximations: How accurate is the "gold standard", CCSD(T) at the complete basis set limit? *J. Chem. Theory Comput.* **2013**, *9*, 2151–2155. [CrossRef] [PubMed]
38. Řezáč, J.; Hobza, P. Benchmark calculations of interaction energies in noncovalent complexes and their applications. *Chem. Rev.* **2016**, *116*, 5038–5071. [CrossRef] [PubMed]
39. Richard, R.M.; Lao, K.U.; Herbert, J.M. Achieving the CCSD(T) basis-set limit in sizable molecular clusters: Counterpoise corrections for the many-body expansion. *J. Phys. Chem. Lett.* **2013**, *4*, 2674–2680. [CrossRef] [PubMed]
40. Riley, K.E.; Platts, J.A.; Řezáč, J.; Hobza, P.; Hill, J.G. Assessment of the performance of MP2 and MP2 variants for the treatment of noncovalent interactions. *J. Phys. Chem. A* **2012**, *116*, 4159–4169. [CrossRef] [PubMed]
41. Sherrill, C.D.; Takatani, T.; Hohenstein, E.G. An assessment of theoretical methods for nonbonded interactions: Comparison to complete basis set limit coupled-cluster potential energy curves for the benzene dimer, the methane dimer, benzene–methane, and benzene–H_2S. *J. Phys. Chem. A* **2009**, *113*, 10146–10159. [CrossRef] [PubMed]
42. Tkatchenko, A.; DiStasio, R.A.; Head-Gordon, M.; Scheffler, M. Dispersion-corrected Møller–Plesset second-order perturbation theory. *J. Chem. Phys.* **2009**, *131*, 94106. [CrossRef] [PubMed]
43. Zhao, Y.; Truhlar, D.G. The M06 suite of density functionals for main group thermochemistry, thermochemical kinetics, noncovalent interactions, excited states, and transition elements: Two new functionals and systematic testing of four M06-class functionals and 12 other function. *Theor. Chem. Acc.* **2008**, *120*, 215–241. [CrossRef]
44. An, Y.; Doney, A.C.; Andrade, R.B.; Wheeler, S.E. Stacking interactions between 9-methyladenine and heterocycles commonly found in pharmaceuticals. *J. Chem. Inf. Model.* **2016**, *56*, 906–914. [CrossRef] [PubMed]
45. Wheeler, S.E.; Bloom, J.W.G. Toward a more complete understanding of noncovalent interactions involving aromatic rings. *J. Phys. Chem. A* **2014**, *118*, 6133–6147. [CrossRef] [PubMed]
46. Wheeler, S.E.; Houk, K.N. Substituent effects in the benzene dimer are due to direct interactions of the substituents with the unsubstituted benzene. *J. Am. Chem. Soc.* **2008**, *130*, 10854–10855. [CrossRef] [PubMed]
47. Podeszwa, R.; Bukowski, R.; Szalewicz, K. Potential energy surface for the benzene dimer and perturbational analysis of π–π interactions. *J. Phys. Chem. A* **2006**, *110*, 10345–10354. [CrossRef] [PubMed]
48. Sinnokrot, M.O.; Sherrill, C.D. Unexpected substituent effects in face-to-face π-stacking interactions. *J. Phys. Chem. A* **2003**, *107*, 8377–8379. [CrossRef]
49. Tsuzuki, S.; Honda, K.; Uchimaru, T.; Mikami, M.; Tanabe, K. Origin of attraction and directionality of the π/π interaction: Model chemistry calculations of benzene dimer interaction. *J. Am. Chem. Soc.* **2002**, *124*, 104–112. [CrossRef] [PubMed]
50. Sinnokrot, M.O.; Valeev, E.F.; Sherrill, C.D. Estimates of the ab initio limit for π–π interactions: The benzene dimer. *J. Am.Chem. Soc.* **2002**, *124*, 10887–10893. [CrossRef] [PubMed]
51. Kroon-Batenburg, L.; Van Duijneveldt, F. The use of a moment-optimized DZP basis set for describing the interaction in the water dimer. *J. Mol. Struct.: THEOCHEM* **1985**, *22*, 185–199. [CrossRef]
52. Hobza, P.; Zahradnik, R. Intermolecular interactions between medium-sized systems. Nonempirical and empirical calculations of interaction energies. Successes and failures. *Chem. Rev.* **1988**, *88*, 871–897. [CrossRef]
53. Hobza, P.; Selzle, H.L.; Schlag, E.W. Potential energy surface for the benzene dimer. Results of ab initio CCSD(T) calculations show two nearly isoenergetic structures: T-shaped and parallel-displaced. *J. Phys. Chem.* **1996**, *100*, 18790–18794. [CrossRef]
54. Sponer, J.; Leszczynski, J.; Hobza, P. Base stacking in cytosine dimer. A comparison of correlated ab initio calculations with three empirical potential models and density functional theory calculations. *J. Comput. Chem.* **1996**, *17*, 841–850. [CrossRef]
55. Šponer, J.; Leszczyński, J.; Hobza, P. Nature of nucleic acid–base stacking: Nonempirical ab initio and empirical potential characterization of 10 stacked base dimers. Comparison of stacked and H-bonded base pairs. *J. Phys. Chem.* **1996**, *100*, 5590–5596. [CrossRef]

56. Hobza, P.; Šponer, J. Toward true DNA base-stacking energies: MP2, CCSD(T), and complete basis set calculations. *J. Am. Chem. Soc.* **2002**, *124*, 11802–11808. [CrossRef] [PubMed]
57. Cacelli, I.; Cinacchi, G.; Prampolini, G.; Tani, A. Computer simulation of solid and liquid benzene with an atomistic interaction potential derived from ab initio calculations. *J. Am. Chem. Soc.* **2004**, *126*, 14278–14286. [CrossRef] [PubMed]
58. Mignon, P.; Loverix, S.; De Proft, F.; Geerlings, P. Influence of stacking on hydrogen bonding: Quantum chemical study on pyridine–benzene model complexes. *J. Phys.Chem. A* **2004**, *108*, 6038–6044. [CrossRef]
59. Řeha, D.; Kabeláč, M.; Ryjáček, F.; Šponer, J.; Šponer, J.E.; Elstner, M.; Suhai, S.; Hobza, P. Intercalators. 1. Nature of stacking interactions between intercalators (ethidium, daunomycin, ellipticine, and 4′,6-diaminide-2-phenylindole) and DNA base pairs. Ab initio quantum chemical, density functional theory, and empirical potential study. *J. Am. Chem. Soc.* **2002**, *124*, 3366–3376. [CrossRef] [PubMed]
60. Sponer, J.; Gabb, H.A.; Leszczynski, J.; Hobza, P. Base–base and deoxyribose–base stacking interactions in B-DNA and Z-DNA: A quantum-chemical study. *Biophys. J.* **1997**, *73*, 76–87. [CrossRef]
61. Hobza, P.; Kabeláč, M.; Šponer, J.; Mejzlík, P.; Vondrášek, J. Performance of empirical potentials (AMBER, CFF95, CVFF, CHARMM, OPLS, POLTEV), semiempirical quantum chemical methods (AM1, MNDO/M, PM3), andab initio Hartree–Fock method for interaction of DNA bases: Comparison with nonempirical beyond Hartree–Fock results. *J. Comput. Chem.* **1997**, *18*, 1136–1150.
62. Cacelli, I.; Prampolini, G.; Tani, A. Atomistic simulation of a nematogen using a force field derived from quantum chemical calculations. *J. Phys. Chem. B* **2005**, *109*, 3531–3538. [CrossRef] [PubMed]
63. Cacelli, I.; Lami, C.F.; Prampolini, G. Force-field modeling through quantum mechanical calculations: Molecular dynamics simulations of a nematogenic molecule in its condensed phases. *J. Comput. Chem.* **2009**, *30*, 366–378. [CrossRef] [PubMed]
64. Cacelli, I.; Cimoli, A.; Livotto, P.R.; Prampolini, G. An automated approach for the parameterization of accurate intermolecular force-fields: Pyridine as a case study. *J. Comput. Chem.* **2012**, *33*, 1055–1067. [CrossRef] [PubMed]
65. Micillo, R.; Panzella, L.; Iacomino, M.; Prampolini, G.; Cacelli, I.; Ferretti, A.; Crescenzi, O.; Koike, K.; Napolitano, A.; D'Ischia, M. Eumelanin broadband absorption develops from aggregation-modulated chromophore interactions under structural and redox control. *Sci. Rep.* **2017**, *7*, 41532. [CrossRef] [PubMed]
66. Prampolini, G.; Greff da Silveira, L.; Jacobs, M.; Livotto, P.R.; Cacelli, I. Interaction energy landscapes of aromatic heterocycles through a reliable yet affordable computational approach. *J. Chem. Theory Comput.* **2017**. submitted.
67. Estévez, L.; Otero, N.S.; Mosquera, R.A. Computational study on the stacking interaction in catechol complexes. *J. Phys. Chem. A* **2009**, *113*, 11051–11058. [CrossRef] [PubMed]
68. Barone, V.; Cacelli, I.; Ferretti, A.; Prampolini, G.; Villani, G. Proton and electron transfer mechanisms in the formation of neutral and charged quinhydrone-like complexes: A multilayered computational study. *J. Chem.Theory Comput.* **2014**, *10*, 4883–4895. [CrossRef] [PubMed]
69. Boys, S.F.; Bernardi, F. The calculation of small molecular interactions by the differences of separate total energies. Some procedures with reduced errors. *Mol. Phys.* **1970**, *19*, 553–566. [CrossRef]
70. Frisch, M.J.; Trucks, G.W.; Schlegel, H.B.; Scuseria, G.E.; Robb, M.A.; Cheeseman, J.R.; Scalmani, G.; Barone, V.; Mennucci, B.; Petersson, G.A.; et al. *Gaussian 09*; Gaussian, Inc.: Wallingford, CT, USA, 2009.
71. Prampolini, G.; Livotto, P.R.; Cacelli, I. Accuracy of quantum mechanically derived force-fields parameterized from dispersion-corrected DFT data: The benzene dimer as a prototype for aromatic interactions. *J. Chem. Theory Comput.* **2015**, *11*, 5182–5196. [CrossRef] [PubMed]

biomimetics

MDPI

Article

Kaxiras's Porphyrin: DFT Modeling of Redox-Tuned Optical and Electronic Properties in a Theoretically Designed Catechol-Based Bioinspired Platform

Orlando Crescenzi *, Marco d'Ischia and Alessandra Napolitano

Department of Chemical Sciences, University of Naples Federico II, I-80126 Naples, Italy;
dischia@unina.it (M.d.I.); alesnapo@unina.it (A.N.)
* Correspondence: orlando.crescenzi@unina.it; Tel.: +39-081-674206

Academic Editor: Josep Samitier
Received: 19 August 2017; Accepted: 26 October 2017; Published: 7 November 2017

Abstract: A detailed computational investigation of the 5,6-dihydroxyindole (DHI)-based porphyrin-type tetramer first described by Kaxiras as a theoretical structural model for eumelanin biopolymers is reported herein, with a view to predicting the technological potential of this unique bioinspired tetracatechol system. All possible tautomers/conformers, as well as alternative protonation states, were explored for the species at various degrees of oxidation and all structures were geometry optimized at the density functional theory (DFT) level. Comparison of energy levels for each oxidized species indicated a marked instability of most oxidation states except the six-electron level, and an unexpected resilience to disproportionation of the one-electron oxidation free radical species. Changes in the highest energy occupied molecular orbital (HOMO)–lowest energy unoccupied molecular orbital (LUMO) gaps with oxidation state and tautomerism were determined along with the main electronic transitions: more or less intense absorption in the visible region is predicted for most oxidized species. Data indicated that the peculiar symmetry of the oxygenation pattern pertaining to the four catechol/quinone/quinone methide moieties, in concert with the NH centers, fine-tunes the optical and electronic properties of the porphyrin system. For several oxidation levels, conjugated systems extending over two or more indole units play a major role in determining the preferred tautomeric state: thus, the highest stability of the six-electron oxidation state reflects porphyrin-type aromaticity. These results provide new clues for the design of innovative bioinspired optoelectronic materials.

Keywords: porphyrin; eumelanin; DFT; *ab initio* calculation

---

1. Introduction

Eumelanins, according to a recent consensus definition, are "a black-to-brown insoluble subgroup of melanin pigments derived at least in part from the oxidative polymerization of L-dopa via 5,6-dihydroxyindole (DHI) intermediates" [1]. Among organic polymers eumelanins occupy a unique position because of (i) their widespread occurrence in nature, from man and mammals to fish, birds and molluscs; (ii) the variety of biological roles, from photoprotection to scavenging of reactive oxygen species and metal chelation; and (iii) a unique set of physical and chemical properties, including broadband absorption throughout the visible range, water-dependent ionic-electronic semiconductor-like behavior, stable free radical character and efficient nonradiative energy dissipation, making them attractive candidates for biomedical and technological applications. Despite growing interest in eumelanin-type functional materials and systems, a major gap hindering progress toward melanin-based technology relates to the highly insoluble and heterogeneous character of these polymers, which has so far hindered detailed insights into the structural basis of their physicochemical properties. In 2006 Kaxiras et al. [2] proposed on a theoretical basis that eumelanin properties could be interpreted in terms of porphyrin-type building blocks derived from the oxidative cyclization of DHI tetramers built via 2,7′-coupling (Scheme 1).

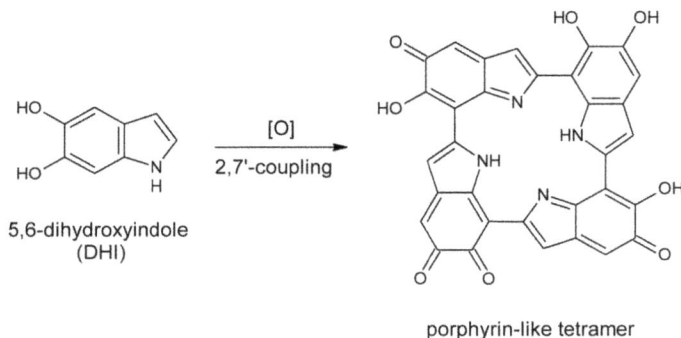

porphyrin-like tetramer

Scheme 1. Formation of a porphyrin-like tetramer (Kaxiras's porphyrin, KP) by oxidation of 5,6-dihydroxyindole (DHI).

Density functional theory (DFT) calculations indicated broadband absorption properties compatible with the optical properties of eumelanins. In a subsequent study, the authors suggested that such tetramers could bind in a covalent manner through interlayer C–C bonds [3].

Inspired by these theoretical studies, several authors adopted Kaxiras's porphyrin (KP) model to explain eumelanin properties [4–7], despite the lack of more than circumstantial or indirect evidence for porphyrin-type structures. Yet, regardless of its actual relevance to eumelanin structure, KP is a noticeable example of polycyclic aromatic system built upon a key eumelanin building block and displaying catechol-controlled electron properties, including aromaticity. This system, which is conceptually related to the basic polycyclic scaffold of electroluminescent DHI-based triazatruxenes [8], displays potential optoelectronic properties that seem worthy of being investigated at a theoretical level in the quest for novel bioinspired functional systems. Of special interest, to this aim, is the impact of the oxidation state on the thermodynamic stability of the porphyrin scaffold and the structural and redox dependence of the highest energy occupied molecular orbital (HOMO)– lowest energy unoccupied molecular orbital (LUMO) gap, which is critical for material design in organic electronics.

The aim of this paper is to integrate and expand previous computational studies on KPs in order to draw a first set of fundamental structure–property relationships. Reported herein is a detailed investigation of the structure and properties of this unique molecular platform as a function of the oxidation state, with special reference to:

(a) The relative stability of the various oxidation states;
(b) The position of tautomeric and acid-base dissociation equilibriums in the various oxidation states;
(c) The spectroscopic (ultraviolet–visible (UV–Vis) and infrared (IR)) properties of the most stable tautomers for each of the possible oxidation states;
(d) The nature of main electronic transitions;
(e) The aromatic character of each oxidation state.

It is expected that elucidation of KPs properties may guide the design of innovative catechol-based redox tunable systems for bioinspired applications.

2. Methods

All calculations were performed with the Gaussian program package [9]. All structures were geometry optimized at the DFT level, with a hybrid functional (PBE0) [10] and a reasonably large basis set (6-31+G(d,p)). For each species, different tautomers/conformers, as well as different protonation states were explored. In those cases where conformational enantiomers exist, a single enantiomeric series has been explored.

Computations were performed either in vacuo, or by adoption of a polarizable continuum model (PCM) [11–14] to account for the influence of the solution environment. In view of the faster convergence, a scaled van der Waals cavity based on universal force field (UFF) radii [15] was used, and polarization charges were modeled by spherical Gaussian functions [16,17]; nonelectrostatic contributions to the solvation free energy were disregarded at this stage; these terms were accounted for in single-point PCM calculations (at the PCM geometries) employing radii and nonelectrostatic terms of the SMD solvation model [18]. Vibrational-rotational contributions to the free energy were also computed. For the preparation of IR spectra, harmonic frequencies were scaled by a factor of 0.9547 [19] and Gaussian line width of 20 cm^{-1} was used.

Ultraviolet–visible spectra of the main species were computed in vacuo or in solution using the time-dependent density functional theory (TD-DFT) approach [20–24], with the PBE0 functional and the 6-311++G(2d,2p) basis set. To produce graphs, transitions below 5.6 eV were selected, and an arbitrary Gaussian line width of 0.25 eV was imposed; the spectra were finally converted to a wavelength scale.

3. Results and Discussion

For the purposes of this study, several different oxidation states were considered, namely: fully reduced (KP-Red), one-electron oxidation (KP-1e, as representative open shell state) and KP-2e, KP-4e, KP-6e and KP-8e as closed shell states.

The generation of the relevant starting structures is straightforward for the KP-Red and KP-1e states. However, for the higher oxidation states explored, the number of possible tautomeric species can become quite significant. A simple procedure to generate them would consist in defining the number of heteroatom-bound hydrogens that corresponds to the target oxidation state (e.g., 10 hydrogens for KP-2e, 8 hydrogens for KP-4e, etc.), and placing them in all possible different ways on the 12 available positions of the porphyrin skeleton (i.e., N1, O5, and O6 for each of the four rings). However, this procedure would not distinguish between closed-shell and open-shell species: these latter are predictably characterized by lower stabilities, and were therefore disregarded in the present exploration. Therefore, a systematic procedure for the generation of relevant closed-shell tautomers was devised, based on combining individual DHI units at the following oxidation stages: (i) fully reduced units, with dangling single bonds at the 2- and 7-position (one type); (ii) one-electron oxidized units, with a dangling double bond at the 2-position and a dangling single bonds at the 7-position (two types), or vice versa (two types); (iii) units at the two-electron oxidation level, with dangling single bonds at the 2- and 7-position (three types, i.e., *o*-quinone, quinone methide, or quinone-imine forms), or else with dangling double bonds at both 2- and 7-position (one type). To build up staring structures for the tetrameric porphyrin system, four building blocks were selected in such a way as to match the desired overall oxidation stage of the porphyrin system, and were combined in all possible ways compatible with the individual connectivity requirements; the resulting pool of structures was examined, and duplicates were discarded. For each tautomer, at least two conformers were generated.

To facilitate discussion and comparison among structures at different oxidation stages, an ad hoc naming scheme was devised, whereby the four individual DHI units of the porphyrin system are labeled as "a", "b", "c", and "d" in the sequence of the 2,7′-connectivities; moreover, mobile hydrogens that have been removed on each unit are explicitly indicated: thus, for example, a four-electron oxidized system featuring an *o*-quinone type unit bound through a 2,7′-single bond to a quinone methide unit, would be dubbed "a56_b16_c0_d0".

In a preliminary screening, presented in a separate paper, it was found that of the various oxidation states accessible to KP, the six-electron state was relatively more stable [25]. Herein, we provide a detailed account of the structural and spectral properties of KP at the aforementioned oxidation states in order to draw possible structure–property relationships that may allow prediction of functionality and prompt the design of KP functional systems for specific applications.

The general computational frame adopted in this study to investigate energies, structures and spectroscopic properties of the various species has proved reliable in several previous investigations of DHI-related oligomers [26–29]. However, in order to confirm the reliability in the specific case of KP, we carried out some ad hoc validation tests, which are reported as Supplementary Materials. First, we compared the relative energies of the main tautomers/conformers of KP-4e (neutral form in vacuo) at the PBE0 and B3LYP level [30], using different basis sets, and we fund quite good reproducibility. Analogous calculations were carried out on the main neutral species in all different oxidation states in vacuo; using this set of data, we then recomputed and compared electronic energy changes for disproportionation processes involving KP (see below, Section 3.7). Again, the results appeared rather robust with respect to functional/basis set choice.

A validation of the computed UV–Vis data was sought by comparing TD-PBE0 vs. TD-B3LYP for two related nonoxygenated porphyrin systems that have been synthesized and characterized by Nakamura and colleagues [31,32]. A similar comparison was extended to the high-wavelength electric dipole-allowed transitions of porphin freebase, and of magnesium porphin. TD-B3LYP calculations were also performed on the most significant tautomers/conformers (neutral forms) of KP in all different oxidation states, and the resulting computed spectra were found to resemble closely their TD-PBE0 counterparts. In this connection it is also worth noting that a recent study employing a post-Hartree–Fock approach (CC2) [33] to investigate the onset of electronic absorption spectra of DHI oligomers reported an overall fairly good agreement with literature TD-DFT results.

3.1. The Fully Reduced State (KP-Red)

Tables 1–3 report structures, energies and relative stabilities of tautomers/conformers in the fully reduced state.

The neutral form has S_4 symmetry both in vacuo and in water; alternative symmetries, including C_4 and C_{4h}, correspond to high-order saddle points, and at any rate are much higher in energy. In principle, many conformational possibilities arise in connection with the different orientations that can be adopted by the OH groups. However, studies on DHI [34] have shown that both OH groups are coplanar with the indole ring, and that "closed" geometries (i.e., those featuring an intramolecular H-bond between O5–H and O6–H) are significantly more stable than the corresponding "open" form. Of the two alternative "closed" forms, the one in which both OH bonds point in the opposite direction with respect to nitrogen is marginally more stable, and has been adopted throughout. Of course, "closed" geometries are invariably preferred when one of the oxygen centers bears a negative charge or a radical.

Table 1. KP-Red, neutral form in vacuo.

Tautomer	Conformer [a]	E (Ha) [b]	H_{RRHO} (Ha) [c]	G_{RRHO} (Ha) [d]
a0_b0_c0_d0	S_4	−2050.155247 (0.0)	−2049.641430 (0.0)	−2049.739870 (0.0)

In parentheses relative energies (kcal mol^{-1}) refer to the most stable form identified at the specified level. [a] For chiral structures, only one enantiomer is listed. [b] Electronic energy. [c] Enthalpy computed at 298.15 K within the rigid-rotor/harmonic-oscillator (RRHO) approximation. [d] Gibbs free energy computed at 298.15 K within the RRHO approximation.

Table 2. KP-Red, neutral form in water.

Tautomer	Conformer [a]	G_{PCM} (Ha) [b]	$H_{PCM,RRHO}$ (Ha) [c]	$G_{PCM,RRHO}$ (Ha) [d]	G_{SMD} (Ha) [e]	$G_{SMD,RRHO}$ (Ha) [f]
a0_b0_c0_d0	S_4	−2050.191047 (0.0)	−2049.678884 (0.0)	−2049.779144 (0.0)	−2050.220499 (0.0)	−2049.808596 (0.0)

In parentheses relative energies (kcal mol^{-1}) refer to the most stable form identified at the specified level. [a] For chiral structures, only one enantiomer is listed. [b] Electronic energy including electrostatic contributions at the polarizable continuum model (PCM) level. [c] Enthalpy computed at 298.15 K within the rigid-rotor/harmonic-oscillator (RRHO) approximation. [d] Gibbs free energy computed at 298.15 K within the RRHO approximation. [e] Electronic energy including nonelectrostatic terms according to the SMD solvation model. [f] $G_{SMD,RRHO} = G_{PCM,RRHO} + G_{SMD} - G_{PCM}$.

Table 3. KP-Red, monoanionic form in water.

Tautomer	Conformer [a]	G_{PCM} (Ha) [b]	$H_{PCM,RRHO}$ (Ha) [c]	$G_{PCM,RRHO}$ (Ha) [d]	G_{SMD} (Ha) [e]	$G_{SMD,RRHO}$ (Ha) [f]
a1_b0_c0_d0	C_1, conf1	−2049.709473 (8.6)	-	-	-	-
a5_b0_c0_d0	C_1, conf1	−2049.719292 (2.4)	−2049.220534 (2.6)	−2049.319341 (2.4)	−2049.749637 (1.7)	−2049.349686 (1.7)
a6_b0_c0_d0	C_1, conf1	−2049.723100 (0.0)	−2049.224673 (0.0)	−2049.323215 (0.0)	−2049.752271 (0.0)	−2049.352386 (0.0)

In parentheses relative energies (kcal mol^{-1}) refer to the most stable form identified at the specified level. [a] For chiral structures, only one enantiomer is listed. [b] Electronic energy including electrostatic contributions at the polarizable continuum model (PCM) level. [c] Enthalpy computed at 298.15 K within the rigid-rotor/harmonic-oscillator (RRHO) approximation. [d] Gibbs free energy computed at 298.15 K within the RRHO approximation. [e] Electronic energy including nonelectrostatic terms according to the SMD solvation model. [f] $G_{SMD,RRHO} = G_{PCM,RRHO} + G_{SMD} - G_{PCM}$.

The preferred site for deprotonation of KP-Red is predicted to be O6. Based on comparison of $\Delta G_{SMD,RRHO}$ values with those of a number of reference acid/base pairs of the AH/A$^-$ type, the pK_a is estimated around 8.2.

Ultraviolet–visible spectra of the most significant species of KP-Red were computed at the TD-PBE0 level both in vacuo and in water (PCM) (Figure 1). Overall, they show reasonable similarity to the spectrum computed at the same level for the fully reduced form of 2,7′-DHI dimer: a moderate bathochromic shift of ca. 10 nm in the spectrum of KP-Red could reflect the influence of additional interring conjugation, but also of a geometric constrain imposing s-*trans* conformation for N–C2–C7′–C6′ interring dihedrals. The rather large bandwidth reflects convolution of several different individual electronic transitions: the longest-wavelength contribution, at 372 nm (in vacuo), is HOMO–LUMO in character and is characterized by a moderate oscillator strength (f = 0.07); more intense transitions connect degenerate HOMO−1/HOMO−2 orbitals to the LUMO (357 nm, f = 0.36), and the HOMO to degenerate LUMO+1 and LUMO+2 orbitals (327 nm, f = 0.38).

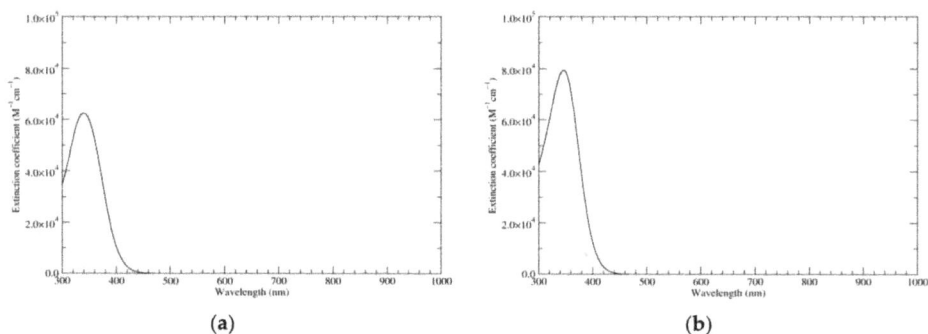

Figure 1. Computed ultraviolet–visible (UV–Vis) spectra of the most significant tautomers/conformers in the reduced state. (**a**) Neutral form in vacuo, a0_b0_c0_d0, S_4. (**b**) Neutral form in water, a0_b0_c0_d0, S_4.

Infrared spectra of the main species are displayed in Figure 2 (Raman spectra have been reported in [25] and will not be discussed here). The spectrum of the reduced species will be useful as a reference point for the discussion of the richer spectra of higher oxidation states.

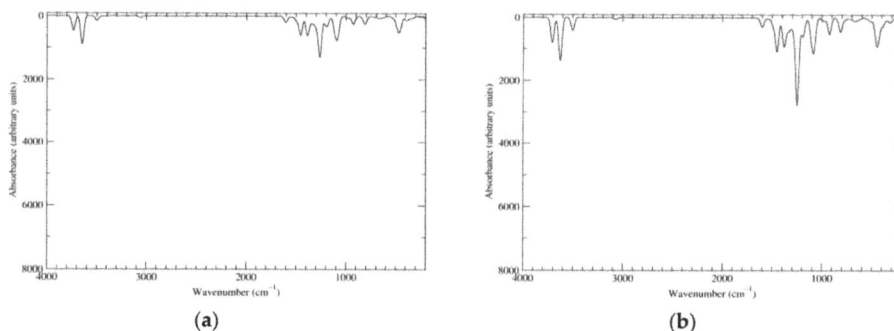

Figure 2. Computed infrared (IR) spectra of the most significant tautomers/conformers in the reduced state. (**a**) Neutral form in vacuo, a0_b0_c0_d0, S_4. (**b**) Neutral form in water, a0_b0_c0_d0, S_4.

3.2. The One-Electron Oxidation State (KP-1e)

Tables 4–6 show structures, energies and relative stabilities of the most stable tautomers/conformers in the one-electron oxidation state, i.e., those within 10 kcal mol^{-1} from the corresponding absolute minimum (complete lists are reported as Supplementary Materials).

While the full S_4 symmetry of the reduced state is unavoidably lost, the overall shape of the molecules is not drastically altered. The radical center is preferentially located at an O6 position, again in close parallel with the case of DHI. Deprotonation is predicted to occur on the O6 position of an adjacent ring, with a pK_a of around 6.2, i.e., more than one pK unit lower with respect to value computed at the same level for the deprotonation of an O6-centered DHI radical. In the same vein, the a56_b0_c0_d0 tautomer, in which both proton and hydrogen atom are removed from the same indole unit, is over 11 kcal mol^{-1} higher in energy than the absolute minimum. Clearly, the possibility to distribute charge and spin over more than one ring favors deprotonation; at the same time, electron delocalization is maximized by the involvement of adjacent indole rings.

Table 4. KP-1e, neutral form in vacuo.

Tautomer	Conformer [a]	E (Ha) [b]	H_{RRHO} (Ha) [c]	G_{RRHO} (Ha) [d]
a5_b0_c0_d0	C_1, conf1	−2049.532339 (5.5)	−2049.031222 (5.4)	−2049.129769 (5.2)
a6_b0_c0_d0	C_1, conf1	−2049.541050 (0.0)	−2049.039878 (0.0)	−2049.138067 (0.0)

In parentheses relative energies (kcal mol^{-1}) refer to the most stable form identified at the specified level. [a] For chiral structures, only one enantiomer is listed. [b] Electronic energy. [c] Enthalpy computed at 298.15 K within the rigid-rotor/harmonic-oscillator (RRHO) approximation. [d] Gibbs free energy computed at 298.15 K within the RRHO approximation.

Table 5. KP-1e, neutral form in water.

Tautomer	Conformer [a]	G_{PCM} (Ha) [b]	$H_{PCM,RRHO}$ (Ha) [c]	$G_{PCM,RRHO}$ (Ha) [d]	G_{SMD} (Ha) [e]	$G_{SMD,RRHO}$ (Ha) [f]
a5_b0_c0_d0	C_1, conf1	−2049.567367 (4.9)	−2049.067779 (4.8)	−2049.167187 (5.1)	−2049.594711 (4.7)	−2049.194531 (4.9)
a6_b0_c0_d0	C_1, conf1	−2049.575106 (0.0)	−2049.075489 (0.0)	−2049.175302 (0.0)	−2049.602176 (0.0)	−2049.202372 (0.0)

In parentheses relative energies (kcal mol^{-1}) refer to the most stable form identified at the specified level. [a] For chiral structures, only one enantiomer is listed. [b] Electronic energy including electrostatic contributions at the polarizable continuum model (PCM) level. [c] Enthalpy computed at 298.15 K within the rigid-rotor/harmonic-oscillator (RRHO) approximation. [d] Gibbs free energy computed at 298.15 K within the RRHO approximation. [e] Electronic energy including nonelectrostatic terms according to the SMD solvation model. [f] $G_{SMD,RRHO} = G_{PCM,RRHO} + G_{SMD} - G_{PCM}$.

Table 6. KP-1e, monoanionic form in water.

Tautomer	Conformer [a]	G_{PCM} (Ha) [b]	$H_{PCM,RRHO}$ (Ha)	$G_{PCM,RRHO}$ (Ha)	G_{SMD} (Ha) [c]	$G_{SMD,RRHO}$ (Ha) [d]
a16_b0_c0_d0	C_1, conf1	−2049.105711 (7.3)	-	-	-	-
a5_b0_c6_d0	C_1, conf1	−2049.103783 (8.5)	-	-	-	-
a5_b6_c0_d0	C_1, conf1	−2049.108387 (5.6)	-	-	-	-
a6_b0_c6_d0	C_1, conf1	−2049.104303 (8.2)	-	-	-	-
a6_b5_c0_d0	C_1, conf1	−2049.106920 (6.5)	-	-	-	-
a6_b6_c0_d0	C_1, conf1	−2049.117326 (0.0)	−2048.630807 (0.0)	−2048.728019 (0.0)	−2049.141803 (0.0)	−2048.752496 (0.0)

In parentheses relative energies (kcal mol^{-1}) refer to the most stable form identified at the specified level. [a] For chiral structures, only one enantiomer is listed. [b] Electronic energy including electrostatic contributions at the polarizable continuum model (PCM) level. [c] Enthalpy computed at 298.15 K within the rigid-rotor/harmonic-oscillator (RRHO) approximation. [d] Gibbs free energy computed at 298.15 K within the RRHO approximation. [e] Electronic energy including nonelectrostatic terms according to the SMD solvation model. [f] $G_{SMD,RRHO} = G_{PCM,RRHO} + G_{SMD} - G_{PCM}$.

Computed UV–Vis spectra of the most significant neutral tautomers of KP-1e are shown in Figure 3. Detailed analysis of the UV–Vis spectra of such open-shell species is difficult, and can be complicated by intrinsic limitations of the TD-DFT model. However, in the case of a6_b0_c0_d0, the predominant tautomer both in vacuo and in water, one can observe that the singly occupied molecular orbital (SOMO) is mostly localized on the "a" ring, with some delocalization on the "d" ring, and to a smaller extent also on the "b" ring. Such limited delocalization is reflected in the appearance of the spectra, in which high-wavelength, low-intensity bands corresponding mostly to transitions into the SOMO are superimposed on a trace resembling the spectrum of KP-Red.

(a)   **(b)**

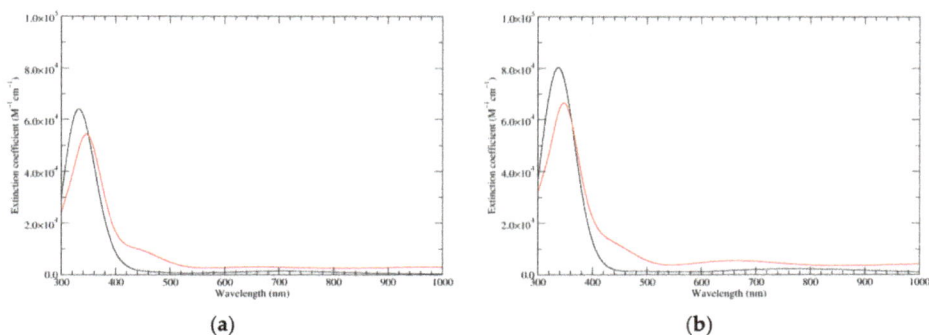

Figure 3. Computed UV–Vis spectra of the most significant tautomers/conformers in the one-electron oxidation state. (**a**) Neutral form in vacuo: black line, a5_b0_c0_d0, C_1, conf1; red line, a6_b0_c0_d0, C_1, conf1. (**b**) Neutral form in water: black line, a5_b0_c0_d0, C_1, conf1; red line, a6_b0_c0_d0 C_1, conf1.

3.3. The Two-Electron Oxidation State (KP-2e)

The most significant structures corresponding to two-electron oxidation state are listed in Tables 7–9.

Table 7. KP-2e, neutral form in vacuo.

Tautomer	Conformer [a]	E (Ha) [b]	H_{RRHO} (Ha) [c]	G_{RRHO} (Ha) [d]
a16_b0_c0_d0	C_1, conf1	−2048.922889 (4.2)	−2048.433678 (4.0)	−2048.531488 (3.1)
a6_b6_c0_d0	C_1, conf1	−2048.929568 (0.0)	−2048.440034 (0.0)	−2048.536382 (0.0)

In parentheses relative energies (kcal mol^{-1}) refer to the most stable form identified at the specified level. [a] For chiral structures, only one enantiomer is listed. [b] Electronic energy. [c] Enthalpy computed at 298.15 K within the rigid-rotor/harmonic-oscillator (RRHO) approximation. [d] Gibbs free energy computed at 298.15 K within the RRHO approximation.

Table 8. KP-2e, neutral form in water.

Tautomer	Conformer [a]	G_{PCM} (Ha) [b]	$H_{PCM,RRHO}$ (Ha) [c]	$G_{PCM,RRHO}$ (Ha) [d]	G_{SMD} (Ha) [e]	$G_{SMD,RRHO}$ (Ha) [f]
a16_b0_c0_d0	C_1, conf1	−2048.952050 (5.1)	-	-	-	-

Table 8. *Cont.*

Tautomer	Conformer [a]	G_{PCM} (Ha) [b]	$H_{PCM,RRHO}$ (Ha) [c]	$G_{PCM,RRHO}$ (Ha) [d]	G_{SMD} (Ha) [e]	$G_{SMD,RRHO}$ (Ha) [f]
a56_b0_c0_d0	C_1, conf1	−2048.947596 (7.9)	-	-	-	-
a6_b6_c0_d0	C_1, conf1	−2048.960200 (0.0)	−2048.472201 (0.0)	−2048.569202 (0.0)	−2048.984556 (0.0)	−2048.593558 (0.0)

In parentheses relative energies (kcal mol^{-1}) refer to the most stable form identified at the specified level. [a] For chiral structures, only one enantiomer is listed. [b] Electronic energy including electrostatic contributions at the polarizable continuum model (PCM) level. [c] Enthalpy computed at 298.15 K within the rigid-rotor/harmonic-oscillator (RRHO) approximation. [d] Gibbs free energy computed at 298.15 K within the RRHO approximation. [e] Electronic energy including nonelectrostatic terms according to the SMD solvation model. [f] $G_{SMD,RRHO} = G_{PCM,RRHO} + G_{SMD} - G_{PCM}$.

Table 9. KP-2e, monoanionic form in water.

Tautomer	Conformer [a]	G_{PCM} (Ha) [b]	$H_{PCM,RRHO}$ (Ha) [c]	$G_{PCM,RRHO}$ (Ha) [d]	G_{SMD} (Ha) [e]	$G_{SMD,RRHO}$ (Ha) [f]
a15_b6_c0_d0	C_1, conf1	−2048.487906 (8.3)	-	-	-	-
a16_b6_c0_d0	C_1, conf1	−2048.501057 (0.0)	−2048.026967 (0.0)	−2048.123602 (0.0)	−2048.520660 (0.0)	−2048.143205 (0.5)
a56_b6_c0_d0	C_1, conf1	−2048.494742 (4.0)	−2048.020572 (4.0)	−2048.118108 (3.4)	−2048.520680 (0.0)	−2048.144046 (0.0)
a6_b16_c0_d0	C_1, conf1	−2048.497643 (2.1)	−2048.023707 (2.0)	−2048.120613 (1.9)	−2048.517642 (1.9)	−2048.140612 (2.2)
a6_b56_c0_d0	C_1, conf1	−2048.487298 (8.6)	-	-	-	-

In parentheses relative energies (kcal mol^{-1}) refer to the most stable form identified at the specified level. [a] For chiral structures, only one enantiomer is listed. [b] Electronic energy including electrostatic contributions at the polarizable continuum model (PCM) level. [c] Enthalpy computed at 298.15 K within the rigid-rotor/harmonic-oscillator (RRHO) approximation. [d] Gibbs free energy computed at 298.15 K within the RRHO approximation. [e] Electronic energy including nonelectrostatic terms according to the SMD solvation model. [f] $G_{SMD,RRHO} = G_{PCM,RRHO} + G_{SMD} - G_{PCM}$.

Already at this oxidation stage, and even restricting the attention to closed-shall forms, the number of alternative tautomers starts to increase significantly. There are three ways to localize the oxidation on a single ring (Figure 4, structures f–h), giving rise to a15_b0_c0_d0, a16_b0_c0_d0, or a56_b0_c0_d0 tautomers; moreover, four tautomers featuring an inter-ring double bond (a1_b1_c0_d0, a1_b6_c0_d0, a6_b1_c0_d0, a6_b6_c0_d0) can be generated by combined two one-electron oxidized units (Figure 4, structures b–e). However, most of these structures are energetically unviable. In practice, the a6_b6_c0_d0 tautomer, featuring a quinone methide structure extended over two adjacent indole rings, is predicted to prevail largely both in vacuo and in water. Thus, the tendency to maximize electron delocalization by involving adjacent indole rings is confirmed in the two-electron oxidation state. The double bond character of the C2 (ring a)-C7 (ring b) connectivity is reflected in the bond length of 1.396 Å (in vacuo; to compare with the other interring bonds of 1.542, 1.457 and 1.447 Å).

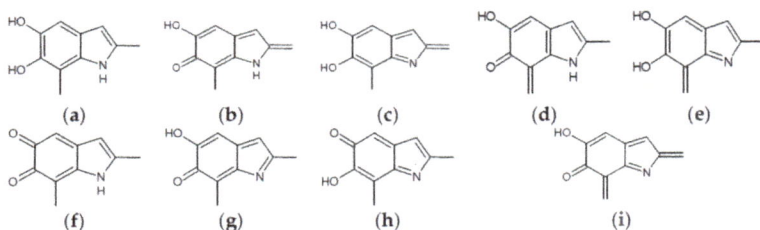

Figure 4. Building blocks for generation of staring structures of closed-shell tautomers of Kaxiras's porphyrin. (**a**) Fully reduced units. (**b–e**) One-electron oxidized units. (**f–i**) Two-electron oxidized units.

Deprotonation of the system is predicted to occur from O5–H of the quinone methide ring, with a pK_a around 6.1, and slightly less probably from the NH (pK_a ca. 6.4).

The two significant tautomeric forms of KP-2e, a16_b0_c0_d0 and a6_b6_c0_d0, display quite similar spectra (Figure 5). Examination of the underlying orbital structure shows that in both cases the high-wavelength region corresponds to an ordered series of several transitions starting from HOMO, HOMO−1, HOMO−2, ... and ending in the LUMO. This latter orbital is rather similar in the two species; however, correspondence of the other orbitals involved is only partial, and moreover some inversions in energy ordering occur. The effect of solvation is also predicted to be comparable in the two cases.

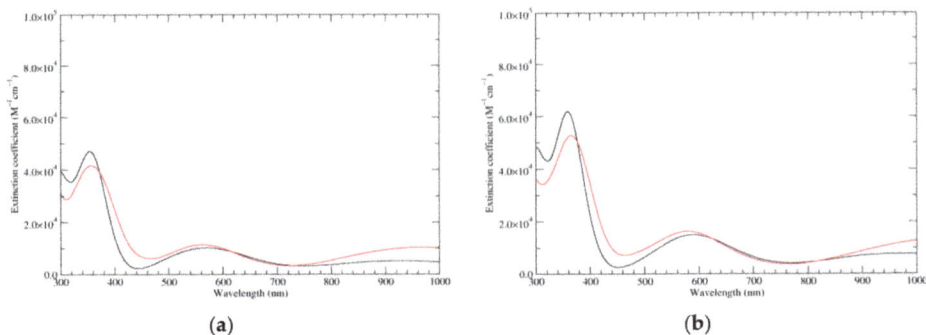

Figure 5. Computed UV–Vis spectra of the most significant tautomers/conformers in the two-electrons oxidation state. (**a**) Neutral form in vacuo: black line, a16_b0_c0_d0, C_1, conf1; red line, a6_b6_c0_d0, C_1, conf1. (**b**) Neutral form in water: black line, a16_b0_c0_d0, C_1, conf1; red line, a6_b6_c0_d0, C_1, conf1.

Figure 6 shows the computed IR spectra of the two main tautomers. With respect to the spectrum of KP-Red, the presence of new features in the carbonyl region is apparent; moreover, these are rather different for the two tautomers. Visual examination of the individual normal modes shows that for a16_b0_c0_d0, a quinone methide localized on ring a, the peak at 1655 cm^{-1} (in vacuo) correspond in essence to an antiphase stretching of the C4–C5 double bond and of the C6–O carbonyl, accompanied however by sizeable deformations of the C5–O–H moiety; the peak at 1562 cm^{-1} can be described as a C6–O carbonyl stretching. In the case of a6_b6_c0_d0, a quinone methide extended over rings a and b, the peak at about 1609 cm^{-1} convolutes two main transitions, separated by some 20 cm^{-1}. The high-wavenumber component corresponds to in-phase stretching of the C6–O (ring a) and C6–O (ring b) carbonyls, which are conjugated through a system of three double bonds; the low-wavenumber component (of higher intensity) represents the corresponding antiphase stretching.

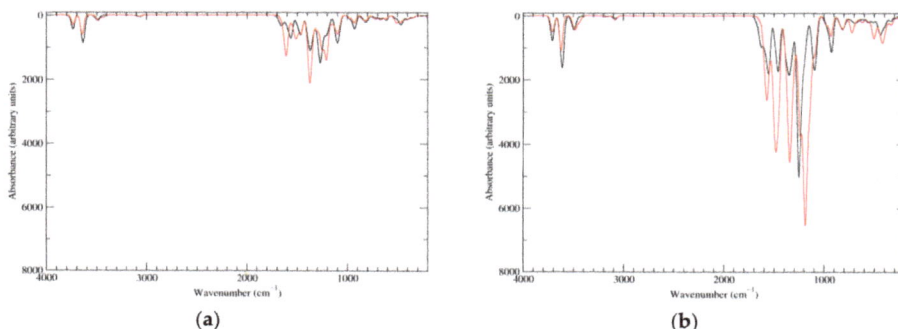

Figure 6. Computed IR spectra of the most significant tautomers/conformers in the two-electrons oxidation state. (**a**) Neutral form in vacuo: black line, a16_b0_c0_d0, C_1, conf1; red line, a6_b6_c0_d0, C_1, conf1. (**b**) Neutral form in water: black line, a16_b0_c0_d0, C_1, conf1; red line, a6_b6_c0_d0, C_1, conf1.

3.4. The Four-Electron Oxidation State (KP-4e)

At this stage, structural complexity is remarkable: 49 closed-shell, neutral tautomers were generated (each in at least two different staring conformation) and fully geometry-optimized. As a rule, each of these structures features many nonequivalent deprotonation sites: an exploration of the monoanionic forms would therefore be extremely challenging, and was not attempted in this case, nor for any of the higher oxidation states.

At this oxidation level, the formation of π-conjugated cyclic systems extending over the four indole units becomes possible, and is obtained in the following tautomers: a1_b1_c1_d1 (28 π-electrons); a1_b1_c1_d6 (26 π-electrons); a1_b1_c6_d6 and a1_b6_c1_d6 (24 π-electrons); a1_b6_c6_d6 (22 π-electrons); a6_b6_c6_d6 (20 π-electrons). The relative energies of these species vary over a large range (see Supplementary Materials); even restricting the attention to species that comply with the $4n + 2$ rule, a1_b1_c1_d6 is 67.6 kcal mol^{-1} above the absolute minimum (in vacuo). Conversely, a6_b6_c6_d6, an antiaromatic system, has a relative energy of just 3.8 kcal mol^{-1}. Overall, it appears that aromatic character by itself is not a major predictor of tautomer stability.

Among all neutral tautomers, six were within 10 kcal mol^{-1} from the absolute minimum in vacuo (and respectively eight in water). These are listed in Tables 10 and 11.

As a matter of fact, a subset of structures is quite close in energy. Tautomers a16_b6_c6_d0 and a6_b6_c16_d0 combine the extended quinone methide structure described for the two-electron oxidation state, with a single two-electron oxidized ring in a localized quinone methide tautomeric arrangement. A new motif appears in a6_b16_c6_d0, in which a two-electron oxidized ring (Figure 4, structure i) is interposed by means of double bonds between two one-electron oxidized building blocks, with formation of an extended system formally delocalized over three indole units. The character of the interring double bonds is confirmed by the lengths of 1.395 Å for C2 (ring a)–C7 (ring b), and 1.400 Å for C2 (ring b)–C7 (ring c) (in vacuo; the interring single bonds have instead lengths of 1.450 and 1.442 Å). In water, the pool of predominant structures is enriched by still another tautomer, a6_b6_c6_d6, formed by a duplication of the two-ring, two-electron building block described above. As a matter of fact, even starting from higher S_4 or C_4 symmetries, the system spontaneously reverts to the C_2 symmetry. Bond lengths reflect the alternation of interring double (1.390 Å, in vacuo) and single bonds (1.448 Å).

Table 10. KP-4e, neutral form in vacuo.

Tautomer	Conformer [a]	E (Ha) [b]	H_{RRHO} (Ha) [c]	G_{RRHO} (Ha) [d]
a16_b0_c16_d0	C_1, conf1	−2047.692169 (6.3)		
	C_2, conf1	−2047.692054 (6.3)	-	-
a16_b6_c6_d0	C_1, conf1	−2047.700139 (1.3)	−2047.235705 (1.2)	−2047.331297 (1.1)
a1_b6_c6_d6	C_1, conf1	−2047.691396 (6.8)	-	-
a6_b16_c6_d0	C_1, conf1	−2047.702153 (0.0)	−2047.237674 (0.0)	−2047.333050 (0.0)
a6_b6_c16_d0	C_1, conf1	−2047.699471 (1.7)	−2047.234863 (1.8)	−2047.330241 (1.8)

26

Table 10. *Cont.*

Tautomer	Conformer [a]	E (Ha) [b]	H_{RRHO} (Ha) [c]	G_{RRHO} (Ha) [d]
	C_1, conf1	−2047.696161 (3.8)	−2047.231453 (3.9)	−2047.325734 (4.6)
a6_b6_c6_d6	C_2, conf1	−2047.696144 (3.8)	−2047.231368 (4.0)	−2047.324833 (5.2)

In parentheses relative energies (kcal mol^{-1}) refer to the most stable form identified at the specified level. [a] For chiral structures, only one enantiomer is listed. [b] Electronic energy. [c] Enthalpy computed at 298.15 K within the rigid-rotor/harmonic-oscillator (RRHO) approximation. [d] Gibbs free energy computed at 298.15 K within the RRHO approximation.

Table 11. KP-4e, neutral form in water.

Tautomer	Conformer [a]	G_{PCM} (Ha) [b]	$H_{PCM,RRHO}$ (Ha) [c]	$G_{PCM,RRHO}$ (Ha) [d]	G_{SMD} (Ha) [e]	$G_{SMD,RRHO}$ (Ha) [f]
a16_b0_c16_d0	C_1, conf1	−2047.716730 (5.1)	-	-	-	-
	C_2, conf1	−2047.716589 (5.2)	-	-	-	-
a16_b6_c6_d0	C_1, conf1	−2047.722869 (1.3)	−2047.260061 (1.3)	−2047.355968 (1.4)	−2047.741845 (1.6)	−2047.374944 (1.5)
a1_b6_c6_d6	C_1, conf1	−2047.714953 (6.2)	-	-	-	-

Biomimetics **2017**, *2*, 21

Table 11. *Cont.*

Tautomer	Conformer [a]	G_{PCM} (Ha) [b]	$H_{PCM,RRHO}$ (Ha) [c]	$G_{PCM,RRHO}$ (Ha) [d]	G_{SMD} (Ha) [e]	$G_{SMD,RRHO}$ (Ha) [f]
a56_b6_c6_d0	C_1, conf1	−2047.714698 (6.4)	-	-	-	-
a6_b16_c6_d0	C_1, conf1	−2047.724901 (0.0)	−2047.262199 (0.0)	−2047.358235 (0.0)	−2047.743929 (0.2)	−2047.377263 (0.0)
a6_b6_c16_d0	C_1, conf1	−2047.722818 (1.3)	−2047.259858 (1.5)	−2047.355768 (1.5)	−2047.742349 (1.2)	−2047.375299 (1.2)
a6_b6_c56_d0	C_1, conf1	−2047.715313 (6.0)	-	-	-	-
a6_b6_c6_d6	C_1, conf1	−2047.723719 (0.7)	−2047.260186 (1.3)	−2047.354624 (2.3)	−2047.744321 (0.0)	−2047.375226 (1.3)
	C_2, conf1	−2047.723546 (0.9)	−2047.260328 (1.2)	−2047.354611 (2.3)	−2047.743800 (0.3)	−2047.374865 (1.5)

In parentheses relative energies (kcal mol^{-1}) refer to the most stable form identified at the specified level. [a] For chiral structures, only one enantiomer is listed. [b] Electronic energy including electrostatic contributions at the polarizable continuum model (PCM) level. [c] Enthalpy computed at 298.15 K within the rigid-rotor/harmonic-oscillator (RRHO) approximation. [d] Gibbs free energy computed at 298.15 K within the RRHO approximation. [e] Electronic energy including nonelectrostatic terms according to the SMD solvation model. [f] $G_{SMD,RRHO} = G_{PCM,RRHO} + G_{SMD} - G_{PCM}$.

The significant degree of structural variation of the predominant tautomers is reflected in the computed UV–Vis spectra (Figure 7). Concerning the high-wavelength region, however, some general trends can be identified. First, the seeming broadband absorptions can be traced back to the partial overlap of a large number of weakly allowed transitions. Moreover, the highest wavelength transition is invariably HOMO–LUMO in character. The transitions that follow display a rather strong orbital mixing: a reasonably clear series of HOMO−1, HOMO−2, . . . –LUMO transitions is interrupted by several transitions into the LUMO+1 orbital, the exact ordering depending on the specific tautomer.

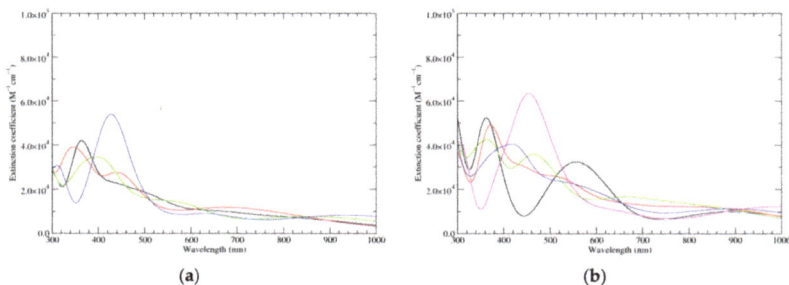

(a)   **(b)**

Figure 7. Computed UV–Vis spectra of the most significant tautomers/conformers in the four-electrons oxidation state. (**a**) Neutral form in vacuo: black line, a16_b6_c6_d0, C_1, conf1; red line, a6_b16_c6_d0, C_1, conf1; green line, a6_b6_c16_d0, C_1, conf1; blue line, a6_b6_c6_d6, C_2, conf1. (**b**) Neutral form in water: black line, a16_b0_c16_d0, C_2, conf1; red line, a16_b6_c6_d0, C_1, conf1; green line, a6_b16_c6_d0, C_1, conf1; blue line, a6_b6_c16_d0, C_1, conf1; magenta line, a6_b6_c6_d6, C_2, conf1.

Figure 8 reports the computed IR spectra of the main tautomers. Concentrating the analysis on the carbonyl/double bond stretching region, one can observe that a16_b6_c6_d0 and a6_b6_c16_d0 display similar features, with main peaks at about 1620 and 1555 cm^{-1} (in vacuo): this reflects structural similarity, both tautomers consisting of localized quinone methide, and an extended quinone methide involving the two following (a16_b6_c6_d0) or preceding (a6_b6_c16_d0) rings. These features are shifted to ca. 1645 and 1527 cm^{-1} in a6_b16_c6_d0, in which inter-ring double bonds form a conjugated system extended over three rings. Inspection of the underlying normal modes shows that the high-energy band results from overlap of several transitions; however, the two main components correspond to motions centered in ring b. The higher-energy component, an antiphase stretching of the C4–C5 double bond and of the C6–O carbonyl, with concomitant deformations of the C5–O–H moiety, is quite similar to that described for the isolated quinone methide (a16_b0_c0_d0 at the two-electron oxidation stage). The lower-energy, more intense contribution involves antiphase stretching of the C6–O (ring a) and C6–O (ring c) carbonyls. Finally, in a6_b6_c6_d6, in which two equivalent two-ring conjugated systems exist, main peaks are at 1616 and 1523 cm^{-1}: although several individual transitions are hidden under the 1616 cm^{-1} peak, the two most intense ones can be described as in-phase (higher energy) and antiphase (lower energy) stretching of the C6–O (ring a) and C6–O (ring b) carbonyls; of course, corresponding motions occur on the symmetry-related rings b and d, with relative phases dictated by the B symmetry species of both modes.

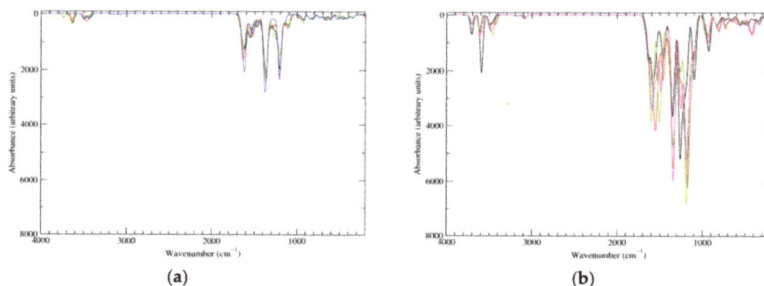

(a)   **(b)**

Figure 8. Computed IR spectra of the most significant tautomers/conformers in the four-electrons oxidation state. (**a**) Neutral form in vacuo: black line, a16_b6_c6_d0, C_1, conf1; red line, a6_b16_c6_d0, C_1, conf1; green line, a6_b6_c16_d0, C_1, conf1; blue line, a6_b6_c6_d6, C_2, conf1. (**b**) Neutral form in water: black line, a16_b0_c16_d0, C_2, conf1; red line, a16_b6_c6_d0, C_1, conf1; green line, a6_b16_c6_d0, C_1, conf1; blue line, a6_b6_c16_d0, C_1, conf1; magenta line, a6_b6_c6_d6, C_2, conf1.

3.5. The Six-Electron Oxidation State (KP-6e)

Seventy-four tautomers of KP-6e were examined. Extended conjugated cyclic systems are possible in the following cases: a16_b16_c1_d1 and a16_b1_c16_d1 (22 π-electrons); a16_b16_c1_d6, a16_b16_c6_d1 and a16_b1_c16_d6 (20 π-electrons); a16_b16_c6_d6 and a16_b6_c16_d6 (18 π-electrons). The 22 π-electrons systems, although predictably aromatic, are high in energy (ca. 100 kcal mol^{-1} above the minimum; see Supplementary Materials), even more than the 20 π-electrons tautomers (relative energies of ca. 45 kcal mol^{-1}). However, the two 18 π-electrons tautomers are indeed the two most stable isomers identified at this oxidation stage. Of them, a16_b6_c16_d6 is strongly favored both in vacuo and in water; a16_b16_c6_d6 is separated from the absolute minimum by ca. 8–9 kcal mol^{-1} (Tables 12 and 13).

Table 12. KP-6e, neutral form in vacuo.

Tautomer	Conformer [a]	E (Ha) [b]	H_{RRHO} (Ha) [c]	G_{RRHO} (Ha) [d]
a16_b16_c6_d6	C_1, conf1	−2046.467842 (9.1)	-	-
	C_1, conf1	−2046.482284 (0.0)	-	-
a16_b6_c16_d6	C_2, conf1	−2046.482242 (0.0)	−2046.043017 (0.0)	−2046.138676 (0.0)

In parentheses relative energies (kcal mol^{-1}) refer to the most stable form identified at the specified level. [a] For chiral structures, only one enantiomer is listed. [b] Electronic energy. [c] Enthalpy computed at 298.15 K within the rigid-rotor/harmonic-oscillator (RRHO) approximation. [d] Gibbs free energy computed at 298.15 K within the RRHO approximation.

Table 13. KP-6e, neutral form in water.

Tautomer	Conformer [a]	G_{PCM} (Ha) [b]	$H_{PCM,RRHO}$ (Ha) [c]	$G_{PCM,RRHO}$ (Ha) [d]	G_{SMD} (Ha) [e]	$G_{SMD,RRHO}$ (Ha) [f]
a16_b16_c6_d6	C_1, conf1	−2046.484846 (7.9)	-	-	-	-
	C_1, conf1	−2046.497644 (−0.1)	-	-	-	-
a16_b6_c16_d6	C_2, conf1	−2046.497479 (0.0)	−2046.059837 (0.0)	−2046.154721 (0.0)	−2046.509330 (0.0)	−2046.166572 (0.0)

In parentheses relative energies (kcal mol^{-1}) refer to the most stable form identified at the specified level. [a] For chiral structures, only one enantiomer is listed. [b] Electronic energy including electrostatic contributions at the polarizable continuum model (PCM) level. [c] Enthalpy computed at 298.15 K within the rigid-rotor/harmonic-oscillator (RRHO) approximation. [d] Gibbs free energy computed at 298.15 K within the RRHO approximation. [e] Electronic energy including nonelectrostatic terms according to the SMD solvation model. [f] $G_{SMD,RRHO} = G_{PCM,RRHO} + G_{SMD} - G_{PCM}$.

This situation may reflect in part an optimal arrangement of NH⋯N hydrogen bonds in the core of the porphyrin ring; however, one should emphasize again the aromatic character of a16_b6_c16_d6 (nine conjugated double bonds in an uninterrupted cyclic arrangement; 18 π-electron system). The molecule features C_2 symmetry: this is brought out clearly by a representation in terms of resonance structures (Scheme 2). The partial double bond character of the interring bonds is reflected in short lengths and in a moderate alternation (in vacuo, C2 (ring a))–C7 (ring b) and C2 (ring c))–C7 (ring d) have a length of 1.417 Å, while C2 (ring b))–C7 (ring c) and C2 (ring d))–C7 (ring a) are 1.409 Å long). The structure is much closer to planarity than in the lower oxidation states, reaching an approximate C_{2h} symmetry.

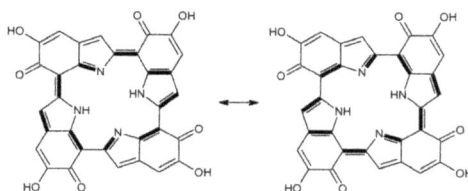

Scheme 2. a16_b6_c16_d6 as a hybrid between two equivalent contributing structures. The pattern of conjugated double bonds forming an 18-electron aromatic system is highlighted in bold.

Taken together, the two factors hinted above (favorable hydrogen bond pattern and extended aromatic π-electron system) confer a remarkable stability to the structure, both with respect to other tautomeric arrangements at the same oxidation level, and in comparison to other oxidation states (see below).

Figure 9 displays the computed UV–Vis spectra of a16_b6_c16_d6. A remarkable extinction coefficient is predicted for the band at ca. 420 nm, higher than in any of the other oxidation states. On account of the symmetry of the species, several electronic transitions are forbidden; many more have low intensity, including notably the highest wavelength transition (1080 nm, $f = 0.03$, mostly HOMO–LUMO in character). In the region above 500 nm, the only other transitions with non-null oscillator strength are the following (in vacuo): 988 nm ($f = 0.06$, HOMO–LUMO+1); 729 nm ($f = 0.02$, HOMO−1–LUMO+1); 678 nm ($f = 0.09$, HOMO−1–LUMO); 569 nm ($f = 0.05$, HOMO−4–LUMO); and 520 nm ($f = 0.04$, HOMO−4–LUMO+1). Orbital mixing is significant: in particular, whenever a given excited state contains a LUMO component, the LUMO+1 is invariably strongly admixed, and vice versa. This behavior is strongly reminiscent of the orbital structure of classical porphyrins, in which, depending on the molecular symmetry, LUMO and LUMO+1 are either exactly or very closely degenerate, and enter as a pair in the composition of the spectroscopically relevant excited states. To be sure, the computed spectrum of a16_b6_c16_d6 displays typical signatures of a porphyrin: the intense peak around 420 nm can be recognized as a Soret (B) band, while the low-intensity peaks at higher wavelength can be classified, at least in part, as Q bands [35]. The 420 nm band corresponds to two distinct transitions, at 431 and 418 nm ($f = 0.56$ and 0.79, respectively): this splitting can be traced back to the lower symmetry of a16_b6_c16_d6 with respect to typical metalloporphyrin complexes endowed with more or less exact D_{4h} symmetry—as a matter of fact, a similar situation is observed in the less symmetric porphyrin free base (D_{2h} in the parent heterocycle).

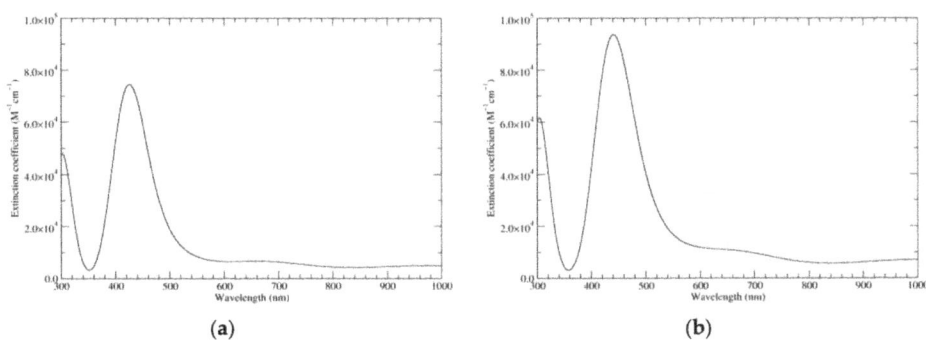

Figure 9. Computed UV–Vis spectra of the most significant tautomers/conformers in the six-electron oxidation state. (**a**) Neutral form in vacuo, a16_b6_c16_d6, C_2, conf1. (**b**) Neutral form in water, a16_b6_c16_d6, C_2, conf1.

As it is well known [36], vibronic features contribute significantly to the appearance of experimental UV–Vis spectra of porphyrins. It is, therefore, expedient to recall that the computational model we adopted does not include any such effects, not even a reproduction of the vibrational structure of the allowed electronic transitions. Much progress has been done recently towards automation and optimization of such computations [37], so that they are becoming feasible even for large molecules in condensed phases; however, they remain by no means trivial. Certainly, the possible involvement of such physical phenomena should be kept in mind if experimental data on the individual KPs become available.

In order to better characterize the nature of the underlying electronic transitions, TD-DFT calculations (Table 14) were repeated on a16_b6_c16_d6 in its C_{2h}-symmetric structure, which corresponds to a first-order saddle point. Use of a more symmetrical structure facilitates classification of the orbitals and of the electronic states in terms of symmetry species; moreover, in the present instance changes in transition wavelengths and intensities are negligible (a more complete version of Table 14, including orbital symmetry labels, is available as Supplementary Materials, alongside corresponding data for the C_2-symmetric structure).

Table 14. Excitations underlying the UV–Vis spectrum of a16_b6_c16_d6, computed at the C_{2h} geometry in vacuo.

Symmetry	Main CI Contributions (Coefficients) [a]	λ (nm) (f) [b]
^1B_u	HOMO−1 → LUMO (−0.12); HOMO → LUMO (0.60); HOMO → LUMO+1 (−0.32)	1080.6 (0.03)
^1B_u	HOMO−4 → LUMO+1 (0.14); HOMO−1 → LUMO (0.16); HOMO−1 → LUMO+1 (−0.15); HOMO → LUMO (0.32); HOMO → LUMO+1 (0.57)	988.5 (0.06)
^1A_g	HOMO−3 → LUMO (0.12); HOMO−2 → LUMO (−0.26); HOMO−2 → LUMO+1 (0.41); HOMO−1 → LUMO+2 (−0.11); HOMO → LUMO+2 (0.49)	779.9 (0.00)
^1B_u	HOMO−2 → LUMO+2 (−0.11); HOMO−1 → LUMO (0.34); HOMO−1 → LUMO+1 (0.59)	729.8 (0.02)
^1A_g	HOMO−2 → LUMO (0.64); HOMO−2 → LUMO+1 (0.22); HOMO → LUMO+2 (0.12)	714.7 (0.00)
^1A_g	HOMO−3 → LUMO (−0.35); HOMO−2 → LUMO+1 (0.46); HOMO−1 → LUMO+2 (−0.14); HOMO → LUMO+2 (−0.35)	679.8 (0.00)
^1B_u	HOMO−4 → LUMO (0.30); HOMO−1 → LUMO (0.51); HOMO−1 → LUMO+1 (−0.31); HOMO → LUMO+1 (−0.20)	678.2 (0.09)
^1A_g	HOMO−3 → LUMO (0.57); HOMO−2 → LUMO+1 (0.20); HOMO → LUMO+2 (−0.34)	638.0 (0.00)
^1A_g	HOMO−3 → LUMO (−0.12); HOMO−3 → LUMO+1 (0.67); HOMO−1 → LUMO+2 (0.14)	579.1 (0.00)
^1B_u	HOMO−5 → LUMO (0.20); HOMO−5 → LUMO+1 (0.17); HOMO−4 → LUMO (0.54); HOMO−3 → LUMO+2 (0.12); HOMO−2 → LUMO+2 (0.24); HOMO−1 → LUMO (−0.21)	568.8 (0.05)
^1B_u	HOMO−5 → LUMO (−0.36); HOMO−5 → LUMO+1 (0.12); HOMO−4 → LUMO+1 (0.49); HOMO−2 → LUMO+2 (0.29); HOMO−1 → LUMO+1 (0.10)	519.9 (0.04)
^1A_g	HOMO−3 → LUMO+1 (−0.16); HOMO−2 → LUMO+1 (0.15); HOMO−1 → LUMO+2 (0.66)	506.4 (0.00)
^1B_u	HOMO−5 → LUMO+1 (−0.36); HOMO−4 → LUMO (−0.15); HOMO−4 → LUMO+1 (−0.19); HOMO−2 → LUMO+2 (0.54)	487.2 (0.10)
^1B_u	HOMO−5 → LUMO (−0.42); HOMO−4 → LUMO+1 (−0.20); HOMO−3 → LUMO+2 (0.51); HOMO−1 → LUMO (−0.10)	458.4 (0.06)
^1B_u	HOMO−5 → LUMO (0.30); HOMO−5 → LUMO+1 (0.35); HOMO−4 → LUMO (−0.27); HOMO−4 → LUMO+1 (0.12); HOMO−3 → LUMO+2 (0.36); HOMO−2 → LUMO+2 (0.11); HOMO−1 → LUMO (0.15); HOMO → LUMO+1 (−0.13); HOMO → LUMO+4 (0.13)	430.7 (0.56)
^1A_g	HOMO−5 → LUMO+2 (0.12); HOMO−4 → LUMO+2 (0.68); HOMO−3 → LUMO (−0.11)	418.5 (0.00)
^1B_u	HOMO−5 → LUMO (−0.19); HOMO−5 → LUMO+1 (0.43); HOMO−4 → LUMO+1 (−0.35); HOMO−3 → LUMO+2 (−0.26); HOMO−2 → LUMO+2 (0.15); HOMO → LUMO (0.16); HOMO → LUMO+4 (0.11); HOMO → LUMO+5 (−0.12)	418.0 (0.79)
^1B_g	HOMO−10 → LUMO (0.24); HOMO−10 → LUMO+1 (0.20); HOMO−9 → LUMO+2 (−0.25); HOMO−8 → LUMO+2 (−0.10); HOMO−6 → LUMO (0.54); HOMO−6 → LUMO+1 (0.15)	407.1 (0.00)
^1A_u	HOMO−10 → LUMO+2 (−0.12); HOMO−9 → LUMO (0.49); HOMO−9 → LUMO+1 (0.14); HOMO−8 → LUMO (0.23); HOMO−8 → LUMO+1 (0.27); HOMO−6 → LUMO+2 (−0.28)	404.4 (0.00)

[a] HOMO: Highest energy occupied molecular orbital; LUMO: Lowest energy unoccupied molecular orbital. [b] Only transitions with λ > 400 nm are listed.

Visual inspection of the molecular orbitals (available as Supplementary Materials) confirms that the LUMO and LUMO+1 b_g orbitals correspond rather well to the degenerate pair of unoccupied e_g orbitals that are involved in the UV–Vis transitions of typical metalloporphyrins. However, the situation becomes more intricate when the occupied orbitals are examined: in the present case, HOMO, HOMO−1, HOMO−4 and HOMO−5 all belong to the same A_u symmetry species, and mix heavily in the transitions leading to the "Soret" band. This complicates a straightforward interpretation in terms of the classical Gouterman "four-orbital" model [38].

Simulated IR spectra of a16_b6_c16_d6 are shown in Figure 10. Main peaks in the carbonyl/double bond stretching region are predicted at about 1655 and 1490 cm^{-1} (in vacuo). Four transitions of comparable intensity underlie the 1655 cm^{-1} peak. The highest energy component, at 1670 cm^{-1}, corresponds to the usual antiphase stretching of the C4–C5 double bond and of the C6–O carbonyl (with concomitant deformations of the C5–O–H moiety), centered on the "b" indole unit (and of course on the symmetry-related "d" indole unit: all four modes belong to the B symmetry species). The transition at 1668 cm^{-1} is similar, but centered on ring a. Conversely, the 1644 cm^{-1} transition corresponds to an in-phase stretching of C4–C5 and C6–O of ring a, with a counterpart at 1639 cm^{-1} for the analogous motion on ring b. The peak at ca. 1490 cm^{-1} is composed by two main transitions, which involve notably interring bonds; in particular, stretching contributions of C7 (ring b))–C7a (ring b) and of C2 (ring b)–C7 (ring c) are prominent in the high-energy (1509 cm^{-1}) component; C2 (ring b)–C7 (ring c) is also well present in the low-energy (1482 cm^{-1}) counterpart, which however involves a substantial contribution from in-plane bending of the N–H. Again, motions in the symmetry-related rings are dictated by the B symmetry species of both modes.

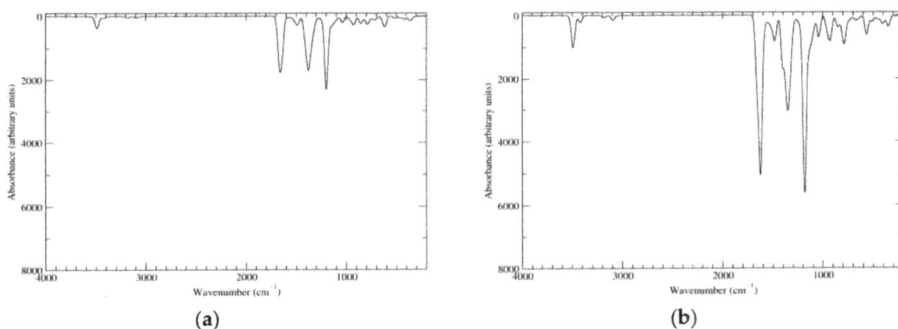

Figure 10. Computed IR spectra of the most significant tautomers/conformers in the six-electrons oxidation state. (**a**) Neutral form in vacuo, a16_b6_c16_d6, C_2, conf1. (**b**) Neutral form in water, a16_b6_c16_d6, C_2, conf1.

3.6. The Eight-Electron Oxidation State (KP-8e)

The highest oxidation state that has been examined in this study corresponds to removing eight electrons and eight protons from the fully reduced tetramer. At such high oxidation states, the number of possible tautomers starts to decrease: 24 tautomers were generated, and several conformers were optimized as usual. A fully π-conjugated cyclic system extending over the four rings is present in a16_b16_c16_d16; however, with 16 π-electrons the system is predicted to be antiaromatic, and at any rate it is high in energy (ca. 36 kcal mol^{-1} above the minimum). Energetically significant species are collected in Tables 15 and 16.

Table 15. KP-8e, neutral form in vacuo.

Tautomer	Conformer [a]	E (Ha) [b]	H_{RRHO} (Ha) [c]	G_{RRHO} (Ha) [d]
	C_1, conf1	−2045.192843 (−0.1)	-	-
	C_1, conf2	−2045.192633 (0.0)	-	-
a16_b56_c16_d56	C_2, conf1	−2045.192633 (0.0)	−2044.778733 (0.0)	−2044.873783 (0.0)

In parentheses relative energies (kcal mol^{-1}) refer to the most stable form identified at the specified level. [a] For chiral structures, only one enantiomer is listed. [b] Electronic energy. [c] Enthalpy computed at 298.15 K within the rigid-rotor/harmonic-oscillator (RRHO) approximation. [d] Gibbs free energy computed at 298.15 K within the RRHO approximation.

Table 16. KP-8e, neutral form in water.

Tautomer	Conformer [a]	G_{PCM} (Ha) [b]	$H_{PCM,RRHO}$ (Ha) [c]	$G_{PCM,RRHO}$ (Ha) [d]	G_{SMD} (Ha) [e]	$G_{SMD,RRHO}$ (Ha) [f]
a16_b16_c56_d56	C_1, conf1	−2045.205247 (9.5)	-	-	-	-
a16_b56_c16_d56	C_1, conf1	−2045.220522 (0.0)	-	-	-	-
	C_2, conf1	−2045.220445 (0.0)	−2044.807661 (0.0)	−2044.902116 (0.0)	−2045.234643 (0.0)	−2044.916314 (0.0)
a16_b56_c56_d56	C_1, conf1	−2045.208959 (7.2)	-	-	-	-

In parentheses relative energies (kcal mol^{-1}) refer to the most stable form identified at the specified level. [a] For chiral structures, only one enantiomer is listed. [b] Electronic energy including electrostatic contributions at the polarizable continuum model (PCM) level. [c] Enthalpy computed at 298.15 K within the rigid-rotor/harmonic-oscillator (RRHO) approximation. [d] Gibbs free energy computed at 298.15 K within the RRHO approximation. [e] Electronic energy including nonelectrostatic terms according to the SMD solvation model. [f] $G_{SMD,RRHO} = G_{PCM,RRHO} + G_{SMD} - G_{PCM}$.

As it turns out, a single form, namely a16_b56_c16_d56, predominates largely both in vacuo and in water. Here all rings are at the same two-electron oxidation state, and localized quinone methide units alternate with localized *o*-quinones so as to create an optimal pattern of transannular NH···N hydrogen bonds. However, porphyrin-type aromaticity is lost at this stage. The structure is C_2-symmetric, but less close to planarity that the main tautomer of KP-6e; interring bonds are longer (1.419 and 1.434 Å, in vacuo).

The electronic spectrum (Figure 11) resembles roughly that of KP-6e; however, the main peak is shifted hypso- and hypochromically. In the low-energy region, transitions of significant intensity (in vacuo) are predicted at 985 ($f = 0.10$; mostly HOMO–LUMO), 788 ($f = 0.18$; main components HOMO–LUMO and HOMO−1–LUMO+1), and 586 nm ($f = 0.05$; HOMO−1–LUMO+2).

The IR spectrum of a16_b56_c16_d56 (Figure 12) shows in the carbonyl region a major peak (1690 cm^{-1} in vacuo, 1657 cm^{-1} in water). In terms of normal modes, this corresponds to a stretching motion of the *o*-quinone carbonyl groups, which is in phase for what concerns the two carbonyls on the same ring, and antiphase for what concerns the symmetry-related b and d rings (B symmetry species). The mode in which the two carbonyls on a same ring move in antiphase is slightly higher in energy, but has much lower intensity. Modes centered on the quinone methide rings (a and c) account for the low-wavenumber shoulder of the peak. Major contributing transitions can also be identified for the bands at 1575–1535 cm^{-1}. The principal contribution to the high-energy band comes from a mode involving in-phase stretching of C7 (ring a)–C7a (ring a) and of C2 (ring a)–C7 (ring b); bending of the NH group is prominent (as usual, motions of the other rings account for a B symmetry species). Analogous stretching motions of C7 (ring a)–C7a (ring a), C7 (ring b)–C7a (ring b), C2 (ring b)–C7 (ring c) is involved in the low-energy transition; however, the NH bending contribution is much smaller in this case.

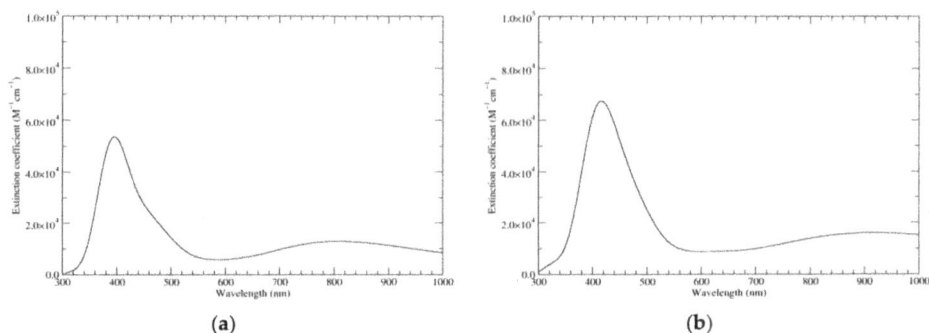

Figure 11. Computed UV–Vis spectra of the most significant tautomers/conformers in the eight-electrons oxidation state. (**a**) Neutral form in vacuo, a16_b56_c16_d56, C_2, conf1. (**b**) Neutral form in water, a16_b56_c16_d56, C_2, conf1.

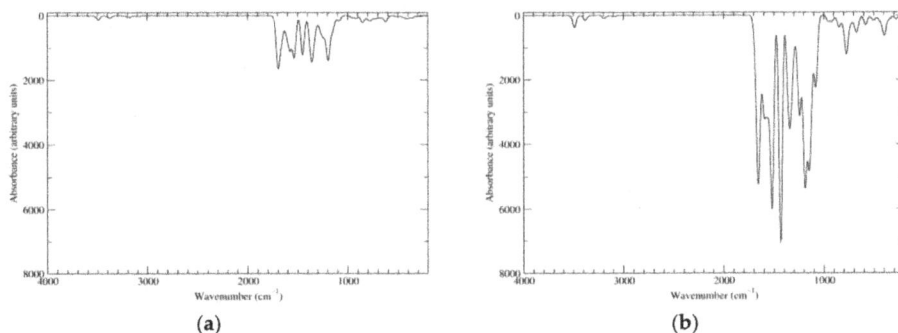

Figure 12. Computed IR spectra of the most significant tautomers/conformers in the eight-electrons oxidation state. (**a**) Neutral form in vacuo, a16_b56_c16_d56, C_2, conf1. (**b**) Neutral form in water, a16_b56_c16_d56, C_2, conf1.

3.7. HOMO–LUMO Gaps

Trends in HOMO and LUMO energies and in the resulting HOMO–LUMO gaps are illustrated in Table 17 for the main neutral species in each oxidation state.

Table 17. HOMO and LUMO energies, and corresponding HOMO–LUMO gaps, computed at the PBE0/6-311++G(2d,2p)//BE0/6-31+G(d,p) level in vacuo and in water.

Oxidation State	Tautomer	Conformer	E_{HOMO} (eV)	E_{LUMO} (eV)	Gap (eV)	E_{HOMO} (eV)	E_{LUMO} (eV)	Gap (eV)
			In Vacuo			In Water		
KP-Red	a0_b0_c0_d0	S_4	−5.074	−1.308	3.767	−5.363	−1.499	3.864
KP-1e	a5_b0_c0_d0	C_1, conf1	−5.341	−1.556	3.784	−5.465	−1.644	3.822
	a6_b0_c0_d0	C_1, conf1	−5.073	−1.512	3.561	−5.314	−1.609	3.705
KP-2e	a16_b0_c0_d0	C_1, conf1	−5.269	−3.449	1.821	−5.480	−3.696	1.783
	a6_b6_c0_d0	C_1, conf1	−5.272	−3.708	1.565	−5.329	−3.834	1.495
KP-4e	a16_b0_c16_d0	C_2, conf1	-	-	-	−5.603	−3.748	1.855
	a16_b6_c6_d0	C_1, conf1	−5.506	−4.114	1.392	−5.443	−4.086	1.357
	a6_b16_c6_d0	C_1, conf1	−5.479	−4.084	1.395	−5.443	−4.057	1.386
	a6_b6_c16_d0	C_1, conf1	−5.451	−4.053	1.398	−5.418	−4.020	1.398
	a6_b6_c6_d6	C_2, conf1	−5.479	−4.371	1.108	−5.280	−4.145	1.135
KP-6e	a16_b6_c16_d6	C_2, conf1	−5.978	−4.280	1.698	−5.890	−4.171	1.719
KP-8e	a16_b56_c16_d56	C_2, conf1	−6.393	−4.787	1.605	−6.132	−4.558	1.574

As expected, the oxidation state plays the largest role; however, even within the range of energetically significant species, changes connected to the specific tautomer/conformer are sometimes significant. HOMO energies display a rather regular decrease as a function of oxidation state; by contrast, LUMO energies shows larger and less regular changes, with a marked downward step of ca. 2.0 eV between KP-Red/KP-1e and KP-2e. The combined effect of these two factors is that the HOMO–LUMO gap changes significantly along the series, reaching a minimum for KP-4e (1.108 eV for a6_b6_c6_d6; ca. 1.4 eV for the other tautomers). Values for KP-2e, KP-6e and KP-8e are in the range 1.5–1.8 eV. Solvent effects on the HOMO–LUMO gap are significant in some cases; however, the overall picture is not substantially modified.

3.8. Disproportionation Equilibria

Table 18 lists reaction free energies computed in vacuo and in water for several disproportionation processes involving KP at different oxidation levels. For ease of comparison, values referred to the conversion of one mole of reactant are also listed (this would entail use of fractional stoichiometric coefficients for the products).

Table 18. Reaction free energies computed for several disproportionation processes involving KP.

Reaction	ΔG_{RRHO} (kcal mol^{-1}) (In Vacuo) [a]	$\Delta G_{SMD,RRHO}$ (kcal mol^{-1}) (In Water) [b]
2 KP-1e ⇌ KP-Red + KP-2e	−0.1 (−0.0$_4$)	1.6 (0.8)
2 KP-2e ⇌ KP-Red + KP-4e	−0.1 (−0.0$_5$)	0.8 (0.4)
2 KP-4e ⇌ KP-2e + KP-6e	−5.6 (−2.8)	−3.5 (−1.8)
2 KP-6e ⇌ KP-4e + KP-8e	44.3 (22.1)	24.8 (12.4)
6 KP-1e ⇌ 5 KP-Red + KP-6e	−6.3 (−1.0)	2.9 (0.5)
3 KP-2e ⇌ 2 KP-Red + KP-6e	−5.8 (−1.9)	−1.9 (−0.6)
3 KP-4e ⇌ KP-Red + 2 KP-6e	−11.3 (−3.8)	−6.2 (−2.1)
4 KP-6e ⇌ KP-Red + 3 KP-8e	121.4 (30.4)	68.2 (17.1)

For each oxidation state and condition (vacuo, water), the free energy of the most stable form identified was used in the calculations. In parentheses values referred to the conversion of one mole of reactant. [a] Reaction free energy computed at 298.15 K within the rigid-rotor/harmonic-oscillator (RRHO) approximation. [b] Reaction free energy computed at 298.15 K at the polarizable continuum model (PCM) level within the RRHO approximation, and including nonelectrostatic terms according to the SMD solvation model ($G_{SMD,RRHO} = G_{PCM,RRHO} + G_{SMD} - G_{PCM}$).

In vacuo, the predicted behavior is very clear: all forms at oxidation stages intermediate between the fully reduced and the six-electron oxidized level are predicted to undergo spontaneous disproportionation, leading eventually to formation of KP-Red and KP-6e.

KP-8e corresponds to the highest oxidation level computed in this study; therefore, its disproportionation could not be modeled. However, disproportionation of KP-6e to KP-Red and KP-8e is so unfavorable (i.e., the opposite comproportionation process is so favorable) that KP-8e could not coexist with any substantial amounts of forms below the six-electron oxidation level; therefore, it could become significant only under specific oxidizing conditions.

Moreover, oxidation levels above KP-8e can be expected to be even less stable. For example, a closed-shell structure for KP-10e requires that at least two indole units reach the three-electron oxidation state. The remaining two units can then contribute a two-electron oxidation level each; in alternative, one of them can be at the one-electron oxidation level, and the other one again at the three-electron level. There is just one way to write a three-electron oxidized building block (Figure 13). Intrinsically, this would be a high-energy site: as a matter of fact, this type of building block was not employed in the generation of starting structures of lower oxidation levels. Moreover, there is no way to combine two such units with two two-electron oxidized building blocks (see Figure 4), nor any way to arrange three three-electron units with a one-electron building block. Analogous considerations hold for KP-12e.

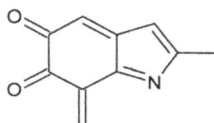

Figure 13. A three-electron oxidized building block, which would be needed for the generation of closed-shell tautomers at oxidation levels above KP-8e.

The one-electron oxidation free radical species displays a remarkably low tendency to disproportionation: however, the process is still predicted to be favorable in vacuo.

In water, disproportionations of KP-6e remain highly endoergic, but less so than in vacuo. All other disproportionation processes are slightly less favorable in water than in vacuo. In particular, the conversion of KP-2e to KP-Red and KP-4e becomes unfavorable, however, this is compensated by subsequent disproportionation of KP-4e, so that the overall conversion of KP-2e to KP-Red and KP-6e remains spontaneous. By contrast, disproportionation of KP-1e (either to KP-Red and KP-2e, or all the way to KP-Red and KP-6e) is predicted to be (slightly) endoergic: this would imply that the free radical should be present in appreciable concentrations in partially oxidized mixtures of KP tetramers in aqueous conditions. This could be taken as an indication of a high reactivity of KP under such conditions [39].

More in general, the comparatively small values of reaction free energies in water would allow for co-existence of species at different oxidation stages. This is particularly relevant in view of the predicted variations of UV–Vis absorption spectra as a function of oxidation level, which is characterized by significant shifts in the absorption properties, especially in the low-energy visible end of the spectrum.

4. Conclusions

The basic variations in the structural features, chemical stability and optical properties of the DHI-based porphyrin system reported by Kaxiras in 2006 have been investigated in a systematic manner by a DFT approach as a function of main accessible oxidation states. Main results can be summarized as follows:

i.   As a rule, oxidation of the tetracatechol system results in a loss of stability, with the noticeable exception of the six-electron level and the unexpectedly low tendency of the one-electron oxidation free radical species to disproportionate.

ii.   Oxidation is associated with variable changes in the HOMO–LUMO gaps leading to the development of more or less intense absorption bands in the visible region.

iii. Tautomerism becomes significant with increasing oxidation states, whereby conjugated systems extending over two or more indole units usually play a major role in determining the preferred tautomeric state.

iv. The highest stability predicted for the six-electron oxidation state would at least in part reflect porphyrin-type aromaticity imparted to the core tetrapyrrole ring system by oxidation at the catechol centers.

Overall, these results allow us to conclude that the co-existence in the planar cyclic platform of a symmetric tetracatechol oxygenation pattern that can interact with the four central NH centers endows this intriguing tetraindole system with highly tunable optical, redox and electronic properties that are governed by reversible and dynamic interconversion of catechol/semiquinone/quinone/quinone methide moieties. These results may guide the design of innovative bioinspired optoelectronic materials of potential technological interest.

Supplementary Materials: The following are available online at www.mdpi.com/2313-7673/2/4/21/s1. Figure S1: Selected molecular orbitals of the a16_b6_c16_d6 tautomer of KP-6e, computed in vacuo at the C_{2h} geometry, Figure S2: Ultraviolet–visible spectrum of the parent non-oxygenated porphyrin ([31], compound **1**) and of a related brominated and oxidized aromatic derivative (*ibid.*, compound **8**) computed at different theory levels, Figure S3: Ultraviolet–visible spectra of the most significant tautomers/conformers (neutral forms) in the different oxidation states, computed at the TD-B3LYP/6-311++G(2d,2p)//PBE0/6-31+G(d,p) level in vacuo, Figure S4: Molecular structures of the most significant tautomers/conformers (neutral forms in water) in the different oxidation states, Tables S1–S15: Structures, energies and relative stabilities of the most stable tautomers/conformers in all oxidation/protonation states explored, in vacuo and in water, Tables S16 and S17: Excitations underlying the UV–Vis spectrum of the a16_b6_c16_d6 tautomer of KP-6e, computed in vacuo at the C_{2h} geometry and at the C_2 geometry, Table S18: Comparison of the electronic energies of the main tautomers/conformers of the neutral form of KP-4e, computed in vacuo at different theory levels, Tables S19 and S20: Electronic energies of main neutral species of KP computed in vacuo at different theory levels, and resulting electronic energy changes for disproportionation processes, Tables S21 and S22: Comparison of the electronic transitions of porphin freebase and of magnesium-porphin, computed in vacuo at different levels.

Acknowledgments: Computational resources were provided by the SCoPE Data Center of the University of Naples Federico II, Naples, Italy.

Author Contributions: The manuscript was written through contributions of all authors.

Conflicts of Interest: The authors declare no conflict of interest.

References

1. d'Ischia, M.; Wakamatsu, K.; Cicoira, F.; Di Mauro, E.; Garcia-Borron, J.C.; Commo, S.; Galva'n, I.; Ghanem, G.; Kenzo, K.; Meredith, P.; et al. Melanins and melanogenesis: From pigment cells to human health and technological applications. *Pigm. Cell Melanoma Res.* **2015**, *28*, 520–544. [CrossRef] [PubMed]

2. Kaxiras, E.; Tsolakidis, A.; Zonios, G.; Meng, S. Structural model of eumelanin. *Phys. Rev. Lett.* **2006**, *97*, 218102. [CrossRef] [PubMed]

3. Meng, S.; Kaxiras, E. Theoretical models of eumelanin protomolecules and their optical properties. *Biophys. J.* **2008**, *94*, 2095–2105. [CrossRef] [PubMed]

4. Liebscher, J.; Mrówczyński, R.; Scheidt, H.A.; Filip, C.; Hadade, N.D.; Turcu, R.; Bende, A.; Beck, S. Structure of polydopamine: A never-ending story? *Langmuir* **2013**, *29*, 10539–10548. [CrossRef] [PubMed]

5. Chen, C.-T.; Ball, V.; de Almeida Gracio, J.J.; Singh, M.K.; Toniazzo, V.; Ruch, D.; Buehler, M.J. Self-assembly of tetramers of 5,6-dihydroxyindole explains the primary physical properties of eumelanin: Experiment, simulation, and design. *ACS Nano* **2013**, *7*, 1524–1532. [CrossRef] [PubMed]

6. Fan, H.; Yu, X.; Liu, Y.; Shi, Z.; Liu, H.; Nie, Z.; Wu, D.; Jin, Z. Folic acid–polydopamine nanofibers show enhanced ordered-stacking via π–π interactions. *Soft Matter* **2015**, *11*, 4621–4629. [CrossRef] [PubMed]

7. Kim, Y.J.; Khetan, A.; Wu, W.; Chun, S.-E.; Viswanathan, V.; Whitacre, J.F.; Bettinger, C.J. Evidence of porphyrin-like structures in natural melanin pigments using electrochemical fingerprinting. *Adv. Mater.* **2016**, *28*, 3173–3180. [CrossRef] [PubMed]

8. Li, X.-C.; Wang, C.-Y.; Lai, W.-Y.; Huanga, W. Triazatruxene-based materials for organic electronics and optoelectronics. *J. Mater. Chem. C* **2016**, *4*, 10574–10587. [CrossRef]

9. Frisch, M.J.; Trucks, G.W.; Schlegel, H.B.; Scuseria, G.E.; Robb, M.A.; Cheeseman, J.R.; Scalmani, G.; Barone, V.; Mennucci, B.; Petersson, G.A.; et al. *Gaussian 09, Revision D.01*; Gaussian, Inc.: Wallingford, CT, USA, 2009.

10. Adamo, C.; Barone, V. Toward reliable density functional methods without adjustable parameters: The PBE0 model. *J. Chem. Phys.* **1999**, *110*, 6158–6169. [CrossRef]

11. Miertus, S.; Scrocco, E.; Tomasi, J. Electrostatic interaction of a solute with a continuum. A direct utilizaion of *ab initio* molecular potentials for the prevision of solvent effects. *J. Chem. Phys.* **1981**, *55*, 117–129. [CrossRef]

12. Cossi, M.; Scalmani, G.; Rega, N.; Barone, V. New developments in the polarizable continuum model for quantum mechanical and classical calculations on molecules in solution. *J. Chem. Phys.* **2002**, *117*, 43–54. [CrossRef]

13. Scalmani, G.; Barone, V.; Kudin, K.N.; Pomelli, C.S.; Scuseria, G.E.; Frisch, M.J. Achieving linear-scaling computational cost for the polarizable continuum model of solvation. *Theor. Chem. Acc.* **2004**, *111*, 90–100.

14. Tomasi, J.; Mennucci, B.; Cammi, R. Quantum mechanical continuum solvation models. *Chem. Rev.* **2005**, *105*, 2999–3093. [CrossRef] [PubMed]

15. Rappé, A.K.; Casewit, C.J.; Colwell, K.S.; Goddard, W.A., III; Skiff, W.M. UFF, a full periodic table force field for molecular mechanics and molecular dynamics simulations. *J. Am. Chem. Soc.* **1992**, *114*, 10024–10035. [CrossRef]

16. York, D.A.; Karplus, M. A smooth solvation potential based on the conductor-like screening model. *J. Phys. Chem. A* **1999**, *103*, 11060–11079. [CrossRef]

17. Scalmani, G.; Frisch, M.J. Continuous surface charge polarizable continuum models of solvation. I. General formalism. *J. Chem. Phys.* **2010**, *132*, 114110. [CrossRef] [PubMed]

18. Marenich, A.V.; Cramer, C.J.; Truhlar, D.G. Universal solvation model based on solute electron density and on a continuum model of the solvent defined by the bulk dielectric constant and atomic surface tensions. *J. Phys. Chem. B* **2009**, *113*, 6378–6396. [CrossRef] [PubMed]

19. Merrick, J.P.; Moran, D.; Radom, L. An evaluation of harmonic vibrational frequency scale factors. *J. Phys. Chem. A* **2007**, *111*, 11683–11700. [CrossRef] [PubMed]

20. Stratmann, R.E.; Scuseria, G.E.; Frisch, M.J. An efficient implementation of time-dependent density-functional theory for the calculation of excitation energies of large molecules. *J. Chem. Phys.* **1998**, *109*, 8218–8224. [CrossRef]

21. Bauernschmitt, R.; Ahlrichs, R. Treatment of electronic excitations within the adiabatic approximation of time dependent density functional theory. *Chem. Phys. Lett.* **1996**, *256*, 454–464. [CrossRef]

22. Casida, M.E.; Jamorski, C.; Casida, K.C.; Salahub, D.R. Molecular excitation energies to high-lying bound states from time-dependent density-functional response theory: Characterization and correction of the time-dependent local density approximation ionization threshold. *J. Chem. Phys.* **1998**, *108*, 4439–4449. [CrossRef]

23. Adamo, C.; Scuseria, G.E.; Barone, V. Accurate excitation energies from time-dependent density functional theory: Assessing the PBE0 model. *J. Chem. Phys.* **1999**, *111*, 2889–2899. [CrossRef]

24. Scalmani, G.; Frisch, M.J.; Mennucci, B.; Tomasi, J.; Cammi, R.; Barone, V. Geometries and properties of excited states in the gas phase and in solution: Theory and application of a time-dependent density functional theory polarizable continuum model. *J. Chem. Phys.* **2006**, *124*, 094107. [CrossRef] [PubMed]

25. Alfieri, M.L.; Micillo, R.; Panzella, L.; Crescenzi, O.; Oscurato, S.L.; Maddalena, P.; Napolitano, A.; Ball, V.; d'Ischia, M. The structural basis of polydopamine film formation: Probing 5,6-dihydroxyindole-based eumelanin type units and the porphyrin issue. *ACS Appl. Mater. Interfaces* **2017**. [CrossRef] [PubMed]

26. Pezzella, A.; Panzella, L.; Crescenzi, O.; Napolitano, A.; Navaratman, S.; Edge, R.; Land, E.J.; Barone, V.; d'Ischia, M. Short-lived quinonoid species from 5,6-dihydroxyindole dimers en route to eumelanin polymers: Integrated chemical, pulse radiolytic, and quantum mechanical investigation. *J. Am. Chem. Soc.* **2006**, *128*, 15490–15498. [CrossRef] [PubMed]

27. d'Ischia, M.; Crescenzi, O.; Pezzella, A.; Arzillo, M.; Panzella, L.; Napolitano, A.; Barone, V. Structural effects on the electronic absorption properties of 5,6-dihydroxyindole oligomers: The potential of an integrated experimental and DFT approach to model eumelanin optical properties. *Photochem. Photobiol.* **2008**, *84*, 600–607. [CrossRef] [PubMed]

28. Pezzella, A.; Panzella, L.; Crescenzi, O.; Napolitano, A.; Navaratman, S.; Edge, R.; Land, E.J.; Barone, V.; d'Ischia, M. Lack of visible chromophore development in the pulse radiolysis oxidation of 5,6-dihydroxyindole-2-carboxylic acid oligomers: DFT investigation and implications for eumelanin absorption properties. *J. Org. Chem.* **2009**, *74*, 3727–3734. [CrossRef] [PubMed]
29. Prampolini, G.; Cacelli, I.; Ferretti, A. Intermolecular interactions in eumelanins: A computational bottom-up approach. I. small building blocks. *RSC Adv.* **2015**, *5*, 38513–38526. [CrossRef]
30. Becke, A.D. Density-functional thermochemistry. III. The role of exact exchange. *J. Chem. Phys.* **1993**, *98*, 5648–5652. [CrossRef]
31. Nakamura, S.; Hiroto, S.; Shinokubo, H. Synthesis and oxidation of cyclic tetraindole. *Chem. Sci.* **2012**, *3*, 524–527. [CrossRef]
32. Nakamura, S.; Kondo, T.; Hiroto, S.; Shinokubo, H. Porphyrin analogues that consist of Indole, benzofuran, and benzothiophene subunits. *Asian J. Org. Chem.* **2013**, *2*, 312–319. [CrossRef]
33. Tuna, D.; Udvarhelyi, A.; Sobolewski, A.L.; Domcke, W.; Domratcheva, T. Onset of the electronic absorption spectra of isolated and π-stacked oligomers of 5,6-dihydroxyindole: An *ab initio* study of the building blocks of eumelanin. *J. Phys. Chem. B* **2016**, *120*, 3493–3502. [CrossRef] [PubMed]
34. Il'ichev, Y.V.; Simon, J.D. Building blocks of eumelanin: Relative stability and excitation energies of tautomers of 5,6-dihydroxyindole and 5,6-indolequinone. *J. Phys. Chem. B* **2003**, *107*, 7162–7171. [CrossRef]
35. Giovannetti, R. The use of spectrophotometry UV–Vis for the study of porphyrins. In *Macro to Nano Spectroscopy*; Uddin, J., Ed.; InTech: Rijeka, Croatia, 2012; pp. 87–108. ISBN 978-953-51-0664-7.
36. Minaev, B.; Wang, Y.H.; Wang, C.K.; Luo, Y.; Agren, H. Density functional theory study of vibronic structure of the first absorption QX band in free-base porphin. *Spectrochim. Acta A Mol. Biomol. Spectrosc.* **2006**, *65*, 308–323. [CrossRef] [PubMed]
37. Baiardi, A.; Bloino, J.; Barone, V. General time dependent approach to vibronic spectroscopy including Franck–Condon, Herzberg–Teller, and Duschinsky effects. *J. Chem. Theory Comput.* **2013**, *9*, 4097–4115. [CrossRef] [PubMed]
38. Gouterman, M.; Wagnière, G.H.; Snyder, L.C. Spectra of porphyrins: Part II. Four orbital model. *J. Mol. Spectrosc.* **1963**, *11*, 108–127. [CrossRef]
39. Pezzella, A.; Crescenzi, O.; Panzella, L.; Napolitano, A.; Land, E.J.; Barone, V.; d'Ischia, M. Free radical coupling of *o*-semiquinones uncovered. *J. Am. Chem. Soc.* **2013**, *135*, 12142–12149. [CrossRef] [PubMed]

biomimetics

MDPI

Article

The Antioxidant Activity of Quercetin in Water Solution

Riccardo Amorati [1,*], Andrea Baschieri [1], Adam Cowden [2] and Luca Valgimigli [1,*]

[1] University of Bologna, Department of Chemistry "G. Ciamician", Via S. Giacomo 11, 40126 Bologna, Italy; andrea.baschieri2@unibo.it

[2] School of Chemistry (Rm 267), University of Edinburgh, West Mains Road, Edinburgh EH9 3FJ, UK; adamcowden@gmail.com

* Correspondence: riccardo.amorati@unibo.it (R.A.); luca.valgimigli@unibo.it (L.V.);
  Tel.: +39-051-209-5689 (R.A.); +39-051-209-5683 (L.V.)

Academic Editor: Daniel Ruiz-Molina
Received: 30 May 2017; Accepted: 22 June 2017; Published: 27 June 2017

Abstract: Despite its importance, little is known about the absolute performance and the mechanism for quercetin's antioxidant activity in water solution. We have investigated this aspect by combining differential oxygen-uptake kinetic measurements and B3LYP/6311+g (d,p) calculations. At pH = 2.1 (30 °C), quercetin had modest activity ($k_{inh} = 4.0 \times 10^3$ M^{-1} s^{-1}), superimposable to catechol. On raising the pH to 7.4, reactivity was boosted 40-fold, trapping two peroxyl radicals in the chromen-4-one core and two in the catechol with k_{inh} of 1.6×10^5 and 7.0×10^4 M^{-1} s^{-1}. Reaction occurs from the equilibrating mono-anions in positions 4' and 7 and involves firstly the OH in position 3, having bond dissociation enthalpies of 75.0 and 78.7 kcal/mol, respectively, for the two anions. Reaction proceeds by a combination of proton-coupled electron-transfer mechanisms: electron–proton transfer (EPT) and sequential proton loss electron transfer (SPLET). Our results help rationalize quercetin's reactivity with peroxyl radicals and its importance under biomimetic settings, to act as a nutritional antioxidant.

Keywords: catechol; peroxyl radicals; proton-coupled electron transfer; kinetics; thermodynamics; mechanisms

1. Introduction

Quercetin is perhaps the most famous flavonoid antioxidant: a member of the flavonols family, bearing a catechol moiety (ring B) linked in position 2 to the polyhydroxylated chromen-4-one core (Scheme 1). It is found in a large variety of dietary vegetables [1], which makes its presence in raw food nearly ubiquitous, and the typical daily intake was estimated as 25 mg/person/day in the U.S. diet [2]. Some sources such as capers, dill, cilantro, radish, carob, fennel, radicchio and onions are particularly rich, with 30–230 mg of quercetin or its glycosides per 100 g of edible portion [1–5], while green and black tea infusions have been reported to contain, respectively, about 480 and 330 mg/L of quercetin glycosides [1]. Considering that oral bioavailability can be as large as 50% (including its phenolic metabolites) [1,3], clearly, under a vegetable-rich dietary regime, quercetin can significantly contribute to the physiological antioxidant defense. Quercetin has also been attributed a range of healthy biological activities, including anti-inflammatory, blood vessel protecting, anti-platelet aggregation, antiviral, anti-cataract, enhancement of cognitive function, and anti-cancer [1,2,6–9]. Although regulatory agencies such as the European Food Safety Authority (EFSA) indicate that there is insufficient clinical evidence to support them [10], it is noteworthy that many such health-related claims are directly or indirectly associated to quercetin's purported antioxidant activity. Its antioxidant activity has also recently been exploited, for instance, in the development of bioinspired synergic nano-antioxidants [11], or of co-delivery vincristine–quercetin nanodrugs for the treatment of lymphoma [12], and it has inspired the design of pH-responsive nanocarriers for drug release [13]. Owing to its importance, the antioxidant activity of quercetin has been the subject of many investigations [14,15]; however, surprisingly little is known about the actual mechanism and relevant quantitative values of its antioxidant activity under biomimetic settings (i.e., on the mechanism and absolute kinetics of reaction with peroxyl radicals [16–20] in water).

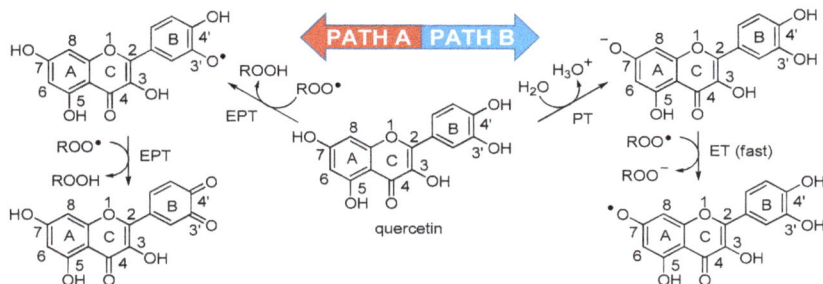

Scheme 1. Possible reaction pathways for quercetin in water. Two mechanistic possibilities, arising from current literature, could account for the antioxidant activity of quercetin in water solution: Path A illustrates the electron–proton transfer (EPT) to peroxyl radicals from the catechol moiety, while Path B depicts the sequential proton loss electron transfer (SPLET = PT/ET) to water/peroxyl radicals. ET: Electron transfer; PT: Proton transfer.

At variance with the assessment of antioxidant activity by rapid assays [20], which has occurred in the majority of studies, and carries no mechanistic information [20], relevant kinetic data on the reaction with peroxyl radicals—namely the absolute rate constant and the stoichiometric factor for the antioxidant, k_{inh} and n, respectively—have recently been obtained by inhibited autoxidation studies in organic solution, specifically in chlorobenzene [11,21–23], in *tert*-butanol [21] and in acetonitrile [11,23]. Those studies converge, indicating that the catechol ring B is the "active" moiety, trapping $n = 2$ peroxyl radicals by formal H-atom transfer (the mechanism can actually be described as a concerted electron–proton transfer, EPT [24]) as depicted in Scheme 1, Path A. The values of k_{inh} recorded in those studies range ~5 × 10^5 M^{-1} s^{-1} in PhCl (30 °C), ~2 × 10^4 M^{-1} s^{-1} in *t*-BuOH (50 °C) to ~1 × 10^4 M^{-1} s^{-1} in MeCN (30 °C), being superimposable to those of catechol itself [22] (See Table 1), and indicating that the polyhydroxy chromen-4-one core has little role in the antioxidant activity. On the other hand, using the persistent 2,2-diphenyl-1-picrylhydrazyl radical (DPPH•) as the model oxidant, Litwinienko and coworkers found that in protic solvents the reactivity of several flavonoids, including quercetin, was related to their pK for acid dissociation (pK_a). Reaction of quercetin was found to be 1000-fold faster in methanol than in dioxane and was suggested to occur upon deprotonation in position 7 (ring A), via a mechanism named sequential proton loss electron transfer (SPLET) [25], and composed of a proton transfer (PT) to the solvent and an electron transfer (ET) to the oxidizing radical, as depicted in Scheme 1, Path B. This indicates that, owing to its acidity, in water solution at pH 7.4, quercetin should react with peroxyl radicals by a similar mechanism, in which the catechol has only a secondary role. Unfortunately, the study of the antioxidant behavior of water-soluble quercetin in water solution has so far been precluded by the lack of suitable methods of investigation. Our recent development and validation of one such method [26] has paved the way to filling this fundamental gap. Therefore, we report here a kinetic study on the antioxidant behavior of quercetin in water solution, accompanied with quanto-mechanical calculations aimed at clarifying its mechanism of reaction with peroxyl radicals. Calculations have also been matched with previously available spectroscopic data, so to compose a hopefully clear picture of the redox chemistry of quercetin in water, under biomimetic settings.

Table 1. Rate constants k_{inh} and stoichiometric factors n for the trapping of peroxyl radicals by quercetin and reference antioxidants, measured in the inhibited autoxidation of tetrahydrofuran (THF) in buffered water at 303 K. Reference values in other solvents are reported for comparison.

| | Water | | | | MeCN | | PhCl | |
| | pH = 2.1 | | pH = 7.4 | | | | | |
	k_{inh} (M^{-1} s^{-1})	n	k_{inh} (M^{-1} s^{-1})	n	k_{inh} (M^{-1} s^{-1})	n	k_{inh} (M^{-1} s^{-1})	n
Quercetin	$(4.0 \pm 0.5) \times 10^3$	-	$(1.6 \pm 0.3) \times 10^5$ $(7.0 \pm 0.5) \times 10^4$	2.1 ± 0.2 2.0 ± 0.2	$1.2 \times 10^{4\,a}$	2.1	$5.6 \times 10^{5\,b}$	2.1
Catechol [c]	3.0×10^3	-	7.0×10^3	-	2.5×10^4	2	$5.5 \times 10^{5\,b}$	2.0
PMHC [c]	1.9×10^5	1.8	2.0×10^5	1.8	6.8×10^5	2	3.2×10^6	2.0

[a] Data from reference [22]. [b] Data from reference [23]. [c] Data from reference [26]. PMHC: 2,2,5,7,8-Pentamethyl-6-chromanol.

2. Materials and Methods

2.1. Materials

All chemicals and solvents were of the highest purity commercially available (Sigma–Aldrich, Milan, Italy). 2,2'-Azobis(2-methylpropion-amidine)dihydrochloride (AAPH) and quercetin were used as received. 2,2,5,7,8-Pentamethyl-6-chromanol (PMHC) was recrystallized from hexane, and catechol was recrystallized from ethyl acetate/hexane. Tetrahydrofuran (THF) was distilled and stored under argon at 5 °C; the content in hydroperoxides was determined by spectrophotometry at 262 nm in isopropanol upon reaction with triphenylphosphine, and found to be <50 ppm (μg g^{-1}) [26]. Buffers were prepared in bidistilled water as previously described [26]: buffer pH 2.1, $NaH_2PO_4 \cdot 2H_2O$ (0.39 g, 0.05 mole) and H_3PO_4 85% (0.17 mL, 0.05 mole) were dissolved in water (50 mL); buffer pH 7.4, Na_2HPO_4 (0.595 g, 0.096 mole) and $NaH_2PO_4 \cdot 2H_2O$ (0.125 g, 0.016 mole) were dissolved in water (50 mL). Buffer solutions were mixed with the desired amount of THF (typically 3:1 *v/v*) after having adjusted the pH to the desired value [26].

2.2. Kinetic Measurements

Autoxidation experiments were performed in a two-channel oxygen-uptake apparatus based on a Validyne DP 15 differential pressure transducer (Validyne Engineering, Northridge, CA, USA) built in our laboratory and described previously [27]. Azo-initiator AAPH was prepared in concentrated stock solutions that were injected into the reaction mixture to the desired final concentrations (typically 12.5–75 mM). AAPH solutions were freshly prepared every 4 h and stored at 5 °C to avoid excessive hydrolysis. In a typical experiment, an air-saturated solution of THF–water (1:3 *v/v*) containing the desired buffer (0.1 M) and the initiator was equilibrated with an identical reference solution containing an excess of PMHC. After equilibration, and when a constant O_2 consumption was reached, a concentrated solution of the antioxidant was injected into the sample flask. The oxygen consumption of the sample was measured, after calibration of the apparatus, from the differential pressure recorded with time between the two channels. Initiation rates, R_i, were determined for each set of conditions by matching autoxidation experiments, using PMHC as the reference antioxidant, by means of Equation (1). Absolute k_{inh} values were determined, after independent assessment of R_i, from 4 to 10 inhibited autoxidation experiments with antioxidant (AH) concentration in the range 10–50 μM by means of Equation (2) and values of n were determined from the same experiments by Equation (1) [26,27].

$$R_i = \frac{n[\text{Antioxidant}]}{\tau} \tag{1}$$

$$-\frac{d[O_2]}{dt} = \frac{k_p[\text{THF}]R_i}{nk_{inh}[\text{AH}]} \tag{2}$$

2.3. Calculations

Geometry optimization and frequencies were computed in the gas phase at the B3LYP/6-31+g (d,p) level; stationary points were confirmed by checking the absence of imaginary frequencies. Thermochemistry was computed at 298 K using the scaling factor of 0.9806 [28]. Relative stabilities in the gas phase and in water were computed by applying the free energy correction at 298 K to single point calculations at the B3LYP/6-311+g (d,p) level without and with the polarizable continuum model (PCM), respectively. Bond dissociation enthalpies (BDE$_{OH}$) values were computed by the isodesmic approach using unsubstituted phenol as the reference, whose BDE$_{OH}$ in water is 88.2 kcal/mol, by using enthalpy corrected total free energy in solution obtained by the PCM method [29,30]. The ultraviolet–visible (UV–vis) spectra of the most stable conformers were calculated by time-dependent density functional theory (TD-DFT), performed at the B3LYP/6-311+g (d,p) level either in the gas phase or in water by using the PCM model [31]. Conformational isomers had very similar calculated transitions in the wavelength range considered (350–700 nm), therefore only the most stable conformers were considered. Calculations were performed by using Gaussian 03 software [32] (see Appendix A in the Supplementary Materials). To visually compare calculated spectra with experimental spectra, each transition obtained by TD-DFT calculation was convoluted by a Gaussian function g_i (x), Equation (3), where e_i, fw_i and σ_i are the energy (in eV), the oscillator strength and the full width at half maximum (FWHM) of the peak, respectively.

$$g_i(x) = 2\sqrt{\frac{\ln 2}{\pi}} \frac{fw_i}{\sigma_i} \exp\left(-\left(\frac{2(x - e_i)\sqrt{\ln 2}}{\sigma_i}\right)^2\right) \tag{3}$$

The UV–vis spectrum $f(x)$ is built as a sum of N bands (Equation (4)), where S is a scale factor and g_i (x) is the Gaussian function defined above.

$$f(x) = S\sum_{i=1}^{N} g_i(x) \tag{4}$$

The scale factor S was adjusted manually, on a trial-and-error basis, to reproduce the intensity of experimental spectra, whereas σ_i was fixed to 0.5 eV. Calculated spectra were finally converted to the wavelength scale (nm) to be compared to experimental spectra [33].

3. Results and Discussion

3.1. Kinetic Measurements with Peroxyl Radicals

The antioxidant activity of quercetin was measured by studying the inhibited autoxidation of THF in buffered water solution initiated at 30 °C by the thermal decomposition of azo-initiator AAPH [26], monitoring oxygen consumption by a differential oxygen-uptake apparatus [26,27] (Figure 1). PMHC, a less lipophilic mimic of α-tocopherol with identical reactivity to peroxyl radicals [26], was used as the reference antioxidant. Measurements were performed both at pH 2.1 and at pH 7.4 and results are collected in Table 1 along with those of reference antioxidants obtained under comparable settings.

At pH 2.1, quercetin exhibited modest antioxidant behavior: at concentrations up to 50 μM, it was only able to slow down the oxygen uptake, without giving a neat inhibition period as observed with reference PMHC, thereby preventing the assessment of n, the number of peroxyl radicals trapped by each antioxidant molecule. The measured rate constant k_{inh}, calculated by assuming $n = 2$, was 4.0×10^3 M^{-1} s^{-1}, very similar to that recorded for simple catechol. This result adds to kinetic measurements in organic solvents (Table 1), showing matched reactivity of quercetin and catechol, and suggesting a superimposable reaction mechanism: a rate-determining concerted EPT from the catechol moiety to the peroxyl radical (Scheme 1, Path A).

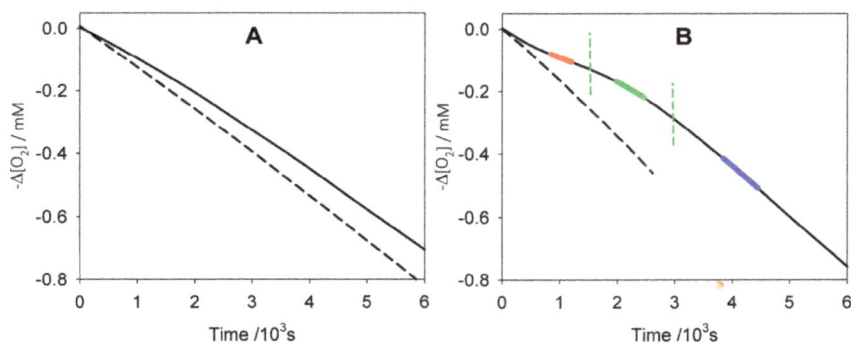

Figure 1. Oxygen consumption during the autoxidation of tetrahydrofuran (THF, 3.1 M) initiated by 2,2'-azobis(2-methylpropion-amidine)dihydrochloride (AAPH, 0.025 M) at 30 °C without inhibitors (dash) or in the presence of quercetin (solid). (**A**) In phosphate buffer 0.1 M pH = 2.1 with quercetin 2.5×10^{-5} M; (**B**) In phosphate buffer 0.1 M pH = 7.4 with quercetin 5.0×10^{-6} M, vertical lines indicate the time lapse corresponding to the trapping of $n = 2$ peroxyl radicals.

Based on this mechanism, the decrease in reactivity recorded on moving from PhCl to MeCN to buffered water as the reaction medium, is perfectly explained by the progressively stronger hydrogen-bonding of the "reactive" OH group to the solvent (Scheme 2), thereby causing a decrease in its reactivity [26,34,35].

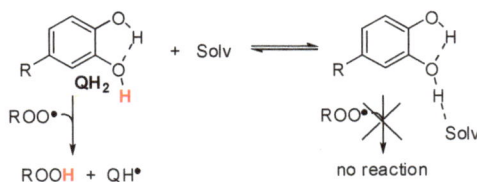

Scheme 2. Kinetic solvent effect explaining the reduced reactivity in H-bond accepting solvents (Solv).

Interestingly, when the pH was raised to 7.4, the reactivity of quercetin increased significantly. Oxygen-uptake plots showed a neat inhibition period corresponding to the trapping of four peroxyl radicals. The first half of the inhibition period provided a k_{inh} of 1.6×10^5 M^{-1} s^{-1} (i.e., 40-fold higher than at pH 2.1) matching the reactivity of reference PMHC. The analysis of the second half of the inhibition period afforded a somewhat reduced k_{inh} of 7×10^4 M^{-1} s^{-1}, yet about 20-fold faster than at pH 2.1. Such a major boost in the antioxidant performance clearly suggests a change in the reaction mechanism, and can be compared to the negligible change in reactivity of monophenolic PMHC and to the modest enhancement of reactivity for catechol, whose k_{inh} grows only by a factor of 2 (Table 1) on raising the pH from 2.1 to 7.4 [26]. Indeed, at pH 7.4, the reactivity of quercetin surpasses that of catechol by over one order of magnitude and the stoichiometric factor is approximately doubled.

Despite its modest magnitude, the enhanced reactivity of catechol upon raising the pH was explained by partial deprotonation which yielded the more electron-rich phenoxide anion [26]; somewhat similarly, the enhanced reactivity of quercetin with DPPH$^\bullet$ radical in ionizing solvents, as compared with non-ionizing solvents of similar H-bond accepting ability, was explained by deprotonation of the OH group in position 7 which yielded an electron-rich phenoxide that would undergo fast ET to the oxidizing radical (Scheme 1, Path B) [25,36]. However, in quercetin, this points toward the major role of the OH in 7 rather than the catechol. On the other hand, previous studies underline the importance of the OH in position 3: the reaction of quercetin toward the radical of a synthetic analogue of α-tocopherol in ethanol was found to be 29-fold faster than that of rutin, where the OH in 3 is glycosylated and unavailable for reaction [37]. To shed some light on these mechanistic possibilities and identify the most stable anions and transient intermediates of quercetin, we turned to quanto-mechanical calculations.

3.2. Quanto-Mechanical Calculations on Quercetin

To help identify the most likely mechanism for reaction of quercetin with peroxyl radicals in water, we first calculated the most stable transient intermediates of quercetin and its main anions in the gas phase and in water solution, based on the relative free energies of formation, by using a PCM, at B3LYP/6-311+g (d,p) level, which had previously been shown to be reliable for phenolic compounds [29,30]. Subsequently, we referred to the time-resolved UV–vis spectra obtained by pulse radiolysis by Jovanovic and co-workers during the reaction of quercetin with $N_3{}^\bullet$ radical in water at various pH [38]. In order to assign the experimental spectra to specific phenoxyl radicals of quercetin, we matched them with the spectra that we calculated for any transient, using TD-DFT methods, which are emerging as powerful tools to investigate radical reactions by allowing the assignment of transient UV–vis spectra [39–42].

3.2.1. Neutral Radicals from Quercetin

Upon removing a H-atom from a quercetin's OH group, five neutral phenoxyl radicals can be formed, whose relative stabilities are reported in Figure 2 (see Figure S1 for all the structures).

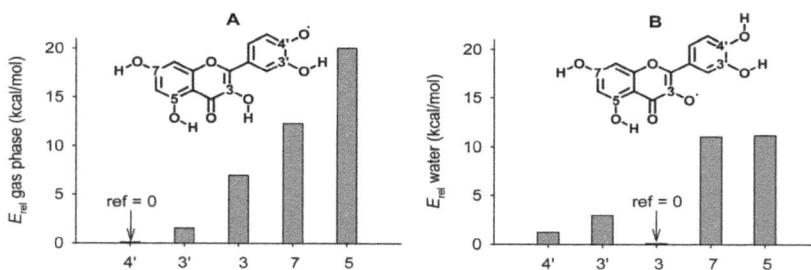

Figure 2. Relative stability of the neutral radicals obtained by H-atom abstraction from quercetin, calculated in the gas phase (**A**) and in water (**B**), determined from the relative energy (E_{rel}) with reference to the most stable species (ref = 0). Numbers on the *x*-axis indicate the position of the radical.

In the gas phase, the most stable radical is in position 4', because the –O$^\bullet$ moiety can accept a relatively strong intramolecular H-bond from the neighboring –OH group and the spin density can be delocalized in ring C (see Figure S2). Calculations performed by using water as the implicit solvent showed that the radical in position 3 is more stable than that in position 4' (+1.2 kcal/mol), while the other three radicals (in positions 3', 7 and 5) have higher energies (by +3.0, +11.0, +11.2 kcal/mol, respectively), in line with previous reports [43]. The inversion of the stabilities of radicals in positions 4' and 3, on moving from the gas phase to water, is due to the high polarity of the catechol (ring B), which is more strongly solvated than the –OH in position 3, which donates an intramolecular H-bond to the neighboring carbonyl. In general, compared to a "free" OH, the cleavage of an intramolecular H-bonded –OH is more energetically costly in the gas phase than in water.

The UV–vis spectra of the radicals of quercetin were measured by pulse radiolysis by Jovanovic et al. [38], by reacting quercetin with the $N_3{}^\bullet$ radical in water at various pH values. The changes of the transient spectra in the pH range 2.6 < pH < 12 showed that the phenoxyl radical from quercetin has two deprotonation equilibria, at pK_a = 4.2 and 9.4, which were attributed, respectively, to the equilibria between the neutral phenoxyl radical and the radical anion and that between the radical anion and the corresponding radical dianion [38]. The spectrum recorded at pH 2.6 is characterized by a λ_{max} of 515 nm and a shoulder at 440 nm (see Figure 3). To assign the structure of the radical originating from the spectrum recorded at pH 2.6, the UV–vis absorption spectra of all neutral radicals which could be formed after H-atom abstraction from quercetin were calculated, and the matching with the experimental spectrum is shown in Figure 3.

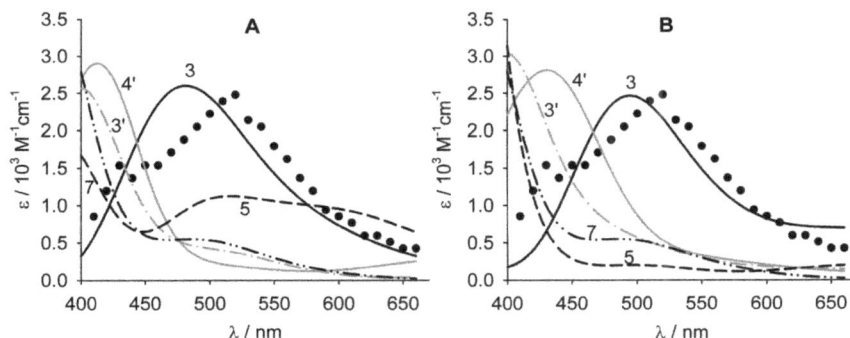

Figure 3. Calculated ultraviolet–visible (UV–vis) spectra of the neutral radicals of quercetin in the gas phase (**A**) or in water (**B**), compared to the experimental spectrum measured at pH 2.6 (●) by Jovanovich et al. [38]. Numbers indicate the position of the radical.

Both the calculations in gas phase and water agree, indicating that the best fit of the experimental spectrum at pH = 2.6 is given by the radical formed in position 3. It should be noted that the quality of the matching of simulated spectra improves in water, since calculations predict a bathochromic shift due to dipolar interaction with the solvent (particularly for radicals in positions 3 and 4'). In water, the radical in position 3 has two strong transitions at 500 and 477 nm, that originate an absorbance peak at 495 nm, that is very near to the maximum of 515 nm determined by pulse radiolysis (Figure 3B). Interestingly, this assignment agrees with the radical stabilities reported in Figure 2. In Figure 3, it is also evident that the radicals in positions 5, 7 and 3' have the most significant transitions below 400 nm, so their spectra do not reproduce the shape of the experimental spectra. Many studies have assumed that, upon H-atom abstraction from quercetin, the most stable radical formed is that in position 4'. As a matter of fact, in water, this radical has two strong transitions at 442 and 396 nm, which create an absorption band at 430 nm, far from the experimental maximum. Nonetheless, small amounts of the radical in position 4', which is calculated to be only 1.2 kcal/mol less stable than that of the radical in position 3, may be present in solution, and it likely originates from the shoulder at 440 nm visible in the experimental spectrum. Assuming that the shoulder at 440 nm can be entirely attributed to the radical in position 4', from the relative areas of the two bands at 515 and 440 nm, it can be estimated that the ratio of the two radicals in positions 3 and 4' is acidic solution is approximately 9:1.

3.2.2. Radical Anions from Quercetin

The stabilities of all the possible ten radical anions were calculated and the results are reported in Figure 4 (see Figure S3 for the structures).

In the gas phase, the most stable tautomer is the radical anion in position 7-4', while the radical anion in position 3-4' is less stable by 0.8 kcal/mol. In water, the stability order is reversed, as the radical anion in position 3-4' is more stable than that in position 7-4' by 6.8 kcal/mol. Similar to what was said in the case of the neutral radicals, upon moving from gas phase to water, the loss of a H-atom from the "free" 7–OH is made more energetically expensive, while that from the 3–OH, that is protected by an intra-molecular H-bond, becomes relatively easier. Interestingly, the radical anion involving the catechol on ring B (i.e., in position 3-4') that has been proposed by Jovanovic as a putative structure for the quercetin radical at pH 5.3 [38], is predicted to be less stable than the radical anion in position 7-4' by 8.6 kcal/mol in the gas phase, while in water it is less stable than the radical anion in position 3-4' by 3.4 kcal/mol.

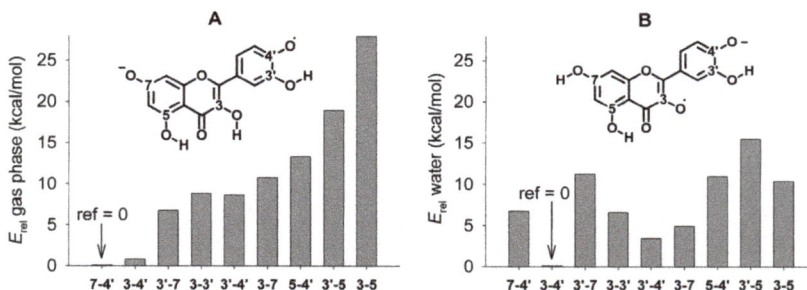

Figure 4. Relative stability of the radical anions obtained by H-atom abstraction from deprotonated quercetin, calculated in the gas phase (**A**) and in water (**B**), determined from the relative energy (E_{rel}) with reference to the most stable species (ref = 0). Numbers on the *x*-axis indicate the position of the radical and of the negative charge. The radical anion in 5-7 is omitted for clarity (E_{rel} is 32.7 and 21.0 kcal/mol in the gas phase and water, respectively).

The experimental UV–vis spectrum of quercetin, recorded by pulse radiolysis at pH 5.8, shows an absorption maximum at 557 nm and a shoulder at about 450 nm. In Figure 5, this spectrum is compared to those calculated for the most stable radical anions.

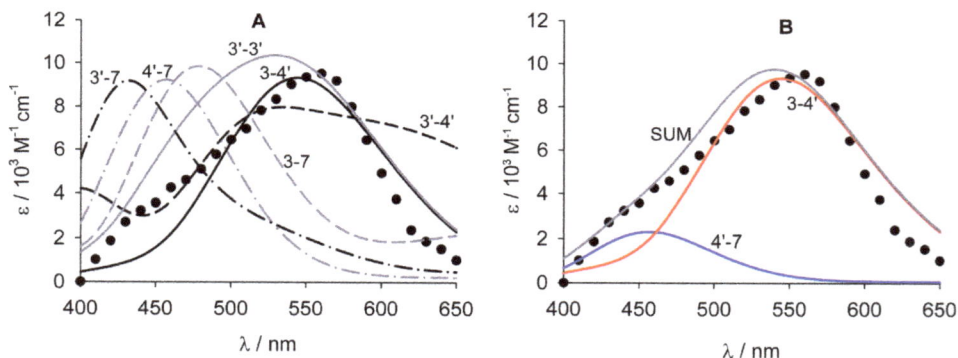

Figure 5. (**A**) Calculated UV–vis spectra of the radical anions of quercetin in water, compared to the experimental spectrum measured at pH 5.8 (●) by Jovanovich et al. [38]. The spectra for the tautomers involving the 5–OH are omitted for clarity. (**B**) Simulation of the spectrum arising from the overlay of radical anions in positions 3-4' and 4'-7 in relative ratio 3.2:1 (SUM). Numbers indicate the position of the radical and of the negative charge.

Considering the spectra calculated in water, which seems the most appropriate approach in the case of charged species, the radical anion that more closely matches the experimental spectrum is that involving the positions 3-4'. Its calculated λ_{max} at 547 nm originates from two transitions at 538 and 561 nm. The assignment of the experimental spectrum to the radical anion in position 3-4' agrees with the relative stabilities of the various species reported in Figure 4. The λ_{max} for the radical anion in position 3-3' (see Figure 5) is at lower wavelength (530 nm) than that in position 3-4', due to the convolution of the main transition at 554 nm with minor transitions between 470 and 512 nm.

This radical anion was calculated to be less stable than that in position 3-4' by 6.6 kcal/mol. The radical anion in position 3-4' has two transitions, with comparable intensity at 513 and 620 nm, originating from a large absorption band which does not reproduce the shape of the experimental spectrum. This radical is less stable than that in position 3-4' by 3.4 kcal/mol. The radical anions involving the 5–OH (not shown in Figure 5) are destabilized by at least 10 kcal/mol with respect to the radical anion in position 3-4', and have the main absorption peaks at about 470 nm. All radical anions involving the 7–OH group show absorption maxima between 435 and 480 nm, that may be responsible for the shoulder at 450 nm visible in the experimental spectrum. Considering the radical anion in position 4'-7 as the one showing the closest electronic transition, we obtained a good matching of the experimental spectrum by simulating the UV–vis spectrum that would arise from the mixture of radical anions in positions 3-4' and 4'-7 in the ration of approximately 3:1 (Figure 5B).

3.3. Bond Dissociation Enthalpies

To rationalize the reactivity toward peroxyl radicals, the dissociation enthalpy of the OH bonds of quercetin and of the quercetin anion was computed in water. Undissociated quercetin reacts preferentially at the 3–OH group, with a BDE_{OH} of 83.1 kcal/mol (Table 2).

Table 2. Bond dissociation enthalpies (BDE_{OH}) of quercetin and of its anions, calculated in water at B3LYP/6-311+g (d,p) level using the isodesmic approach.

Starting Compound	Abstracted OH	BDE_{OH} (kcal/mol)
Neutral Quercetin	4'	83.3
	3'	86.0
	3	83.1
	7	94.0
	5	94.7
Anion 4'	7	81.2
	3'	79.1
	3	75.0
	5	86.4
Anion 7	4'	79.5
	3'	84.8
	3	78.7
	5	95.7

Concerning the BDE_{OH} of dissociated quercetin, calculations indicated that the most acidic –OH group is that in position 4', followed by those in positions 7 (+1.6 kcal/mol), 3' (3.2 kcal/mol), 3 (+3.9 kcal/mol) and 5 (+6.9 kcal/mol), in agreement with previous calculations [44]. The BDE of all –OH groups of the two most stable anions were then calculated as reported in Table 2. The results indicate that both anions in positions 4' and 7 preferentially react in position 3, its BDE_{OH} being 75.0 and 78.7 kcal/mol, respectively. The results in Table 2 justify that deprotonation boosts the reactivity of quercetin by lowering the BDE_{OH}.

3.4. A Proposed Mechanism for the Antioxidant Activity of Quercetin in Water

The reactivity of quercetin in acidic medium is superimposable to that of simple catechol, hence it most likely involves the catechol moiety. Calculations indicated that the most stable transient is the radical in position 3, which is more stable than the radical in position 4' by as little as 1.2 kcal/mol: a difference of marginal relevance considering the difficulty to reproduce the aqueous medium with a PCM.

Considering that both the radicals in positions 3 and 4' coexist in the transient UV spectrum obtained by pulse radiolysis and that the modest difference in BDE_{OH} (0.2 kcal/mol, see Table 2) is unlikely to be visible in kinetic measurements, our data indicate that both reaction pathways summarized in Scheme 3 could coexist and contribute to the observed reactivity.

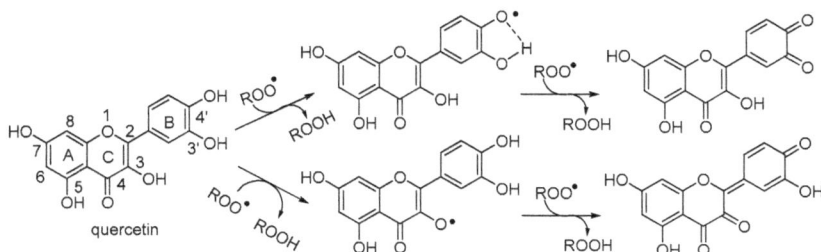

Scheme 3. Mechanistic proposal for the observed antioxidant activity of quercetin in water at pH = 2.1.

At close to neutral pH, instead, the reactivity of quercetin is far less simple than Scheme 1 would suggest. The trapping of four peroxyl radicals with different rate constants indicates that two distinct moieties are independently reacting with two radicals each: namely, the chromen-4-one core (rings A + C) and the catechol moiety (ring B). Clearly, deprotonation has a major role and a rate-determining reaction with peroxyl radicals must occur from the mono-anion, as was previously recognized [25]. Litwinienko and coworkers reported that the pK_a of quercetin in water/methanol is 8.45, indicating that a significant portion would be dissociated at pH 7.4, and identified position 7 as the most acidic [25]. Our calculations, as well as others [44], suggest instead that the anion in position 4' would be marginally more stable (again, some approximation in the calculations' model is to be considered). Interestingly, both the anions in positions 4' and 7 have the lowest BDE_{OH} in position 3, which matches the assignment of pulse radiolysis transients as a mixture of radical anions in positions 4'-3 and 7-3 in different ratios (main signal and shoulder, respectively). On this basis, we suggest that both equilibrating anions contribute to the observed reactivity as tentatively illustrated in Scheme 4. For each route and for the trapping of each of the four peroxyl radicals, current data do not allow to clearly distinguish whether the reaction occurs by concerted EPT or stepwise PT/ET, and most likely a combination of the two mechanisms is operating.

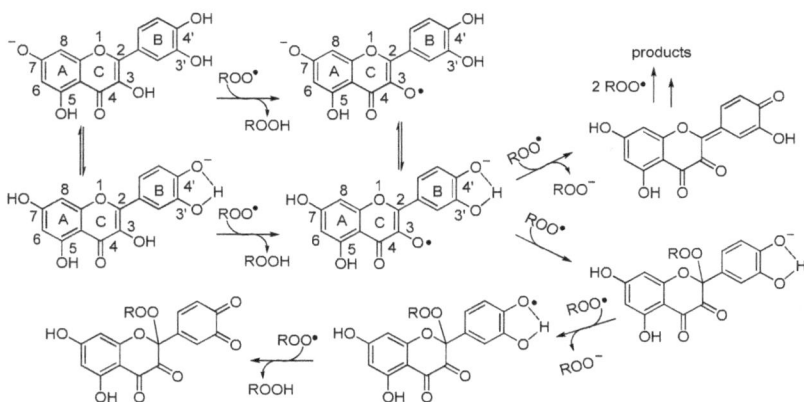

Scheme 4. Mechanistic proposal for the observed antioxidant activity of quercetin in water at pH = 7.4.

4. Conclusions

The antioxidant activity of quercetin in water strongly depends on pH, which affects both its absolute performance and its mechanism. Under acidic conditions (pH = 2.1), the antioxidant performance is modest and comparable to simple catechol: its mechanism appears to involve the trapping of two peroxyl radicals by the catechol moiety or by the catechol moiety (in position 4′) and by the –OH in position 3, which are the most reactive sites. Under more biomimetic settings (pH = 7.4), however, the antioxidant performance is boosted 40-fold, approaching that of α-tocopherol mimic PMHC and trapping twice as many (i.e., four) peroxyl radicals: two in the chromen-4-one core and two in the catechol moiety. The fastest rate-controlling reaction with peroxyl radicals comes from both equilibrating mono-anions in position 4′ (in the catechol) and position 7 (in the chromen-4-one core) and involves the –OH in position 3. Although the mono-anion may account for only about 10% of the quercetin in solution at pH 7.4, based on the reported pK_a of 8.45 [25] (lower pK_a values have also been reported [45]), the much higher reactivity of such an electron-rich structure would overwhelm that of the undissociated form, driving the whole process and progressively shifting the deprotonation equilibrium. Formal H-atom transfer to peroxyl radicals is likely to occur by a combination of concerted EPT and of stepwise PT/ET (i.e., SPLET) mechanisms, both being favored under such settings. The obtained results help explain and support the relevance of quercetin as a nutritional antioxidant. They will also prove useful in the rational design of novel catechol-based bioinspired antioxidants.

Supplementary Materials: The following are available online at www.mdpi.com/2313-7673/2/3/9/s1, Figure S1: Structures of the anions and of the neutral phenoxyl radicals of quercetin, Figure S2: Spin distribution of quercetin radicals, Figure S3: Structures of the radical anions of quercetin, Appendix A: Details on quanto-mechanical calculations.

Acknowledgments: This work was supported by the University of Bologna (FARB Project FFBO123154). A.C thanks the ERASMUS+ mobility program.

Author Contributions: R.A. and L.V. conceived and designed the experiments; R.A. also performed the DFT calculations; A.B. and A.C. performed the kinetic studies and A.B. also analyzed the data; L.V. wrote the paper, with the help of R.A.

Conflicts of Interest: The authors declare no conflict of interest. The founding sponsors had no role in the design of the study; in the collection, analyses, or interpretation of data; in the writing of the manuscript, and in the decision to publish the results.

References

1. Crozier, A.; Jaganath, I.J.; Clifford, M.N. Dietary phenolics: Chemistry, bioavailability and effects on health. *Nat. Prod. Rep.* **2009**, *26*, 1001–1043. [CrossRef] [PubMed]
2. Formica, J.V.; Regelson, W. Review of the biology of quercetin and related bioflavonoids. *Food Chem. Toxicol.* **1995**, *33*, 1061–1080. [CrossRef]
3. Erlund, I. Review of the flavonoids quercetin, hesperetin, and naringenin. Dietary sources, bioactivities, bioavailability, and epidemiology. *Nut. Res.* **2004**, *24*, 851–874. [CrossRef]
4. Bhagwat, S.; Haytowitz, D.B.; Holden, J.M. *USDA Database for the Flavonoid Content of Selected Foods*; Release 3; U.S. Department of Agriculture: Quilcene, WA, USA, 2011. Available online: https://www.ars.usda.gov/ARSUserFiles/80400525/Data/Flav/Flav_R03.pdf (accessed on 03 March 2017).
5. Justesen, U.; Knuthsen, P. Composition of flavonoids in fresh herbs and calculation of flavonoid intake by use of herbs in traditional Danish dishes. *Food Chem.* **2001**, *73*, 245–250. [CrossRef]
6. D'Andrea, G. Quercetin: A flavonol with multifaceted therapeutic applications? *Fitoterapia* **2015**, *106*, 256–271. [CrossRef] [PubMed]
7. Sharmila, G.; Bhat, F.A.; Arunkumar, R.; Elumalai, P.; Raja Singh, P.; Senthilkumar, K.; Arunakaran, J. Chemopreventive effect of quercetin, a natural dietary flavonoid on prostate cancer in in vivo model. *Clin. Nutr.* **2014**, *33*, 718–726. [CrossRef] [PubMed]
8. Murakami, A.; Ashida, H.; Terao, J. Multitargeted cancer prevention by quercetin. *Cancer Lett.* **2008**, *269*, 315–325. [CrossRef] [PubMed]

9.   Miles, S.I.; McFarland, M.; Niles, R.M. Molecular and physiological actions of quercetin: Need for clinical trials to assess its benefits in human disease. *Nutr. Rev.* **2014**, *72*, 720–734. [CrossRef] [PubMed]

10. EFSA Panel on Dietetic Products, Nutrition and Allergies (NDA). Scientific Opinion on the substantiation of health claims related to quercetin and protection of DNA, proteins and lipids from oxidative damage (ID 1647), "cardiovascular system" (ID 1844), "mental state and performance" (ID 1845), and "liver, kidneys" (ID 1846) pursuant to Article 13(1) of Regulation (EC) No 1924/2006. *EFSA J.* **2011**, *9*, 2067. [CrossRef]

11. Massaro, M.; Riela, S.; Guernelli, S.; Parisi, F.; Lazzara, G.; Baschieri, A.; Valgimigli, L.; Amorati, R. A synergic nanoantioxidant based on covalently modified halloysite-trolox nanotubes with intra-lumen loaded quercetin. *J. Mater. Chem. B* **2016**, *4*, 2229–2241. [CrossRef]

12. Zhu, B.; Yu, L.; Yue, Q.C. Co-delivery of vincristine and quercetin by nanocarriers for lymphoma combination chemotherapy. *Biomed. Pharmacother.* **2017**, *91*, 287–294. [CrossRef] [PubMed]

13. Wu, M.; Cao, Z.; Zhao, Y.; Zeng, R.; Tu, M.; Zhao, J. Novel self-assembled pH-responsive biomimetic nanocarriers for drug delivery. *Mater. Sci. Eng. C* **2016**, *64*, 346–353. [CrossRef] [PubMed]

14. Leopoldini, M.; Russo, N.; Toscano, M. The molecular basis of working mechanism of natural polyphenolic antioxidants. *Food Chem.* **2011**, *125*, 288–306. [CrossRef]

15. Galano, A.; Mazzone, G.; Alvarez-Diduk, R.; Marino, T.; Alvarez-Idaboy, J.R.; Russo, N. Food antioxidants: Chemical insights at the molecular level. *Annu. Rev. Food Sci. Technol.* **2016**, *7*, 335–352. [CrossRef] [PubMed]

16. Valgimigli, L.; Amorati, R.; Petrucci, S.; Pedulli, G.F.; Hu, D.; Hanthorn, J.J.; Pratt, D.A. Unexpected acid catalysis in reactions of peroxyl radicals with phenols. *Angew. Chem. Int. Ed.* **2009**, *48*, 8348–8351. [CrossRef] [PubMed]

17. Kumar, S.; Engman, L.; Valgimigli, L.; Amorati, R.; Fumo, M.G.; Pedulli, G.F. Antioxidant profile of ethoxyquin and some of its S, Se, and Te analogues. *J. Org. Chem.* **2007**, *72*, 6046–6055. [CrossRef] [PubMed]

18. Amorati, R.; Lynett, P.T.; Valgimigli, L.; Pratt, D.A. The reaction of sulfenic acids with peroxyl radicals: Insights into the radical-trapping antioxidant activity of plant-derived thiosulfinates. *Chem. Eur. J.* **2012**, *18*, 6370–6379. [CrossRef] [PubMed]

19. Matera, R.; Gabbanini, S.; Berretti, S.; Amorati, R.; De Nicola, G.R.; Iori, R.; Valgimigli, L. Acylated anthocyanins from sprouts of *Raphanus sativus* cv. Sango: Isolation, structure elucidation and antioxidant activity. *Food Chem.* **2015**, *166*, 397–406. [CrossRef] [PubMed]

20. Amorati, R.; Valgimigli, L. Advantages and limitations of common testing methods for antioxidants. *Free Radic. Res.* **2015**, *49*, 633–649. [CrossRef] [PubMed]

21. Pedrielli, P.; Pedulli, G.F.; Skibsted, L.H. Antioxidant mechanism of flavonoids. Solvent effect on rate constant for chain-breaking reaction of quercetin and epicatechin in autoxidation of methyl linoleate. *J. Agric. Food Chem.* **2001**, *49*, 3034–3040. [CrossRef] [PubMed]

22. Tarozzi, A.; Bartolini, M.; Piazzi, L.; Valgimigli, L.; Amorati, R.; Bolondi, C.; Djemil, A.; Mancini, F.; Andrisano, V.; Rampa, A. From the dual function lead AP2238 to AP2469, a multi-target-directed ligand for the treatment of Alzheimer's disease. *Pharmacol. Res. Perspect.* **2014**, *2*, e00023. [CrossRef] [PubMed]

23. Amorati, R.; Valgimigli, L.; Panzella, L.; Napolitano, A.; d'Ischia, M. 5-S-lipoylhydroxytyrosol, a multidefense antioxidant featuring a solvent-tunable peroxyl radical-scavenging 3-thio-1,2-dihydroxybenzene motif. *J. Org. Chem.* **2013**, *78*, 9857–9864. [CrossRef] [PubMed]

24. Zielinski, Z.; Presseau, N.; Amorati, R.; Valgimigli, L.; Pratt, D.A. Redox chemistry of selenenic acids and the insight it brings on transition state geometry in the reactions of peroxyl radicals. *J. Am. Chem. Soc.* **2014**, *136*, 1570–1578. [CrossRef] [PubMed]

25. Musialik, M.; Kuzmicz, R.; Pawłowski, T.S.; Litwinienko, G. Acidity of hydroxyl groups: An overlooked influence on antiradical properties of flavonoids. *J. Org. Chem.* **2009**, *74*, 2699–2709. [CrossRef] [PubMed]

26. Amorati, R.; Baschieri, A.; Morroni, G.; Gambino, R.; Valgimigli, L. Peroxyl radical reactions in water solution: A gym for proton-coupled electron-transfer theories. *Chem. Eur. J.* **2016**, *22*, 7924–7934. [CrossRef] [PubMed]

27. Amorati, R.; Baschieri, A.; Valgimigli, L. Measuring antioxidant activity in bioorganic samples by the differential oxygen uptake apparatus: Recent advances. *J. Chem.* **2017**, *2017*, 6369358. [CrossRef]

28. Scott, A.P.; Radom, L. Harmonic vibrational frequencies: An evaluation of Hartree–Fock, Møller–Plesset, quadratic configuration interaction, density functional theory, and semiempirical scale factors. *J. Phys. Chem.* **1996**, *100*, 16502–16513. [CrossRef]

29. Guerra, M.; Amorati, R.; Pedulli, G.F. Water effect on the O–H dissociation enthalpy of *para*-substituted phenols: A DFT study. *J. Org. Chem.* **2004**, *69*, 5460–5467. [CrossRef] [PubMed]

30. Lind, J.; Shen, T.; Eriksen, E.; Merenyi, G. The one-electron reduction potential of 4-substituted phenoxyl radicals in water. *J. Am. Chem. Soc.* **1990**, *112*, 479–482. [CrossRef]

31. Amat, A.; Clementi, C.; De Angelis, F.; Sgamellotti, A.; Fantacci, S. Absorption and emission of the apigenin and luteolin flavonoids: A TDDFT investigation. *J. Phys. Chem. A* **2009**, *113*, 15118–15126. [CrossRef] [PubMed]

32. Frisch, M.J.; Trucks, G.W.; Schlegel, H.B.; Scuseria, G.E.; Robb, M.A.; Cheeseman, J.R.; Montgomery, J.A., Jr.; Vreven, T.; Kudin, K.N.; Burant, J.C.; et al. *Gaussian 03, Revision D.02*; Gaussian, Inc.: Wallingford, CT, USA, 2004.

33. Bremond, E.A.G.; Kieffer, J.; Adamo, C. A reliable method for fitting TD-DFT transitions to experimental UV–visible spectra. *J. Mol. Struct. THEOCHEM* **2010**, *954*, 52–56. [CrossRef]

34. Amorati, R.; Valgimigli, L. Modulation of the antioxidant activity of phenols by non-covalent interactions. *Org. Biomol. Chem.* **2012**, *10*, 4147–4158. [CrossRef] [PubMed]

35. Lucarini, M.; Pedulli, G.F.; Valgimigli, L. Do peroxyl radicals obey the principle that kinetic solvent effects on H-Atom abstraction are independent of the nature of the abstracting radical? *J. Org. Chem.* **1998**, *63*, 4497–4499. [CrossRef]

36. Foti, M.C.; Daquino, C.; Di Labio, G.A.; Ingold, K.U. Kinetics of the oxidation of quercetin by 1,1-diphenyl-2-picrylhydrazyl (DPPH). *Org. Lett.* **2011**, *13*, 4826–4829. [CrossRef] [PubMed]

37. Mukai, K.; Oka, W.; Watanabe, K.; Egawa, Y.; Nagaoka, S.; Terao, J. Kinetic study of free-radical-scavenging action of flavonoids in homogeneous and aqueous Triton X-100 micellar solutions. *J. Phys. Chem. A* **1997**, *101*, 3746–3753. [CrossRef]

38. Jovanovic, S.V.; Steenken, S.; Hara, Y.; Simic, M.G. Reduction potentials of flavonoid and model phenoxyl radicals. Which ring in flavonoids is responsible for antioxidant activity? *J. Chem. Soc. Perkin Trans. 2* **1996**, 2497–2504. [CrossRef]

39. Shen, L.; Zhang, H.-Y.; Ji, H.-F. Successful application of TD-DFT in transient absorption spectra assignment. *Org. Lett.* **2005**, *7*, 243–246. [CrossRef] [PubMed]

40. Lalevee, J.; Allonas, X.; Fouassier, J.-P.; Ingold, K.U. Absolute rate constants for some intermolecular reactions of aminoalkylperoxyl radicals. Comparison with alkylperoxyls. *J. Org. Chem.* **2008**, *73*, 6489–6496. [CrossRef] [PubMed]

41. Chatgilialoglu, C.; D'Angelantonio, M.; Guerra, M.; Kaloudis, P.; Mulazzani, Q.G. A Reevaluation of the ambident reactivity of the guanine moiety towards hydroxyl radicals. *Angew. Chem. Int. Ed.* **2009**, *48*, 2214–2217. [CrossRef] [PubMed]

42. Kaloudis, P.; D'Angelantonio, M.; Guerra, M.; Spadafora, M.; Cismas, C.; Gimisis, T.; Mulazzani, Q.G.; Chatgilialoglu, C. Comparison of isoelectronic 8-HO-G and 8-NH2-G derivatives in redox processes. *J. Am. Chem. Soc.* **2009**, *131*, 15895–15902. [CrossRef] [PubMed]

43. Leopoldini, M.; Marino, T.; Russo, N.; Toscano, M. Density functional computations of the energetic and spectroscopic parameters of quercetin and its radicals in the gas phase and in solvent. *Theor. Chem. Acc.* **2004**, *111*, 210–216. [CrossRef]

44. Fiorucci, S.; Golebiowski, J.; Cabrol-Bass, D.; Antonczak, S. DFT study of quercetin activated forms involved in antiradical, antioxidant, and prooxidant biological processes. *J. Agric. Food Chem.* **2007**, *55*, 903–911. [CrossRef] [PubMed]

45. Milane, H.A.; Ubeaud, G.; Vandamme, T.F.; Jung, L. Isolation of quercetin's salts and studies of their physicochemical properties and antioxidant relationships. *Bioorg. Med. Chem.* **2004**, *12*, 3627–3635. [CrossRef] [PubMed]

Chapter 2:
Catechol-Based Biomechanisms and Bioactivity

biomimetics

Article

Examining Potential Active Tempering of Adhesive Curing by Marine Mussels

Natalie A. Hamada [1], Victor A. Roman [1], Steven M. Howell [1] and Jonathan J. Wilker [1,2,*]

[1] Department of Chemistry, Purdue University, 560 Oval Drive, West Lafayette, IN 47907-2084, USA; nhamada@purdue.edu (N.A.H.); roman6@purdue.edu (V.A.R.); showell2@niu.edu (S.M.H.)
[2] School of Materials Engineering, Purdue University, 701 West Stadium Avenue, West Lafayette, IN 47907-2045, USA
*   Correspondence: wilker@purdue.edu; Tel.: +1-765-496-3382

Academic Editors: Marco d'Ischia and Daniel Ruiz-Molina
Received: 26 July 2017; Accepted: 17 August 2017; Published: 21 August 2017

Abstract: Mussels generate adhesives for staying in place when faced with waves and turbulence of the intertidal zone. Their byssal attachment assembly consists of adhesive plaques connected to the animal by threads. We have noticed that, every now and then, the animals tug on their plaque and threads. This observation had us wondering if the mussels temper or otherwise control catechol chemistry within the byssus in order to manage mechanical properties of the materials. Here, we carried out a study in which the adhesion properties of mussel plaques were compared when left attached to the animals versus detached and exposed only to an aquarium environment. For the most part, detachment from the animal had almost no influence on the mechanical properties on low-energy surfaces. There was a slight, yet significant difference observed with attached versus detached adhesive properties on high energy surfaces. There were significant differences in the area of adhesive deposited by the mussels on a low- versus a high-energy surface. Mussel adhesive plaques appear to be unlike, for example, spider silk, for which pulling on the material is needed for assembly of proteinaceous fibers to manage properties.

Keywords: adhesion; adhesive; byssus; catechol; DOPA; mussel; plaque; thread

---

1. Introduction

1.1. Catechols in the Sea

Mussels, sandcastle worms, and tube worms may be the most famous proponents of catechol chemistry [1]. These animals attach themselves to rocks using protein-based adhesives containing 3,4-dihydroxyphenylalanine (DOPA), for which the amino acid sidechain is a pendant catechol group. The surface adhesive properties of DOPA groups arise when the catechol ring is in the reduced (i.e., not oxidized) state [2,3]. Although some evidence does exist for oxidation when bonding at organic surfaces [4]. Cohesive strength for the glues is derived from one electron (to semiquinone) or two-electron (to quinone) oxidation of DOPA to then generate covalent cross-links, often with iron beginning such reactivity [5,6].

As our understanding of these natural systems has expanded, so too has there been a blossoming of biomimetic systems [7–10]. In a typical scenario, synthetic polymer backbones are synthesized to substitute for the protein matrix [11,12]. Derivatives of catechol are then appended to the polymer chain. In doing so, these efforts have given rise to new functional materials, including hydrogels, coatings, and adhesives. With greater understanding of the animals' biology, chemistry, and mechanics will come the design ideas for new biomimetic systems.

1.2. Animals Managing Their Glue with Mechanical Forces

Such a chemical perspective is helpful for materials design, although beyond the grasp of mussels, themselves. We have been wondering how the animals manage their adhesive. They are well known to attach atop substrates with the byssal plaque and thread structure visible in Figure 1. Our research group has been working with these shellfish for several years now [13–16]. During these studies, we have noticed that the animals are not completely passive after deposition of adhesive plaques. Mussels do not apply the glue and then simply sit around. Rather, upon closer observation, it appears as if the animals are pulling on their threads. This byssal tugging can be observed with movement of the animal while the threads remain relatively stationary, appearing as if the animals are shaking around a bit. Such motion is quite distinct from the constant opening and closing of their shells (i.e., valves) or movement resulting from turbulent/high water flow.

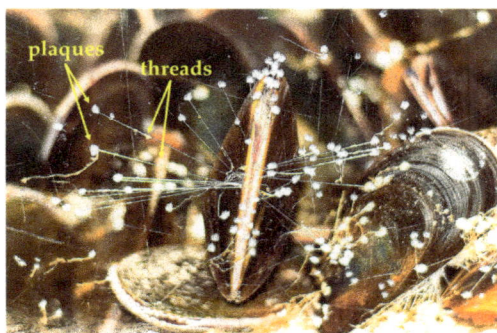

Figure 1. Marine mussel adhered to a glass wall of an aquarium tank. The byssal adhesive assemblies consist of adhesive plaques and threads which remain tethered to the byssal retractor muscles inside the shells.

From the surface up, there is the adhesive plaque and the thread, which is connected to the byssal retractor muscle hidden inside their shells. This muscle, controlled by a decentralized nervous system of paired ganglia, can contract and relax, thereby changing position of the animal relative to the substrate, as well as balance tension on the adhesive threads [17]. Perhaps an occasional tug induces mechanical strain on the thread and glue after deposition, helps to order the molecules, and creates a material well suited for living amongst the crushing waves of the intertidal zone.

1.3. Mussel Byssus and Mechanics

Mussels have created several strategies for mitigating mechanical forces from their surroundings. The byssal threads are often spread out in many directions, thereby aiding the ability to deal with forces from waves and turbulent waters [18–20]. These threads are made of a solid, non-porous material. The distal portion of threads, closest to the adhesive plaque on rocky substrates, is crystalline [18,21]. The proximal thread, nearest to the retractor muscles within the shell, is more elastic [18,21]. Generating such gradients of stiff to flexible, the threads may be tuned to dampen shocks. By contrast, the plaques have a microporous structure, potentially dissipating mechanical forces throughout the material [22]. Quite interestingly, this multi-component material changes in mechanical properties (e.g., tenacity) with seasons, wave action, food supply, pH, and temperature [23–26]. We wonder if forces, be they from the surroundings or the animals, are influencing the nature of the mussel's attachment system. There has been some evidence suggesting that byssal threads can increase in material hardness with physical agitation or over time as the material ages [27,28]. Such motion, hydrated conditions, and the presence of dissolved oxygen in seawater could promote cross-linking based upon catechol chemistry. Perhaps internal forces could also play a role in curing mussel adhesives.

1.4. Animal Mechanics May Influence the Performance of Mimics

Our current, state of the art mimics of mussel adhesive proteins are often designed with a bottom-up approach. We start with small molecule insights, progress to larger molecules, such as plaques, and then examine bulk properties. When designing the next generation of biomimetic materials, we may wish to consider a more top-down approach from the animals' perspective.

Examples of linking molecular with macroscopic design have been shown in the production of biomimetic silk [29–31]. Spiders tailor silk protein properties by specific processing with their spinnerets. Mechanical stresses induced by pulling on the silk helps to promote structural transitions from a semi-amorphous material to crystalline β-sheet and α-helical protein structures, giving silk the highly-prized properties of a strong, yet elastic, material [30,32].

It took years of effort to learn how such mechanical forces were essential for obtaining properties of real silk from mimics [30–33]. For example, silk proteins can be expressed in solution, isolated, and used to make materials, but the properties were always lacking when compared to genuine silk from spiders. Only when spinneret mechanical forces were incorporated could the desired, fully mimetic material properties be achieved.

Spider silk is composed of repeat glycine-, proline-, and alanine-rich peptide domains forming hierarchical structures for strength and elasticity [34]. Similarly, mussels have several mussel foot proteins (mfp) rich in glycine and uncharged residues [35]. Mussels might also use a pulling technique to modify the mechanical properties of their proteinaceous materials. This idea was proposed recently with regard to the mussel byssal threads [36]. These threads begin from secretory vesicles and end up being semi-crystalline structures. Mechanical forces from the animal may help organize the molecules into regular structures.

What about the adhesive plaques, themselves? In contrast to mussel and spider threads, the plaques are not crystalline. Rather, the adhesive has a foam-like structure [22]. Might mechanical forces help form the bulk material? We do know that formation of mussel plaques requires a large degree of cross-linking chemistry [1]. The DOPA residues in mussel foot proteins become oxidized [1,6,13,37]. Subsequent reactions with nucleophiles, such as reduced DOPA residues, amines, and thiols can bring about covalent cross-links to cure the material. However, this chemistry takes place within the solid plaque, or perhaps a precursor that is a foam or gel. In any case, mobility of reactive groups is low when compared to typical synthetic solution chemistry in which reagents are combined in solvents.

How do the reactive groups such as an electrophilic semiquinone and a nucleophilic, unoxidized DOPA find each other to couple within a solid material? We do know that such reactions are slow within this solid matrix given that radicals from iron-induced oxidation can be observed hours after the animals deposits this glue [13]. Most often in chemistry, radical species are so short-lived that they cannot be isolated and observed the way that they can for mussel adhesive. Given that these reactive groups appear to be at least somewhat trapped within a solid or semi-solid, we wonder if applied mechanical forces may aid curing.

Figure 2 depicts a scenario in which reactive groups are physically separated from each other. Similar to spider silk, application of mechanical forces can shift orientation or placement of molecules with respect to each other. Bringing the electrophiles and nucleophiles closer together could then enable covalent coupling and curing. This chemistry could help to account for our observations of the animals tugging on their glue. If such interplay exists between curing chemistry and animal mechanics, we will then have an additional parameter to consider when processing mussel-mimicking synthetic polymers.

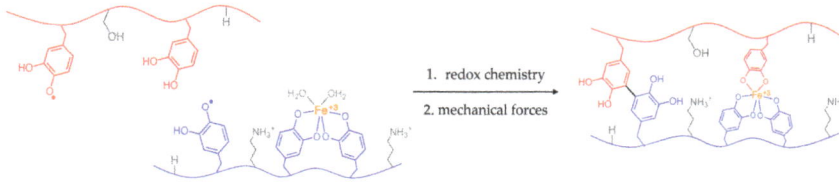

Figure 2. Possible scenario depicting how reactive species within a solid could be brought closer together via mechanical forces. Tugging from byssal retractor mussels could better align reactive groups for adhesive curing.

1.5. Testing the Influence of Forces from Mussels

After noticing consistent byssal 'tugging' by the animals, we became compelled to examine the potential effects of this behavior on material performance. We generated two sets of mussels and their adhesive: attached and detached. In the attached case, mussels deposited their glue onto surfaces and, after three days, the adhesive performance was measured. In the detached case, newly forged threads were immediately severed from the animal, resulting in samples that were connected to the animal for no more than 12 h.

Threads connecting animal and adhesive plaque were cut from the animal at the point providing the maximum thread length. These detached plaques were maintained in the same aquarium system as the animals to complete the three-day period (Figure 3). Animals were periodically checked and fresh adhesive was immediately detached from the animal ($\leq$12 h) until completion of the three-day experiment. Adhesion performance was then assessed.

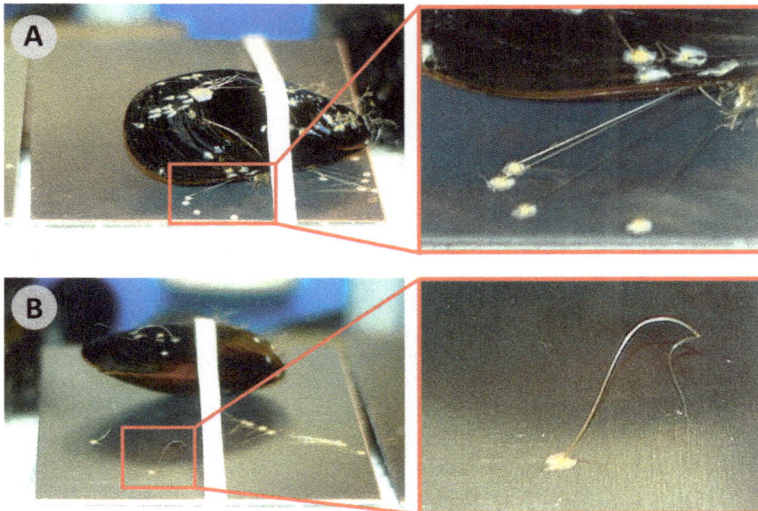

Figure 3. Experimental setup of marine mussels banded to aluminum substrates where: (**A**) Threads remain attached to the animal, taught and controlled by byssal retractor muscles; (**B**) Threads have been detached from the mussel at the closest point near their shell using a razor blade. The soft proximal portion of the threads can be seen where the threads start to curl and wave. The crystalline distal portions remain relatively straight.

These experiments were carried out with both high surface energy 6061-T6 aluminum and low surface energy poly(methyl methacrylate) (PMMA, referred to as acrylic) substrate panels. Typical surface energy values fall around 41 and 169 mJ/m^2 for acrylic and aluminum, respectively [38]. Exploiting low and high surface energy substrates gives insights on the nature in which these adhesives fail, which is further discussed in Section 3.1. In all, four datasets were created—attached on aluminum, detached on aluminum, attached on acrylic, and detached on acrylic. In the end, property differences between attached and detached plaques were found to be quite minimal, but there were observed differences in plaque areas on the two different surfaces.

2. Materials and Methods

2.1. Animal Handling

Live blue mussels (*Mytilus edulis*) were obtained from fishermen in Maine, USA, and shipped to our laboratory. The animals were maintained in an aquarium system that has been described previously [14]. Briefly, the system entails water at 4 °C, with a surge system for periodic turbulent flow to mimic periodic shore breaks along a coast, and artificial day/night cycles to represent Maine in February. Mussels used for this study were all of 50–70 mm in length. Animal feeding remained constant throughout the data collection period, which consisted of an enriched phytoplankton diet (Phyto-Feast, Reef Nutrition, Campbell, CA) every two days.

2.2. Adhesive Deposition

If mussels are placed onto a new surface, they tend to "walk" away, finding neighbors to aggregate with and stick together. Consequently, we placed the animals onto substrate sheets while held loosely in place with rubber bands. The substrates used here were 10 × 10 cm sheets of 6061-T6 aluminum (Farmers Copper Ltd., Texas City, TX, USA) and PMMA (United States Plastic Corp., Lima, OH, USA). Aluminum substrates were cleaned with detergent, ethanol, and acetone rinses followed by 3× rinses with deionized water. Acrylic substrates were prepared in the same manner, excluding the acetone wash. Each mussel/substrate/rubber band assembly was connected to a large plastic grate via zip ties in order to ensure exposure to consistent flow rates in a turbulent environment. Attached and detached samples were alternately arranged along the grate to minimize any deviations in flow rate along the sample set [24]. All mussels were oriented in the same manner in order to reduce variations in drag produced from their shell, with their long axis perpendicular to the upcurrent and the narrower region of valves facing into the current [20]. We did note that the aluminum substrates became slightly darker after three days of exposure in salt water.

2.3. Attached versus Detached Adhesive

Mussels were divided into four groups: attached plaques and detached plaques, each on aluminum or acrylic substrates. For the attached data, animals were placed atop substrates and they formed plaques and threads. Mussels were kept in the turbulent aquarium and adhesion measurements were made after three days. After years of working with these animals, we found that a period of three days is a sufficient amount of time for the mussels to anchor themselves securely to a given substrate. For the detached data, animals were placed onto the sheets of aluminum or acrylic. Once the adhesive was deposited, threads were cut at a point adjacent to the shell. The threads were cut as soon as they were visible or, at the longest, within 12 h of deposition. If further adhesive was produced later, similar cuts were made. These substrates with plaques were maintained in the aquarium, without the respective animals, for a total of three days. Thus, all adhesion data were collected three days after placing the animals within the tank. The attached plaques remained with the mussels for the whole three days. The detached plaques away from mussels for at least 2.5 days.

2.4. Adhesion Measurements

At the end of the experiment, all remaining threads were cut from the animal and immediately tested for adhesive strength using an Instron testing system (Instron 5544, Instron, Norwood, MA, USA). Details of pulling plaques until failure were reported previously [14]. Briefly, clamps surrounded the threads, were tightened, and pulled normal from the surface until failure. The mussel byssus tends to be fanned out into several or all directions from the animal (Figure 1). Adhesion testing can involve pulling each thread/plaque at different angles from the surface [39]. In practice, we tend to prefer using a 90° pull, normal from the surface, for all measurements [14,15]. Use of this approach allows for rapid collection of large data volumes in a consistent manner [14,15]. Such considerations are especially important in studies like this current one in which we have examined almost 1000 plaques. All plaque areas were measured using digital photography and ImageJ software [40].

2.5. Statistics

A total of 36 mussels were used in this experiment, resulting in 18 mussels used for each attached and detached group. Rather than running experiments on 18 mussels per group at one time, two data runs were carried out. Each data run contained nine mussels for collection of attached and nine mussels for collection of detached adhesive. Data from the two runs were then pooled. This approach helps to minimize variabilities, such as animal behavior within a given time period [15,26]. Error bars provided show 95% confidence intervals.

Each adhesive plaque was treated as an individual replicate. Pooled data included an average of all plaques, which we tend to refer to as 'average of all'. We have also examined the average values for each mussel, and pooled those data separately, which we refer to as 'average of the average'. There were no significant differences between these calculation methods, except in the magnitude of the error bars.

Statistics were calculated using SPSS software (IBM Corp., Armonk, NY, USA). An independent Student's *t*-test examined significant differences within adhesion, force of removal, plaque area, and work of adhesion variables. The assumption of homogeny of variances were assessed using Laverne's test, where the a priori α was set at a value of 0.5. Alternate hypotheses were accepted at *p*-values less than 0.05. Significance (*) is indicated in the figures. No statistical differences were found within acrylic variables. Conversely, the aluminum group found significant differences between attached and detached values within adhesion, force, and work of adhesion variables. The different areas of adhesive plaques deposited on acrylic and aluminum substrates were also statistically significant.

3. Results and Discussion

3.1. Adhesive Production and Failure Modes

For each dataset (attached aluminum, detached aluminum, attached acrylic, detached acrylic), 18 animals were used. Table 1 provides the pooled data for the plaques produced by the animals in each case. In all, 923 plaques were deposited. There are several ways in which mussel adhesive may fail under these conditions [14,39]. The plaque can be completely removed off the surface ("adhesive failure"), the plaque itself may tear apart ("cohesive failure"), the junction of plaque and thread might break ("thread–plaque failure") or the thread may become severed ("thread breakage"). In practice, thread breakage is rare given that we cover the entire thread with the clamp providing the forces.

Table 1. Adhesion results of attached versus detached byssus on acrylic and aluminum.

	Acrylic		Aluminum	
	Detached	Attached	Detached	Attached
Number of plaques	231	134	314	244
Average adhesion (kPa)	65 ± 3	66 ± 4	127 ± 4	135 ± 5
Force of detachment (N)	0.34 ± 0.02	0.34 ± 0.02	0.53 ± 0.02	0.58 ± 0.03
Plaque area ($\times 10^{-6}$ m^2)	5.4 ± 0.3	5.3 ± 0.3	4.3 ± 0.1	4.4 ± 0.2
Work of adhesion (N/m)	148 ± 14	176 ± 25	401 ± 22	441 ± 30
Adhesive failure %	90.0	84.2	26.8	23.4
Cohesive failure %	4.8	7.5	57.3	58.6
Thread break failure %	3.5	3.0	10.8	8.6
Thread–plaque interface failure %	1.7	5.3	5.1	9.4

Speaking generally for all adhesives, high energy surfaces, such as metals or rocks, yield strong adhesion, whereas bonds are less robust on low-energy plastics. Translated to the conditions used here, mussel plaques are likely to exhibit significant degrees of cohesive failure on aluminum. Table 1 shows both the attached and detached plaques on aluminum failed cohesively just over half the time. Adhesive failure comprised the majority of remaining events. By contrast, weaker bonding to acrylic was manifested in a high percentage of adhesive failure, at 84% or more (Table 1). When comparing attached versus detached plaques on either surface, there were no major differences in distribution of failure modes. In at least this regard, the animals do not appear to be exerting any influence over the material properties.

3.2. Overall Adhesion

For the low-energy acrylic substrate, Figure 4A shows the adhesion results. Performance of plaques that remained attached to the animals showed no significant difference. Adhesion is often expressed as the force of detachment divided by the overlap area. Hence, Figure 4B,C separate out these values. The force at failure, in Newtons, was very similar for attached and detached plaques on acrylic. Similarly, the plaque areas upon a given substrate remained quite constant between these two datasets.

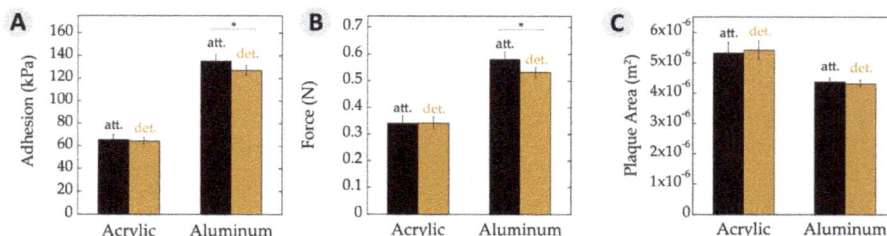

Figure 4. Adhesive properties. (**A**) Average adhesion of attached (att.) and detached (det.) plaques and threads on acrylic and aluminum substrates measured in kilopascals. Adhesion was calculated as a function of removal force divided by plaque area. Note overall increase in adhesion seen on a high surface energy substrate (aluminum) relative to a low energy surface (acrylic); (**B**) Force of byssal removal on acrylic and aluminum substrates measured in newtons; (**C**) Average plaque area deposited on acrylic and aluminum substrates. Note the change in plaque areas on acrylic versus aluminum. Asterisks (*) indicate statistically significant differences ($p \leq 0.05$). All error bars show 95% confidence intervals.

The results slightly changed when looking at mussels on aluminum (Figure 4). The attached plaques showed a small, yet significant, increase in adhesion to the detached counterparts. Likewise, the maximum force at failure was very similar and the plaque areas were nearly identical for those attached versus detached. With aluminum being a high-energy surface and giving rise to generally high adhesion, potential differences might be seen more easily here versus with the acrylic surfaces. Nonetheless, the performance was very similar for adhesion, force at failure, and area.

There is often a direct relationship between adhesive strength and the substrate surface energy [38,41,42]. High-energy surfaces increase the spreadability of the adhesive due to enhanced adsorption and surface binding, yielding an overall stronger bond relative to a low-energy surface. This phenomenon was demonstrated by the changes to overall force and adhesion, where values increased by nearly 100% on aluminum versus acrylic. Interestingly, the adhesive area increased by ≈30% on low-energy acrylic compared to high-energy aluminum. There has been research investigating the relationship between the area of plaques and surface energy, but the reported results are somewhat contradictory [14,43–46]. In the case of adhesive plaques, the mfp-5, mfp-3f, and mfp-3s proteins are deposited along the adhesive–substrate interface [10,47]. The mpf-5 and mfp-3 proteins have the highest DOPA contents at 30% and 19%, respectively. Perhaps the mussel is able to control the spatial deposition and differentiate between proteins. They may then be able to maximize substrate binding via increasing mfp-5 and -3 at the interface of a low-energy substrate by upregulation of these proteins, using mechanical forces to increase the adhesive area, or simply the spreadability of the adhesive proteins behaves differently as the surface energy is changed.

Similarly, a larger plaque area would increase the overall protein as well as catechols in contact with the surface. These changes could thus aid animal binding to low-energy surfaces and explain the larger plaques observed versus upon high-energy aluminum. Placing such observations within the context of designing biomimetic adhesives may teach us two lessons. First, greater overlap area will enhance bonding to plastics. This idea is neither surprising nor unprecedented. More useful, however, may be the second prediction. Mussel mimicking polymers may exhibit higher adhesion between low-energy substrates when the catechol content is greater than those used to bind high-energy surfaces. In other words, polymers designed with a low catechol content make better adhesives on metals and other high-energy surfaces, whereas polymers with a higher catechol content may be used for bonding low-energy plastics.

3.3. Work of Adhesion

Measuring the maximum load at failure of an adhesive material is a common way to express the relevant force. An alternative view on the strength of a bond is the energy required to rupture the joint. This work of adhesion is determined by integrating the area under force-versus-extension curves and factoring in overlap area [38,41,42]. Figure 5A,B shows such plots for the cases of adhesive failure on acrylic and aluminum for typical attached and detached plaques, which generally looked similar in shape. It is interesting to note that force–extension graphs from aluminum samples initiated with a steep slope within the first 1 mm of extension, followed by a more gradual slope until failure. This yield has been observed in mechanical tests of mussel adhesive threads, suggesting that the change in slope corresponds with transition of forces between several constituents and/or phases within the material [28,48]. Our testing system is designed to reduce factors from the threads as much as possible by covering almost the entire thread with the clamps. Nearly all of the observed forces are, thus, to deform the adhesive plaque. These plots provide what may be the first evidence that yielding phenomena could also be at play with the plaques.

The work of adhesion values for attached and detached plaques, both on acrylic and aluminum are depicted in Figure 5C. Here, we chose to present data averaged over all failure modes. Nearly all failure was adhesive in nature when on acrylic. Some differences may be found when separating out, for example, adhesive versus cohesive failure on aluminum [49]. However, the general trends remained the same. The work of adhesion on aluminum substrates showed a slight, yet significant, decrease with detachment relative to the attached adhesive. Similarly, adhesive on acrylic substrates decreased a little when moving from being attached to detached, but the potential changes here remained within the error limits. The observed differences here may be statistically significant, but are, nonetheless, small.

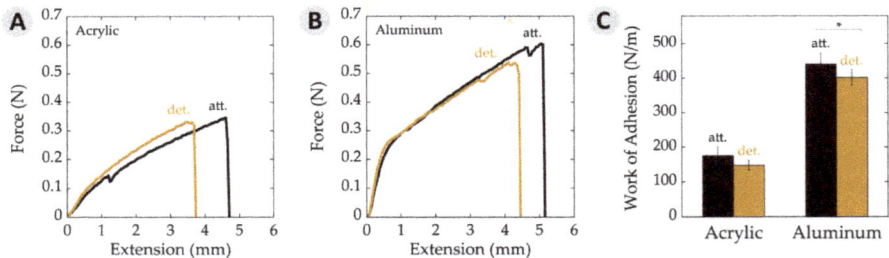

Figure 5. Work of adhesion data per unit area. (**A**) Typical force–extension curves on acrylic for attached (att.) and detached (det.) samples. Curves represent average plaque area, force, and extension values seen on acrylic; (**B**) Typical force–extension curves on aluminum substrates for attached and detached samples. Curves represent average plaque area, force, and extension values seen on aluminum; (**C**) Work of adhesion per unit area showing attached and detached values on acrylic and aluminum substrates. The asterisk (*) indicates a statistically significant difference ($p \leq 0.05$). Error bars depict 95% confidence intervals.

4. Conclusions

Our growing understanding of bioadhesives has given rise to an array of biomimetic materials and applications development. While we make such new systems, we are still teasing out the details contained within the parent systems under the seas. Data presented here help to examine the degree to which mussels influence the performance of their glue after deposition onto the substrate. On both a low- and high-energy surface, bonding was more or less equivalent, whether or not the animal had access to the adhesive. Measured plaque areas increased on a low- versus high-energy surface, suggesting that the animals have means to maximize bonding to a variety of surfaces. The observed tugging of mussels upon their byssus had very little influence on adhesion. These results indicate that future adhesive mimics may slightly benefit from mechanical pre-stressing when on high-energy surfaces. Mimics of mussel threads and spider silks, by contrast, are likely to improve properties with applied mechanical forces during processing. Perhaps the tugging we are seeing is the animals behaving in a manner akin to stretching when waking up in the morning.

Acknowledgments: We appreciate funding for this project provided by the Office of Naval Research (grants N000141612709, N000141310245) and the National Science Foundation (grant CHE-0952928). We would also like to thank Bradley McGill and Lee Huntington (Purdue University, IN, USA) for keeping the mussels alive and happy as clams, and to Trevor Meyer (Purdue University) for photo edits.

Author Contributions: N.A.H., V.A.R., and S.M.H. carried out all of the experiments. N.A.H. and J.J.W. wrote the paper. J.J.W. oversaw the project.

Conflicts of Interest: The authors declare no conflicts of interest.

References

1.	Hagenau, A.; Suhre, M.H.; Scheibel, T.R. Nature as a blueprint for polymer material concepts: protein fiber-reinforced composites as holdfasts of mussels. *Prog. Polym. Sci.* **2014**, *39*, 1564–1583. [CrossRef]
2.	Lee, H.; Scherer, N.F.; Messersmith, P.B. Single-molecule mechanics of mussel adhesion. *Proc. Natl. Acad. Sci. USA* **2006**, *103*, 12999–13003. [CrossRef] [PubMed]

3. Yu, J.; Wei, W.; Danner, E.; Ashley, R.K.; Israelachvili, J.; Waite, J.W. Mussel protein adhesion depends on interprotein thiol-mediated redox modulation. *Nat. Chem. Biol.* **2011**, *7*, 588–590. [CrossRef] [PubMed]

4. Leng, C.; Liu, Y.; Jenkins, C.; Meredith, H.; Wilker, J.J.; Chen, Z. Interfacial structure of a DOPA-inspired adhesive polymer studied by sum frequency generation vibrational spectroscopy. *Langmuir* **2013**, *29*, 6659–6664. [CrossRef] [PubMed]

5. Sagert, J.; Sun, C.; Waite, J.H. Chemical subtleties of mussel and polychaete holdfasts. In *Biological Adhesives*; Smith, A.M., Callow, J.A., Eds.; Springer: Berlin, Germany, 2006; pp. 125–143.

6. Wilker, J.J. The iron-fortified adhesive system of marine mussels. *Angew. Chem. Int. Ed.* **2010**, *49*, 8076–8078. [CrossRef] [PubMed]

7. Faure, E.; Falentin-Daudré, C.; Jérôme, C.; Lyskawa, J.; Fournier, D.; Woisel, P.; Detrembleur, C. Catechols as versatile platforms in polymer chemistry. *Prog. Polym. Sci.* **2013**, *38*, 236–270. [CrossRef]

8. Sedó, J.; Saiz-Poseu, J.; Busqué, F.; Ruiz-Molina, D. Catechol-based biomimetic functional materials. *Adv. Mater.* **2013**, *25*, 653–701. [CrossRef] [PubMed]

9. Ye, Q.; Zhou, F.; Liu, W. Bioinspired catecholic chemistry for surface modification. *Chem. Soc. Rev.* **2011**, *40*, 4244–4258. [CrossRef] [PubMed]

10. Kord Forooshani, P.; Lee, B.P. Recent approaches in designing bioadhesive materials inspired by mussel adhesive protein. *J. Polym. Sci. Part A Polym. Chem.* **2017**, *55*, 9–33. [CrossRef] [PubMed]

11. Moulay, S. Polymers with dihydroxy/dialkoxybenzene moieties. *Comptes Rendus Chim.* **2009**, *12*, 577–601. [CrossRef]

12. Moulay, S. DOPA/Catechol-tethered polymers: bioadhesives and biomimetic adhesive materials. *Polym. Rev.* **2014**, *54*, 436–513. [CrossRef]

13. Sever, M.J.; Weisser, J.T.; Monahan, J.; Srinivasan, S.; Wilker, J.J. Metal-mediated cross-linking in the generation of a marine mussel adhesive. *Angew. Chem. Int. Ed.* **2004**, *43*, 448–450. [CrossRef] [PubMed]

14. Burkett, J.R.; Wojtas, J.L.; Cloud, J.L.; Wilker, J.J. A Method for measuring the adhesion strength of marine mussels. *J. Adhes.* **2009**, *85*, 601–615. [CrossRef]

15. Del Grosso, C.A.; McCarthy, T.W.; Clark, C.L.; Cloud, J.L.; Wilker, J.J. Managing redox chemistry to deter marine biological adhesion. *Chem. Mater.* **2016**, *28*, 6791–6796. [CrossRef]

16. North, M.A.; Del Grosso, C.A.; Wilker, J.J. High strength underwater bonding with polymer mimics of mussel adhesive proteins. *ACS Appl. Mater. Interfaces* **2017**, *9*, 7866–7872. [CrossRef] [PubMed]

17. Waite, J.H. The Formation of Mussel Bysses: Anatomy of a natural manufacturing process. In *Structure, Cellular Synthesis and Assembly of Biopolymers*; Case, S.T., Ed.; Springer: Berlin, Germany, 1992; pp. 55–74. ISBN 3540555498.

18. Carrington, E.; Waite, J.H.; Sara, G.; Sebens, K.P. Mussels as a model system for integrative ecomechanics. *Ann. Rev. Mar. Sci.* **2015**, *7*, 443–469. [CrossRef] [PubMed]

19. Bell, E.C.; Gosline, J.M. Strategies for life in flow: Tenacity, morphometry, and probability of dislodgment of two *Mytilus* species. *Mar. Ecol. Prog. Ser.* **1997**, *159*, 197–208. [CrossRef]

20. Dolmer, P.; Svane, I. Attachment and orientation of *Mytilus edulis* L. in flowing water. *Ophelia* **1994**, *40*, 63–74. [CrossRef]

21. Qin, X.-X.; Coyne, K.J.; Waite, J.H. Tough tendons mussel byssus has collagen with silk-like domains. *J. Biol. Chem.* **1997**, *272*, 32623–32627. [CrossRef] [PubMed]

22. Filippidi, E.; DeMartini, D.G.; de Molina, P.M.; Danner, E.W.; Kim, J.; Helgeson, M.E.; Waite, J.H.; Valentine, M.T. The microscopic network structure of mussel (*Mytilus*) adhesive plaques. *J. R. Soc. Interface* **2015**, *12*, 20150827. [CrossRef] [PubMed]

23. O'Donnell, M.J.; George, M.N.; Carrington, E. Mussel byssus attachment weakened by ocean acidification. *Nat. Clim. Chang.* **2013**, *3*, 587–590. [CrossRef]

24. Carrington, E.; Moeser, G.M.; Thompson, S.B.; Coutts, L.C.; Craig, C.A. Mussel attachment on rocky shores: The effect of flow on byssus production. *Integr. Comp. Biol.* **2008**, *48*, 801–807. [CrossRef] [PubMed]

25. Moeser, G.M.; Leba, H.; Carrington, E. Seasonal influence of wave action on thread production in *Mytilus edulis*. *J. Exp. Biol.* **2006**, *209*, 881–890. [CrossRef] [PubMed]

26. Carrington, E. Seasonal variation in the attachment strength of blue mussels: Causes and consequences. *Limnol. Oceanogr.* **2002**, *47*, 1723–1733. [CrossRef]

27. Sun, C.; Vaccaro, E.; Waite, J.H. Oxidative stress and the mechanical properties of naturally occurring chimeric collagen-containing fibers. *Biophys. J.* **2001**, *81*, 3590–3595. [CrossRef]

28. Aldred, N.; Wills, T.; Williams, D.N.; Clare, A.S. Tensile and dynamic mechanical analysis of the distal portion of mussel (*Mytilus edulis*) byssal threads. *J. R. Soc. Interface* **2007**, *4*, 1159–1167. [CrossRef] [PubMed]
29. Lazaris, A.; Arcidiacono, S.; Huang, Y.; Zhou, J.-F.; Duguay, F.; Chretien, N.; Welsh, E.A.; Soares, J.W.; Karatzas, C.N. Spider silk fibers spun from soluble recombinant silk produced in mammalian cells. *Science* **2002**, *295*, 472LP–476LP. [CrossRef] [PubMed]
30. Rising, A.; Johansson, J. Toward spinning artificial spider silk. *Nat. Chem. Biol.* **2015**, *11*, 309–315. [CrossRef] [PubMed]
31. Rammensee, S.; Slotta, U.; Scheibel, T.; Bausch, A.R. Assembly mechanism of recombinant spider silk proteins. *Proc. Natl. Acad. Sci. USA* **2008**, *105*, 6590–6595. [CrossRef] [PubMed]
32. Becker, N.; Oroudjev, E.; Mutz, S.; Cleveland, J.P.; Hansma, P.K.; Hayashi, C.Y.; Makarov, D.E.; Hansma, H.G. Molecular nanosprings in spider capture-silk threads. *Nat. Mater.* **2003**, *2*, 278–283. [CrossRef] [PubMed]
33. Römer, L.; Scheibel, T. The elaborate structure of spider silk: Structure and function of a natural high performance fiber. *Prion* **2008**, *2*, 154–161. [CrossRef] [PubMed]
34. Holland, G.P.; Jenkins, J.E.; Creager, M.S.; Lewis, R.V.; Yarger, J.L. Quantifying the fraction of glycine and alanine in β-sheet and helical conformations in spider dragline silk using solid-state NMR. *Chem. Comm.* **2008**, 5568–5570. [CrossRef] [PubMed]
35. Harrington, M.J.; Waite, J.H. Holdfast heroics: Comparing the molecular and mechanical properties of *Mytilus californianus* byssal threads. *J. Exp. Biol.* **2007**, *210*, 4307–4318. [CrossRef] [PubMed]
36. Priemel, T.; Degtyar, E.; Dean, M.N.; Harrington, M.J. Rapid self-assembly of complex biomolecular architectures during mussel byssus biofabrication. *Nat. Commun.* **2017**, *8*, 14539. [CrossRef] [PubMed]
37. Harrington, M.J.; Masic, A.; Holten-Andersen, N.; Waite, J.H.; Fratzl, P. Iron-clad fibers: A metal-based biological strategy for hard flexible coatings. *Science* **2010**, *328*, 216–220. [CrossRef] [PubMed]
38. Kinloch, A. *Adhesion and Adhesives: Science and Technology*; Springer Science & Business Media: Berlin, Germany, 2012; ISBN 9401577641.
39. Desmond, K.W.; Zacchia, N.A.; Waite, J.H.; Valentine, M.T. Dynamics of mussel plaque detachment. *Soft Matter* **2015**, *11*, 6832–6839. [CrossRef] [PubMed]
40. Abràmoff, M.D.; Magalhães, P.J.; Ram, S.J. Image processing with ImageJ. *Biophotonics Int.* **2004**, *11*, 36–42.
41. Gutowski, W. Thermodynamics of Adhesion. In *Fundamentals of Adhesion*; Lee, L.-H., Ed.; Springer: Boston, MA, USA, 1991; pp. 87–135. ISBN 978-1-4899-2073-7.
42. Pocius, A.V.; Dillard, D.A. *Adhesion Science and Engineering: Surfaces, Chemistry and Applications*; Elsevier: Amsterdam, The Netherlands, 2002; ISBN 0080525989.
43. Young, G.A.; Crisp, D.J. Marine animals and adhesion. *Adhesion* **1982**, *6*, 19–39.
44. Crisp, D.J.; Walker, G.; Young, G.A.; Yule, A.B. Adhesion and substrate choice in mussels and barnacles. *J. Colloid Interface Sci.* **1985**, *104*, 40–50. [CrossRef]
45. Aldred, N.; Ista, L.K.; Callow, M.E.; Callow, J.A.; Lopez, G.P.; Clare, A.S. Mussel (*Mytilus edulis*) byssus deposition in response to variations in surface wettability. *J. R. Soc. Interface* **2006**, *3*, 37–43. [CrossRef] [PubMed]
46. Zhang, W.; Yang, H.; Liu, F.; Chen, T.; Hu, G.; Guo, D.; Hou, Q.; Wu, X.; Su, Y.; Wang, J. Molecular interactions between DOPA and surfaces with different functional groups: A chemical force microscopy study. *RSC Adv.* **2017**, *7*, 32518–32527. [CrossRef]
47. Qin, Z.; Buehler, M.J. Molecular mechanics of mussel adhesion proteins. *J. Mech. Phys. Solids* **2014**, *62*, 19–30. [CrossRef]
48. Vaccaro, E.; Waite, J.H. Yield and post-yield behavior of mussel byssal thread: A self-healing biomolecular materials. *Biomacromolecules* **2001**, *2*, 906–911. [CrossRef] [PubMed]
49. Hamada, N.A.; Roman, V.A.; Howell, S.M.; Wilker, J.J. Purdue University, Lafayette, IN, USA. Potential differences in adhesive properties as a function of failure modes, 2017.

biomimetics

MDPI

Article

2-*S*-Lipoylcaffeic Acid, a Natural Product-Based Entry to Tyrosinase Inhibition via Catechol Manipulation

Raffaella Micillo [1], **Valeria Pistorio** [2], **Elio Pizzo** [2], **Lucia Panzella** [1,*], **Alessandra Napolitano** [1] and **Marco d'Ischia** [1]

[1]   Department of Chemical Sciences, University of Naples "Federico II", Via Cintia 4, I-80126 Naples, Italy; raffaella.micillo@unina.it (R.M.); alesnapo@unina.it (A.N.); dischia@unina.it (M.d.I.)
[2]   Department of Biology, University of Naples "Federico II", Via Cintia 4, I-80126 Naples, Italy; valeria.pistorio@gmail.com (V.P.); elipizzo@unina.it (E.P.)
*   Correspondence: panzella@unina.it; Tel.: +39-081-674131

Academic Editor: Ille C. Gebeshuber
Received: 24 July 2017; Accepted: 9 August 2017; Published: 10 August 2017

Abstract: Conjugation of naturally occurring catecholic compounds with thiols is a versatile and facile entry to a broad range of bioinspired multifunctional compounds for diverse applications in biomedicine and materials science. We report herein the inhibition properties of the caffeic acid- dihydrolipoic acid *S*-conjugate, 2-*S*-lipoylcaffeic acid (LC), on mushroom tyrosinase. Half maximum inhibitory concentration (IC_{50}) values of 3.22 ± 0.02 and 2.0 ± 0.1 µM were determined for the catecholase and cresolase activity of the enzyme, respectively, indicating a greater efficiency of LC compared to the parent caffeic acid and the standard inhibitor kojic acid. Analysis of the Lineweaver–Burk plot suggested a mixed-type inhibition mechanism. LC proved to be non-toxic on human keratinocytes (HaCaT) at concentrations up to 30 µM. These results would point to LC as a novel prototype of melanogenesis regulators for the treatment of pigmentary disorders.

Keywords: depigmenting agents; L-DOPA; dopachrome; tyrosinase; melanin; caffeic acid; dihydrolipoic acid; lipoic acid; keratinocytes

1. Introduction

Several pigmentary disorders, such as melasma or lentigo, are associated with the overproduction or accumulation of melanin as the result of inflammatory responses or abnormal function of melanocytes inducing a local excess of pigmentation known as "hypermelanosis" [1–3]. The medical and aesthetical unfavorable impact of such disorders has prompted a constant search for new non-toxic depigmenting agents [4–6].

Since skin complexion is under control of several factors, including activity, expression, and stability of tyrosinase and related enzymes, melanocytes homeostasis, and melanosome transfer to the keratinocytes, commercially available depigmenting agents and melanogenesis regulators usually act through different mechanisms [7,8].

One of the most common approaches for control of pigmentation involves the inhibition of tyrosinase (EC 1.14.18.1) [9,10], a copper-containing enzyme exhibiting cresolasic or monophenolasic activity (hydroxylation of monophenols to *o*-diphenols) and catecholasic or diphenolasic activity (dehydrogenation of catechols to *o*-quinones) by a mechanism involving electron exchange with the copper atoms. In particular, tyrosinase catalyzes the key steps of melanogenesis, namely the hydroxylation and oxidation of L-tyrosine to dopaquinone [11,12].

When considering a new tyrosinase inhibitor several factors should be taken into account, such as product efficacy, cytotoxicity, solubility, cutaneous absorption, and stability. Increasing attention has been paid to the adverse effects of depigmenting agents in vitro and in vivo, especially further to the case of rhododendrol, a phenolic skin whitening agent that has been recently withdrawn from the market because of its cytotoxic effects on melanocytes and the consequent induction of leukoderma [13–16]. This has been ascribed to the tyrosinase-catalyzed oxidation of rhododendrol producing toxic metabolites [14–17].

Several catecholic compounds have raised interest due to their tyrosinase inhibition properties [18–21]. Among the catecholic compounds of natural origin, a prominent position is occupied by caffeic acid (3,4-dihydroxycinnamic acid) due to its health-beneficial properties [22–24]. A number of caffeic acid derivatives, mostly amides, have been described as tyrosinase inhibitors, while caffeic acid is not, and this ability has been attributed to both the structural similarity of the caffeic acid moiety to the substrate 3,4-dihydroxy-L-phenylalanine (L-DOPA) [20] and to the hydrophobicity and copper-chelating properties imparted by the particular nitrogen substituents [25–28].

Recently, with a view to synthesizing multifunctional antioxidants inspired to bioactive thiol-conjugates of naturally occurring phenolic compounds [29–35], we have focused our attention to dihydrolipoic acid (DHLA), the reduced form of lipoic acid (LA). The LA/DHLA system is known to be a powerful antioxidant being able to reduce reactive oxygen species, scavenge hydroxyl radicals, hypochlorous acid, and peroxynitrite, and chelate Fe^{2+} ions; moreover, LA and its reduced form can exert their functions both in membrane and in cytoplasm, because of their solubility in fats and water [36,37]. DHLA has also been reported to react with dopaquinone to give lipoyl-DOPA conjugates, and such a kind of reaction is able to affect melanin production resulting in depigmentation [38,39].

On this basis, we report herein the tyrosinase inhibition properties of a conjugate of DHLA with caffeic acid, namely 2-*S*-lipoylcaffeic acid (LC) (Figure 1).

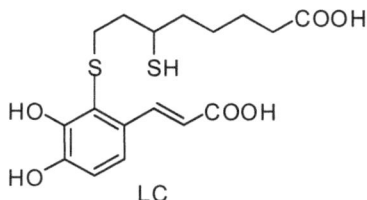

Figure 1. Structure of 2-*S*-lipoylcaffeic acid (LC).

2. Materials and Methods

2.1. Materials

2-Iodobenzoic acid, oxone®, (±)-LA, sodium borohydride, sodium dithionite, L-DOPA, mushroom tyrosinase (EC 1.14.18.1), *p*-coumaric acid, caffeic acid, kojic acid, and 3-(4,5-dimethyl-2-thiazolyl)-2,5-diphenyl-2H-tetrazolium bromide (MTT) were purchased from Sigma-Aldrich (Milan, Italy). Dulbecco's modified Eagle medium (DMEM), L-glutamine, penicillin/streptomycin, and fetal bovine serum (FBS) were purchased from Euroclone (Milan, Italy). All solvents were high-performance liquid chromatography (HPLC) grade. Double-distilled deionized water was used throughout the study.

2-Iodoxybenzoic acid (IBX) [40] and DHLA [41] were synthetized as reported.

2.2. Methods

Ultraviolet–visible (UV–Vis) spectra were recorded on a Jasco V-730 spectrophotometer (Lecco, Italy).

Nuclear magnetic resonance (NMR) spectra were recorded at 400 MHz on a Bruker instrument (Milan, Italy).

HPLC analyses were performed on an Agilent 1100 binary pump instrument (Agilent Technologies, Milan, Italy) equipped with a UV–Vis detector, using an octadecylsilane-coated column, 250 mm × 4.6 mm,

5 μm particle size (Phenomenex SphereClone ODS, Bologna, Italy) at 0.7 mL/min, and the following gradient: 0.1% formic acid (eluent a)/methanol (eluent b): 40% b, 0–10 min; from 40 to 80% b, 10–47.5 min. The detection wavelength was set at 280 nm.

Liquid chromatography–mass spectrometry (LC–MS) analysis was performed on an HPLC 1100 VL series instrument (Agilent Technologies) with an electrospray ionization source in positive ion mode (ESI+). An Agilent Eclipse XDB-C18, 150 mm × 4.60 mm, 5 μm (Agilent Technologies) was used, with the same eluent used for the HPLC analysis at a flow rate of 0.4 mL/min. Mass spectra were registered under the following conditions: nebulizer pressure 50 psi; drying gas (nitrogen) flow 10 L/min, at 350 °C; and capillary voltage 4000 V.

2.3. Synthesis of 2-S-Lipoylcaffeic Acid

A solution of *p*-coumaric acid (215 mg, 1.3 mmol) in methanol (18 mL) was treated with IBX (561 mg, 2 mmol) under vigorous stirring at room temperature. After 7 min a solution of DHLA (1.12 g, 5.3 mmol) in methanol (18 mL) was added dropwise, and after additional 15 min the reaction mixture was diluted with water and acidified to pH 1 with 6 M HCl. The mixture was then washed with hexane/toluene 8:2 *v*/*v* (10 × 200 mL) and extracted with chloroform (7 × 200 mL). The combined chloroform layers were dried over sodium sulfate and taken to dryness to afford pure LC (101 mg, 20% yield) as a yellow oil.

ESI+/MS: *m*/*z* 387 ([M + H]$^+$), 409 ([M + Na]$^+$); UV: λ_{max} (CH$_3$OH) 252, 320 nm; ^{1}H-NMR (CD$_3$OD): δ (ppm) 1.38 (m, 1H), 1.54 (m, 1H), 1.54 (m, 2H), 1.42 (m, 1H), 1.62 (m, 1H), 1.78 (m, 1H), 1.81 (m, 1H), 2.26 (m, 2H), 2.88 (m, 1H), 2.90 (m, 1H), 2.96 (m, 1H), 6.29 (d, *J* = 16 Hz, 1H), 6.86 (d, *J* = 8.4 Hz, 1H), 7.22 (d, *J* = 8.4 Hz, 1H), 8.40 (d, *J* = 16 Hz, 1H); ^{13}C-NMR (CD$_3$OD): δ (ppm) 25.4 (CH$_2$), 27.3 (CH$_2$), 34.1 (2 × CH$_2$), 39.3 (CH$_2$), 39.4 (CH$_2$), 40.2 (CH), 117.2 (CH), 118.1 (CH), 119.9 (CH), 121.7 (C), 130.7 (C), 144.2 (CH), 147.7 (C), 147.8 (C), 168.2 (C), 174.8 (C).

2.4. Mushroom Tyrosinase Inhibition Assay

One hundred microliters of a methanolic solution of LC were incubated in 2 mL (0.001–1 mM final concentration) of 50 mM phosphate buffer (pH 6.8) at room temperature in the presence of mushroom tyrosinase (20 U/mL). After 10 min 20 μL of a 100 mM solution of L-DOPA or L-tyrosine in 0.6 M HCl (1 mM final concentration) were added and the course of the reaction was followed spectrophotometrically measuring the absorbance at 475 nm for 10 min at 2 min intervals. In control experiments the reaction was run in the absence of LC. When required, the assay was performed as described but by adding L-DOPA to the reaction mixture soon after the addition of LC (3 μM).

In separate experiments, the assay was run as above with LC at 250 μM, in the presence or absence of L-DOPA, and after 10 min the mixture was analyzed by HPLC.

2.5. Investigation of the Mechanism of Inhibition of Mushroom Tyrosinase Activity

The assay was run as above, using different concentrations of L-DOPA (0.125, 0.25, 0.5, 1, and 2 mM) and LC (0, 2, 3 and 5 μM). Data were elaborated to build the Lineweaver–Burk plot.

2.6. Cell Viability Assay

Cytotoxic effects on immortalized human keratinocytes (HaCaT) were determined using the cell proliferation reagent MTT. Briefly, 5×10^3 cells were seeded into a 96-well plate and were incubated overnight at 37 °C with 5% CO$_2$. Medium was then replaced with 100 μL of fresh media containing LC at 0–30 μM and cells were incubated at 37 °C with 5% CO$_2$. After 24, 48, or 72 h the LC-containing medium was removed, and 100 μL of fresh medium without red phenol, containing 10% MTT reagent, were added to each well and cells were incubated for 4 h at 37 °C in the dark. Subsequently, the absorbance at 570 nm was measured in a microtiter plate reader (SINERGY H4, BioTek, AHSI S.P.A., Milan, Italy) and cell viability was expressed as the mean ± standard deviation (SD) percentage compared to control.

3. Results and Discussion

3.1. Preparation of 2-S-Lipoylcaffeic Acid

The synthesis of LC was carried out by adapting a procedure previously reported for the preparation of the conjugation product of hydroxytyrosol with DHLA [30]. This involved the generation of the *o*-quinone of caffeic acid by the regioselective hydroxylation of *p*-coumaric acid with 2-iodoxybenzoic acid (IBX) [42], followed by addition of DHLA. The product was obtained in pure form in ca. 20% yield by a sequential extraction with solvents of increasing polarity, without the need for any chromatographic purification step. NMR, MS, and UV-Vis analysis confirmed the identity of the compound as the conjugation product of DHLA with caffeic acid quinone via the C-2 of the aromatic ring [43,44].

3.2. Inhibition of the Catecholase Activity of Mushroom Tyrosinase by 2-S-Lipoylcaffeic Acid

The enzyme inhibition properties of LC were investigated using mushroom tyrosinase, which is routinely used for preliminary assessment of the activity of potential tyrosinase inhibitors [45–47]. For the assay of catecholase activity, L-DOPA was used as the substrate. The assay is based on the spectrophotometric monitoring of dopachrome formation (wavelength of maximum absorbance (λ_{max}) 475 nm), following oxidative cyclization of dopaquinone produced by tyrosinase-induced oxidation of the substrate, in the presence and in the absence of the inhibitor [48] (Scheme 1).

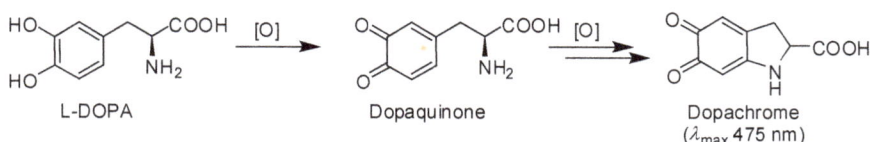

Scheme 1. Dopachrome formation by tyrosinase-catalyzed oxidation of 3,4-dihydroxy-L-phenylalanine (L-DOPA). λ_{max}: Wavelength of maximum absorbance.

LC was incubated in 50 mM phosphate buffer (pH 6.8) in the presence of mushroom tyrosinase (20 U/mL) at room temperature. After 10 min L-DOPA (1 mM final concentration) was added and the absorbance at 475 nm was measured at different times (Figure 2).

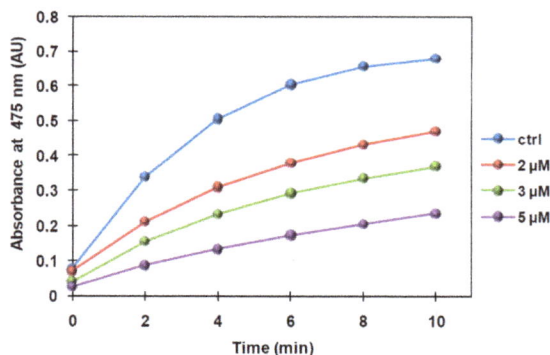

Figure 2. Time course of the absorbance change at 475 nm in the oxidation mixture of L-DOPA (1 mM) with mushroom tyrosinase in the absence (control (ctrl)) or presence of different concentrations of LC. Reported are the mean values of at least three experiments (standard deviation (SD) < 5%). AU: Arbitrary units.

The percentage of inhibition was calculated using the following Equation:

$$\% \text{ inhibition} = \left(1 - \frac{\Delta A_{475/\min} \text{ in the presence of the inhibitor}}{\Delta A_{475/\min} \text{ in the absence of the inhibitor}}\right) \times 100 \qquad (1)$$

As reported in Figure 3, a maximum 80% inhibition was observed with 5 μM LC. A half maximum inhibitory concentration (IC$_{50}$) value of 3.22 ± 0.02 μM was determined.

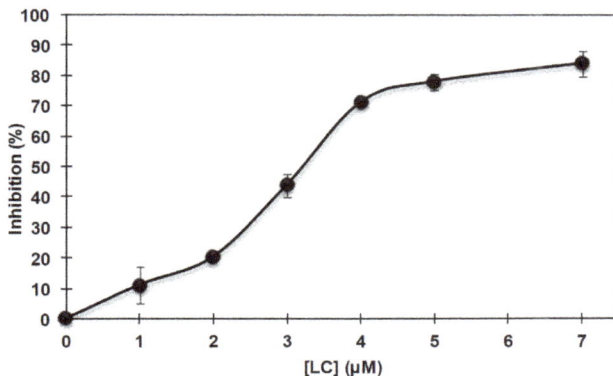

Figure 3. Percent of inhibition of mushroom tyrosinase activity vs. LC concentration using L-DOPA (1 mM) as substrate. Reported are the mean ± SD values of at least three experiments.

Figure 4 shows the inhibition effect of 10 μM LC on the formation of dopachrome when L-DOPA is incubated with mushroom tyrosinase under the conditions previously described. Notably, at the same concentration, caffeic acid and the well-established tyrosinase inhibitor kojic acid [49,50] did not induce any inhibition.

Figure 4. Tyrosinase-catalyzed oxidation mixtures of L-DOPA (1 mM) in the absence (B) or in the presence of 10 μM inhibitor (LC: 2-*S*-lipoylcaffeic acid; CAF: Caffeic acid; KOJ: Kojic acid).

These results not only show the superior inhibition properties of LC, but also underline the importance of the functionalization with DHLA in imparting tyrosinase inhibition properties to the parent catechol caffeic acid. It is well known that insertion of a chalcogen can affect the properties of catechol systems by lowering the O–H bond dissociation enthalpy [29,51,52]. However, the possibility that the higher inhibitory activity observed following conjugation with DHLA is not due to the presence of the sulfur substituent, per se, but rather to the hydrophobicity acquired by the compound cannot be excluded. Actually, the tyrosinase inhibition properties of caffeoyl-amino acidyl-hydroxamic acid derivatives have been ascribed in part to their hydrophobicity, making them suitable for binding to the active site of tyrosinase [26–28].

On the other hand, the primary sulfhydryl group of DHLA has been reported to react with dopaquinone produced by the tyrosinase-catalyzed oxidation of L-DOPA, leading to the formation of covalent lipoyl adducts and inhibiting dopachrome formation [38,39]. Given the presence of a free, although secondary, SH group, in separate experiments the possibility that LC could react with dopaquinone in the same way was investigated: the assay was run under the usual conditions and after 10 min the mixture was analyzed by HPLC which, however, did not reveal any consumption of the inhibitor nor formation of new products, ruling out a possible reaction of LC with the oxidation products of L-DOPA.

These results also suggested that LC is not an alternative substrate of the enzyme, since no consumption was observed even in the absence of L-DOPA. This is an issue of considerable importance, since the toxicity of some depigmenting agents has been mostly attributed to their acting as substrates of tyrosinase and, as such, being oxidized by the enzyme, leading to the formation of highly reactive, cytotoxic *o*-quinones [47,53,54].

Finally, the effect of pre-incubation on the inhibitory activity was investigated, by performing the spectrophotometric assay as above but with the addition of L-DOPA to the reaction mixture immediately after addition of LC. The time course of the absorbance change at 475 nm was comparable to that observed when LC was preincubated with the enzyme for 10 min before addition of L-DOPA, ruling out any role of pre-incubation in the inhibition effects exerted by LC.

3.3. Inhibition of the Cresolase Activity of Mushroom Tyrosinase by 2-S-Lipoylcaffeic Acid

The ability of LC to inhibit the monophenolasic activity of mushroom tyrosinase was investigated as described above using L-tyrosine instead of L-DOPA as substrate (Figure 5). An IC_{50} value of 2.0 ± 0.1 µM was determined.

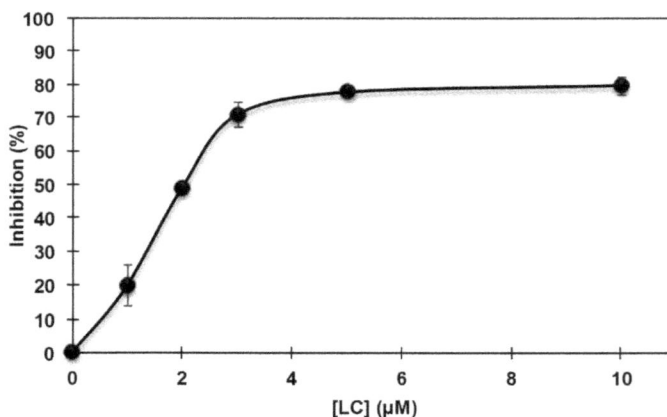

Figure 5. Percent of inhibition of mushroom tyrosinase activity vs. LC concentration using L-tyrosine (1 mM) as the substrate. Reported are the mean ± SD values of at least three experiments.

3.4. Investigation of the Mechanism of Inhibition of Mushroom Tyrosinase Activity by 2-S-Lipoylcaffeic Acid

Lineweaver–Burk plot analysis was used to determine the mode of tyrosinase inhibition by LC. In Figure 6 the double-reciprocal plots of tyrosinase inhibition using L-DOPA as substrate is reported for different concentrations (0, 3, and 5 µM) of LC.

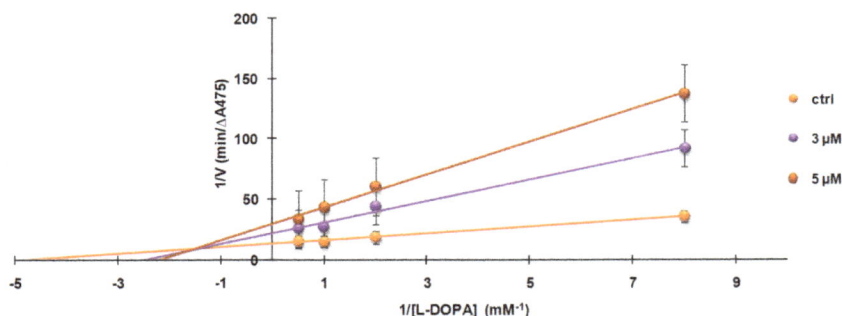

Figure 6. Lineweaver–Burk plot for the inhibition of mushroom tyrosinase-catalyzed L-DOPA oxidation by LC at 0 (control (ctrl)), 3 or 5 µM. Data were obtained as mean ± SD values of $1/V$, inverse of the increase of absorbance at 475 nm per min (ΔA_{475}/min), of three independent experiments with different concentrations of L-DOPA.

The results were a family of straight lines with different slopes and different *x*- and *y*-intercepts, suggestive of a mixed inhibitor, lowering the maximum rate (V_{max}) and increasing the Michaelis constant K_m in a dose dependent-manner (Figure 7).

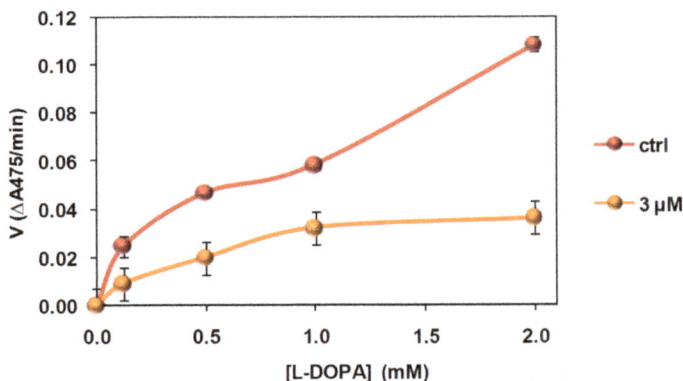

Figure 7. The effect of LC on the enzymatic kinetics for the mushroom tyrosinase-induced oxidation of L-DOPA. Data were obtained as mean ± SD values of the increase of absorbance at 475 nm per min (ΔA_{475}/min) (V) of three independent experiments with different concentrations of L-DOPA.

Several mixed-type inhibitors of mushroom tyrosinase have been described in the literature and, in most cases, complex kinetics are involved and the phenomena have been left unexplained. Recently, non-specific binding sites have been invoked to explain the mixed-type inhibition in mushroom tyrosinase activities [55]. However, in our case, available data do not allow discussion in more detail of how the ternary complex of substrate–enzyme–inhibitor is formed, to assess whether the free thiol group participates in the inhibition mechanism and by what mechanism, and what the role of the hydrophobic aliphatic chain of the DHLA residue is.

3.5. Cytotoxicity Evaluation

With the aim of evaluating the possible use of LC as a tyrosinase inhibitor in vivo, its cytotoxicity was preliminarily evaluated on human keratinocyte cells (HaCaT) by performing the MTT assay [52,56].

As shown in Figure 8, HaCat cells did not exhibit any significant reduction in proliferation rate when incubated with increasing amounts of LC over 72 h.

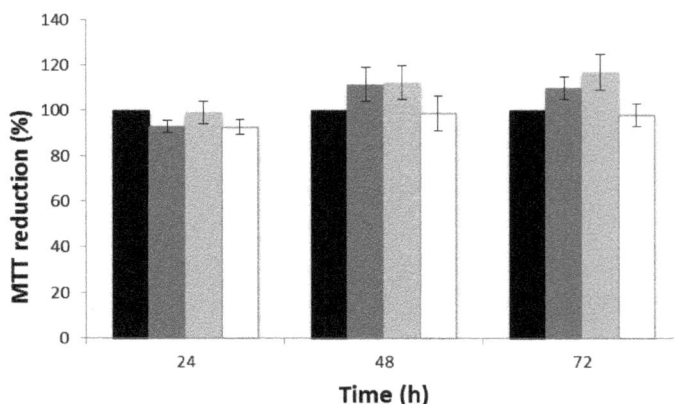

Figure 8. Effect of LC on HaCaT cell viability determined by 3-(4,5-dimethyl-2-thiazolyl)-2,5-diphenyl-2H-tetrazolium bromide (MTT) assay. Cells were cultured in normal growth medium and then subjected to treatment with LC (black: Control; dark grey: 0.3 μM; grey: 3 μM; white: 30 μM) for 24, 48, and 72 h. Cell viability was evaluated by measuring the A_{570nm}. Results are expressed as the percentage (means ± SD from at least three experiments) compared to the control.

4. Conclusions

The use of natural catechols and derivatives as tyrosinase inhibitors for the treatment of pigmentary disorders associated with the overproduction or accumulation of melanin is well documented. We have reported herein that 2-*S*-lipoylcaffeic acid (LC), the *S*-conjugation product of caffeic acid and dihydrolipoic acid, is a promising lead structure for the development of catechol-based natural product-like tyrosinase inhibitors. LC was found to be able to inhibit both the catecholase and cresolase activity of mushroom tyrosinase with IC_{50} values as low as 3 μM, whereas under the same conditions caffeic acid did not show any effect, pointing to insertion of the dihydrolipoyl chain as an effective means of potentiating the inhibitory activity.

Whether this effect is due to the *S*-substituted catechol moiety alone or reflects the cooperative effect of the adjacent SH group cannot be assessed on the basis of the present data. It seems likely, however, that the potent inhibitory effects on tyrosinase reflect a synergic combination of caffeic acid as a substrate (L-DOPA)-like scaffold on which an efficient copper-binding arm (DHLA) is installed modulating catechol redox and chelating properties and conferring to the conjugate a higher degree of lipophilicity.

Although further experiments are needed to assess the actual potential of LC as a depigmenting agent in vivo, it is worth noting that the compound was non-toxic on immortalized human keratinocytes at concentrations up to 30 μM. Moreover, control experiments revealed that LC is not a substrate of tyrosinase, a critical issue for the toxicity of depigmenting agents in vivo. Although we are aware that additional, more cogent experiments are necessary to corroborate the lack of toxicity, the results of this paper can provide the necessary background for further studies on mammalian melanocyte cell lines, which will be directed to confirm activity on mammalian enzyme and pigment cells.

Overall, these results further expand the framework of the practical opportunities offered by the thiol-quinone coupling reactions, with particular reference to the dihydrolipoic/lipoic acid chemistry.

Author Contributions: L.P., A.N., and M.d.I. conceived and designed the experiments; R.M. and V.P. performed the experiments; E.P. and L.P. analyzed the data; and R.M., E.P., L.P., A.N., and M.d.I. wrote the paper.

Conflicts of Interest: The authors declare no conflict of interest.

References

1. Slominski, A. Melanin pigmentation in mammalian skin and its hormonal regulation. *Physiol. Rev.* **2004**, *84*, 1155–1228. [CrossRef] [PubMed]
2. Yamaguchi, Y.; Hearing, V.J. Melanocytes and their diseases. *Cold Spring Harb. Perspect. Med.* **2014**, *4*. [CrossRef] [PubMed]
3. Cardinali, G.; Kovacs, D.; Picardo, M. Mechanisms underlying post-inflammatory hyperpigmentation: Lessons from solar lentigo. *Ann. Dermatol. Venereol.* **2012**, *139* (Suppl. S4), S148–S152. [CrossRef]
4. Smit, N.; Vicanova, J.; Pavel, S. The hunt for natural skin whitening agents. *Int. J. Mol. Sci.* **2009**, *10*, 5326–5349. [CrossRef] [PubMed]
5. Picardo, M.; Carrera, M. New and experimental treatments of cloasma and other hypermelanoses. *Dermatol. Clin.* **2007**, *25*, 353–362. [CrossRef] [PubMed]
6. Gunia-Krzyżak, A.; Popiol, J.; Marona, H. Melanogenesis inhibitors: Strategies for searching for and evaluation of active compounds. *Curr. Med. Chem.* **2016**, *23*, 3548–3574. [CrossRef] [PubMed]
7. Ebanks, J.P.; Wickett, R.R.; Boissy, R.E. Mechanisms regulating skin pigmentation: The rise and fall of complexion coloration. *Int. J. Mol. Sci.* **2009**, *10*, 4066–4087. [CrossRef] [PubMed]
8. Slominski, A.; Zmijewski, M.A.; Pawelek, J. L-Tyrosine and L-dihydroxyphenylalanine as hormone-like regulators of melanocyte functions. *Pigment Cell Melanoma Res.* **2012**, *25*, 14–27. [CrossRef] [PubMed]
9. Pillaiyar, T.; Manickam, M.; Namasivayam, V. Skin whitening agents: Medicinal chemistry perspective of tyrosinase inhibitors. *J. Enzym. Inhib. Med. Chem.* **2017**, *32*, 403–425. [CrossRef] [PubMed]
10. Chang, T.S. An updated review of tyrosinase inhibitors. *Int. J. Mol. Sci.* **2009**, *10*, 2440–2475. [CrossRef] [PubMed]
11. Riley, P.A. Mechanistic aspects of the control of tyrosinase activity. *Pigment Cell Res.* **1993**, *6*, 182–185. [CrossRef] [PubMed]
12. Ito, S.; Wakamatsu, K. Chemistry of mixed melanogenesis—Pivotal roles of dopaquinone. *Photochem. Photobiol.* **2008**, *84*, 582–592. [CrossRef] [PubMed]
13. Lee, C.S.; Joo, Y.H.; Baek, H.S.; Park, M.; Kim, J.H.; Shin, H.J.; Park, N.H.; Lee, J.H.; Park, Y.H.; Shin, S.S.; et al. Different effects of five depigmentary compounds, rhododendrol, raspberry ketone, monobenzone, rucinol and AP736 on melanogenesis and viability of human epidermal melanocytes. *Exp. Dermatol.* **2016**, *25*, 44–49. [CrossRef] [PubMed]
14. Ito, S.; Hinoshita, M.; Suzuki, E.; Ojika, M.; Wakamatsu, K. Tyrosinase-catalyzed oxidation of the leukoderma-inducing agent raspberry ketone produces (E)-4-(3-oxo-1-butenyl)-1,2-benzoquinone: Implications for melanocyte toxicity. *Chem. Res. Toxicol.* **2017**, *30*, 859–868. [CrossRef] [PubMed]
15. Okura, M.; Yamashita, T.; Ishii-Osai, Y.; Yoshikawa, M.; Sumikawa, Y.; Wakamatsu, K.; Ito, S. Effects of rhododendrol and its metabolic products on melanocytic cell growth. *J. Dermatol. Sci.* **2015**, *80*, 142–149. [CrossRef] [PubMed]
16. Ito, S.; Ojika, M.; Yamashita, T.; Wakamatsu, K. Tyrosinase-catalyzed oxidation of rhododendrol produces 2-methylchromane-6,7-dione, the putative ultimate toxic metabolite: Implications for melanocyte toxicity. *Pigment Cell Melanoma Res.* **2014**, *27*, 744–753. [CrossRef] [PubMed]
17. Sasaki, M.; Kondo, M.; Sato, K.; Umeda, M.; Kawabata, K.; Takahashi, Y.; Suzuki, T.; Matsunaga, K.; Inoue, S. Rhododendrol, a depigmentation-inducing phenolic compound, exerts melanocyte cytotoxicity via a tyrosinase-dependent mechanism. *Pigment Cell Melanoma Res.* **2014**, *27*, 754–763. [CrossRef] [PubMed]
18. Chen, C.Y.; Lin, L.C.; Yang, W.F.; Bordon, J.; Wang, H.M.D. An updated organic classification of tyrosinase inhibitors on melanin biosynthesis. *Curr. Org. Chem.* **2015**, *19*, 4–18. [CrossRef]
19. Pillaiyar, T.; Manickam, M.; Jung, S.H. Inhibitors of melanogenesis: A patent review (2009–2014). *Expert Opin. Ther. Pat.* **2015**, *25*, 775–788. [CrossRef] [PubMed]
20. Munoz-Munoz, J.L.; Berna, J.; Garcia-Molina, F.; Garcia-Ruiz, P.A.; Tudela, J.; Rodriguez-Lopez, J.N.; Garcia-Canovas, F. Unravelling the suicide inactivation of tyrosinase: A discrimination between mechanisms. *J. Mol. Catal. B Enzym.* **2012**, *75*, 11–19. [CrossRef]

21. Xue, Y.L.; Miyakawa, T.; Hayashi, Y.; Okamoto, K.; Hu, F.; Mitani, N.; Furihata, K.; Sawano, Y.; Tanokura, M. Isolation and tyrosinase inhibitory effects of polyphenols from the leaves of persimmon, *Diospyros kaki*. *J. Agric. Food Chem.* **2011**, *59*, 6011–6017. [CrossRef] [PubMed]

22. Silva, T.; Oliveira, C.; Borges, F. Caffeic acid derivatives, analogs and applications: A patent review (2009–2013). *Expert Opin. Ther. Pat.* **2014**, *24*, 1257–1270. [CrossRef] [PubMed]

23. Magnani, C.; Isaac, V.L.B.; Correa, M.A.; Salgado, H.R.N. Caffeic acid: A review of its potential use in medications and cosmetics. *Anal. Methods* **2014**, *6*, 3203–3210. [CrossRef]

24. Touaibia, M.; Jean-Francois, J.; Doiron, J. Caffeic acid, a versatile pharmacophore: An overview. *Mini Rev. Med. Chem.* **2011**, *11*, 695–713. [CrossRef] [PubMed]

25. Kuo, Y.H.; Chen, C.C.; Lin, P.; You, Y.J.; Chiang, H.M. *N*-(4-Bromophenethyl) caffeamide inhibits melanogenesis by regulating AKT/glycogen synthase kinase 3 β/microphthalmia-associated transcription factor and tyrosinase-related protein 1/tyrosinase. *Curr. Pharm. Biotechnol.* **2015**, *16*, 1111–1119. [CrossRef] [PubMed]

26. Kwak, S.Y.; Yang, J.K.; Choi, H.R.; Park, K.C.; Kim, Y.B.; Lee, Y.S. Synthesis and dual biological effects of hydroxycinnamoyl phenylalanyl/prolyl hydroxamic acid derivatives as tyrosinase inhibitor and antioxidant. *Bioorg. Med. Chem. Lett.* **2013**, *23*, 1136–1142. [CrossRef] [PubMed]

27. Kwak, S.Y.; Lee, S.; Choi, H.R.; Park, K.C.; Lee, Y.S. Dual effects of caffeoyl-amino acidyl-hydroxamic acid as an antioxidant and depigmenting agent. *Bioorg. Med. Chem. Lett.* **2011**, *21*, 5155–5158. [CrossRef] [PubMed]

28. Tada, T.; Ohnishi, K.; Komiya, T.; Imai, K. Synthetic search for cosmetic ingredients: Preparations, tyrosinase inhibitory and antioxidant activities of caffeic amides. *J. Oleo Sci.* **2002**, *51*, 19–27. [CrossRef]

29. Amorati, R.; Valgimigli, L.; Panzella, L.; Napolitano, A.; d'Ischia, M. 5-*S*-Lipoylhydroxytyrosol, a multidefense antioxidant featuring a solvent-tunable peroxyl radical-scavenging 3-thio-1,2-dihydroxybenzene motif. *J. Org. Chem.* **2013**, *78*, 9857–9864. [CrossRef] [PubMed]

30. Panzella, L.; Verotta, L.; Goya, L.; Ramos, S.; Martin, M.A.; Bravo, L.; Napolitano, A.; d'Ischia, M. Synthesis and bioactivity profile of 5-*S*-lipoylhydroxytyrosol-based multidefense antioxidants with a sizeable (poly)sulfide chain. *J. Agric. Food Chem.* **2013**, *61*, 1710–1717. [CrossRef] [PubMed]

31. Greco, G.; Panzella, L.; Pezzella, A.; Napolitano, A.; d'Ischia, M. Reaction of dihydrolipoic acid with juglone and related naphthoquinones: Unmasking of a spirocyclic 1,3-dithiane intermediate en route to naphtho[1,4]dithiepines. *Tetrahedron* **2010**, *66*, 3912–3916. [CrossRef]

32. De Lucia, M.; Panzella, L.; Pezzella, A.; Napolitano, A.; d'Ischia, M. Plant catechols and their *S*-glutathionyl conjugates as antinitrosating agents: Expedient synthesis and remarkable potency of 5-*S*-glutathionylpiceatannol. *Chem. Res. Toxicol.* **2008**, *21*, 2407–2413. [CrossRef] [PubMed]

33. Panzella, L.; De Lucia, M.; Napolitano, A.; d'Ischia, M. The first expedient entry to the human melanogen 2-*S*-cysteinyldopa exploiting the anomalous regioselectivity of 3,4-dihydroxycinnamic acid-thiol conjugation. *Tetrahedron Lett.* **2007**, *48*, 7650–7652. [CrossRef]

34. Panzella, L.; Napolitano, A.; d'Ischia, M. Oxidative conjugation of chlorogenic acid with glutathione. Structural characterization of addition products and a new nitrite-promoted pathway. *Bioorg. Med. Chem.* **2003**, *11*, 4797–4805. [CrossRef]

35. Panzella, L.; Napolitano, A.; d'Ischia, M. Nitrite-mediated decarboxylative conjugation of caffeic acid with glutathione under mildly acidic conditions. *Bioorg. Med. Chem. Lett.* **2002**, *12*, 3547–3550. [CrossRef]

36. Packer, L.; Witt, E.H.; Tritschler, H.J. α-Lipoic acid as a biological antioxidant. *Free Radic. Biol. Med.* **1995**, *19*, 227–235. [CrossRef]

37. Rochette, L.; Ghibu, S.; Richard, C.; Zeller, M.; Cottin, Y.; Vergely, C. Direct and indirect antioxidant properties of α-lipoic acid and therapeutic potential. *Mol. Nutr. Food Res.* **2013**, *57*, 114–122. [CrossRef] [PubMed]

38. Tsuji-Naito, K.; Hatani, T.; Okada, T.; Tehara, T. Evidence for covalent lipoyl adduction with DOPAquinone following tyrosinase-catalyzed oxidation. *Biochem. Biophys. Res. Commun.* **2006**, *343*, 15–20. [CrossRef] [PubMed]

39. Tsuji-Naito, K.; Hatani, T.; Okada, T.; Tehara, T. Modulating effects of a novel skin-lightening agent, α-lipoic acid derivative, on melanin production by the formation of DOPA conjugate products. *Bioorg. Med. Chem.* **2007**, *15*, 1967–1975. [CrossRef] [PubMed]

40. Frigerio, M.; Santagostino, M.; Sputore, S. A user-friendly entry to 2-iodoxybenzoic acid (IBX). *J. Org. Chem.* **1999**, *64*, 4537–4538. [CrossRef]

41. Gunsalus, I.C.; Barton, L.S.; Gruber, W. Biosynthesis and structure of lipoic acid derivatives. *J. Am. Chem. Soc.* **1956**, *78*, 1763–1766. [CrossRef]
42. Bernini, R.; Fabrizi, G.; Pouysegu, L.; Deffieux, D.; Quideau, S. Synthesis of biologically active catecholic compounds via ortho-selective oxygenation of phenolic compounds using hypervalent iodine(V) reagents. *Curr. Org. Synth.* **2012**, *9*, 650–669. [CrossRef]
43. Guerriero, E.; Sorice, A.; Capone, F.; Costantini, S.; Palladino, P.; d'Ischia, M.; Castello, G. Effects of lipoic acid, caffeic acid and a synthesized lipoyl-caffeic conjugate on human hepatoma cell lines. *Molecules* **2011**, *16*, 6365–6377. [CrossRef] [PubMed]
44. Ferreira-Lima, N.; Vallverdu-Queralt, A.; Meudec, E.; Mazauric, J.-P.; Sommerer, N.; Bordignon-Luiz, M.T.; Cheynier, V.; Le Guerneve, C. Synthesis, identification, and structure elucidation of adducts formed by reactions of hydroxycinnamic acids with glutathione or cysteinylglycine. *J. Nat. Prod.* **2016**, *79*, 2211–2222. [CrossRef] [PubMed]
45. Solano, F.; Briganti, S.; Picardo, M.; Ghanem, G. Hypopigmenting agents: An updated review on biological, chemical and clinical aspects. *Pigment Cell Res.* **2006**, *19*, 550–571. [CrossRef] [PubMed]
46. Jones, K.; Hughes, J.; Hong, M.; Jia, Q.; Orndorff, S. Modulation of melanogenesis by aloesin: A competitive inhibitor of tyrosinase. *Pigment Cell Res.* **2002**, *15*, 335–340. [CrossRef] [PubMed]
47. Ito, S.; Wakamatsu, K. A convenient screening method to differentiate phenolic skin whitening tyrosinase inhibitors from leukoderma-inducing phenols. *J. Dermatol. Sci.* **2015**, *80*, 18–24. [CrossRef] [PubMed]
48. Mason, H.S. The chemistry of melanin: III. Mechanism of the oxidation of dihydroxyphenylalanine by tyrosinase. *J. Biol. Chem.* **1948**, *172*, 83–99. [PubMed]
49. Battaini, G.; Monzani, E.; Casella, L.; Santagostini, L.; Pagliarin, R. Inhibition of the catecholase activity of biomimetic dinuclear copper complexes by kojic acid. *J. Biol. Inorg. Chem.* **2000**, *5*, 262–268. [CrossRef] [PubMed]
50. Kahn, V. Effect of kojic acid on the oxidation of DL-DOPA, norepinephrine, and dopamine by mushroom tyrosinase. *Pigment Cell Res.* **1995**, *8*, 234–240. [CrossRef] [PubMed]
51. Amorati, R.; Fumo, M.G.; Menichetti, S.; Mugnaini, V.; Pedulli, G.F. Electronic and hydrogen bonding effects on the chain-breaking activity of sulfur-containing phenolic antioxidants. *J. Org. Chem.* **2006**, *71*, 6325–6332. [CrossRef] [PubMed]
52. Tanini, D.; Panzella, L.; Amorati, R.; Capperucci, A.; Pizzo, E.; Napolitano, A.; Menichetti, S.; d'Ischia, M. Resveratrol-based benzoselenophenes with an enhanced antioxidant and chain breaking capacity. *Org. Biomol. Chem.* **2015**, *13*, 5757–5764. [CrossRef] [PubMed]
53. Riley, P.A.; Cooksey, C.J.; Johnson, C.I.; Land, E.J.; Latter, A.M.; Ramsden, C.A. Melanogenesis-targeted anti-melanoma pro-drug development: Effect of side-chain variations on the cytotoxicity of tyrosinase-generated ortho-quinones in a model screening system. *Eur. J. Cancer* **1997**, *33*, 135–143. [CrossRef]
54. Cooksey, C.J.; Land, E.J.; Ramsden, C.A.; Riley, P.A. Tyrosinase-mediated cytotoxicity of 4-substituted phenols: Quantitative structure-thiol-reactivity relationships of the derived *o*-quinones. *Anticancer Drug Des.* **1995**, *10*, 119–122. [CrossRef] [PubMed]
55. Hassani, S.; Haghbeen, K.; Fazli, M. Non-specific binding sites help to explain mixed inhibition in mushroom tyrosinase activities. *Eur. J. Med. Chem.* **2016**, *122*, 138–148. [CrossRef] [PubMed]
56. Mosmann, T. Rapid colorimetric assay for cellular growth and survival: Application to proliferation and cytotoxicity assays. *J. Immunol. Methods* **1983**, *65*, 55–63. [CrossRef]

biomimetics

MDPI

Article

Catechol-Containing Hydroxylated Biomimetic 4-Thiaflavanes as Inhibitors of Amyloid Aggregation

Matteo Ramazzotti [1,*]**, Paolo Paoli** [1]**, Bruno Tiribilli** [3]**, Caterina Viglianisi** [2]**,**
Stefano Menichetti [2,*] **and Donatella Degl'Innocenti** [1]

[1]   Dipartimento di Scienze Biomediche, Sperimentali e Cliniche, Università degli Studi di Firenze,
     viale G.B. Morgagni 50, 50134 Firenze, Italy; paolo.paoli@unifi.it (P.P.);
     donatella.deglinnocenti@unifi.it (D.D.)
[2]   Dipartimento di Chimica "Ugo Schiff", Polo Scientifico e Tecnologico, Università degli Studi di Firenze,
     via della Lastruccia 3-13, 50019 Sesto Fiorentino, Firenze, Italy; caterina.viglianisi@unifi.it
[3]   Consiglio Nazionale delle Ricerche (CNR), Istituto dei Sistemi Complessi, Via Madonna del Piano, 10,
     50019 Sesto Fiorentino, Firenze, Italy; bruno.tiribilli@isc.cnr.it
*   Correspondence: matteo.ramazzotti@unifi.it (M.R.); stefano.menichetti@unifi.it (S.M.);
     Tel.: +39-055-275-1248 (M.R.); Tel.: +39-055-457-3535 (S.M.)

Academic Editor: Daniel Ruiz-Molina
Received: 22 February 2017; Accepted: 4 May 2017; Published: 9 May 2017

Abstract: The study of compounds able to interfere in various ways with amyloid aggregation is of paramount importance in amyloid research. Molecules characterized by a 4-thiaflavane skeleton have received great attention in chemical, medicinal, and pharmaceutical research. Such molecules, especially polyhydroxylated 4-thiaflavanes, can be considered as structural mimickers of several natural polyphenols that have been previously demonstrated to bind and impair amyloid fibril formation. In this work, we tested five different 4-thiaflavanes on the hen egg-white lysozyme (HEWL) amyloid model for their potential anti-amyloid properties. By combining a thioflavin T assay, atomic force microscopy, and a cell toxicity assay, we demonstrated that such compounds can impair the formation of high-order amyloid aggregates and mature fibrils. Despite this, the tested 4-thiaflavanes, although non-toxic per se, are not able to prevent amyloid toxicity on human neuroblastoma cells. Rather, they proved to block early aggregates in a stable, toxic conformation. Accordingly, 4-thiaflavanes can be proposed for further studies aimed at identifying blocking agents for the study of toxicity mechanisms of amyloid aggregation.

Keywords: catechol; hydroxylated 4-thiaflavanes; inhibition; amyloid aggregation; hen egg white lysozyme; antioxidant activity

1. Introduction

Amyloid aggregation is a degenerative process characterized by deposition at tissue levels of organized insoluble super-molecular protein assemblies with a typical cross-β secondary structure. Such degeneration gives rise to amyloidosis, a composite range of diseases classically divided into neurodegenerative (e.g., Alzheimer's disease, Parkinson's disease, etc.) and systemic (e.g., cystic fibrosis, light chain amyloidosis) amyloidosis. More than 20 different human proteins, intact or fragmented, proved their amyloidogenicity in vivo, among which we may count amyloid β (Aβ) peptide (in Alzheimer's disease), α-synuclein (in Parkinson's disease), islet amyloid polypeptide (in type II-diabetes), light chains of immunoglobulins, variants of human lysozyme [1,2], and transthyretin (TTR) [3]. It is nowadays widely accepted that amyloid aggregation is a general tendency of polypeptide chains [4–6] that, in fact, may be induced to form amyloid aggregation in appropriate conditions [7].

Lysozyme, a 130-residue-long bacteriolytic enzyme largely distributed in different tissues, organs, and external secretions, has been highlighted as an interesting model for the study of amyloid aggregation. Although wild-type lysozyme is not directly involved in amyloid diseases, several naturally occurring single point mutations (e.g., Ile56Thr, Phe57Ile, Trp64Arg, and Asp67His) are connected with familial non-neuropathic systemic amyloidosis [8]. In addition, the wild-type lysozyme either from humans, horses, or hens, under appropriate conditions, is able to form amyloid fibrils in vitro [9–11].

In this work, we used the hen egg-white lysozyme (HEWL—14.3 kDa, 129 amino acids, 40% identity with the human enzyme) inducing its aggregation through a heat treatment in acidic conditions [12]. Despite the fact that HEWL is not associated with in vivo diseases [13], it has been demonstrated that high temperatures and low pH induce the breakage of X-Asp peptide bonds, leading to the formation of peptide fragments (among which one contains the residues corresponding to those mutated in human familiar diseases Ile56Thr and Asp67His). Such fragments have a high tendency to form amyloid aggregates [11] and amyloid-like fibrils in a few days. In addition, a direct toxic effect of HEWL aggregates added to cell cultures or injected in rat brains, mimicking the toxic effect of Aβ peptide, has been demonstrated [14].

A substantial body of literature over the years documents that extracts from natural herbs and plants or common dietary elements such as wine [15] or green-tea [16] are of great benefit to general human health, mainly due to their antioxidant power [17,18]. Among the molecules proposed to be essential for achieving such effects are polyphenols [19], a wide and heterogeneous group of substances well characterized in terms of structure [20]. Apart from protecting cells from oxidative stress, several phenolic compounds have been shown to be effective in inhibiting amyloid aggregation in various protein models such as transthyretin, microglobulin, α-syunclein, Aβ peptide and lysozyme, with supposed specific action mechanisms independent of their antioxidant properties [21–25]. Of particular relevance for this work, catecholic- and hydroquinone-containing phenols were reported to act as inhibitors of amyloid aggregation for their ability to induce quinoprotein formation [26].

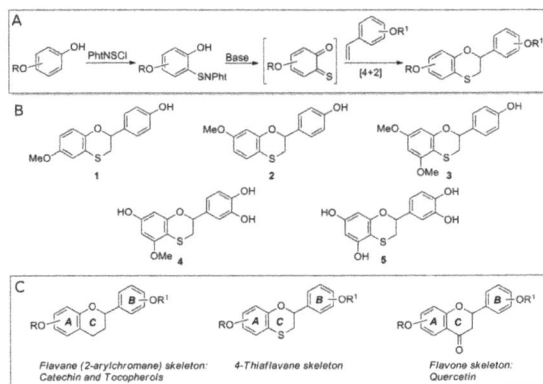

Figure 1. Design of hydroxylated 4-thiaflavanes and their biomimetisms. (**A**) Synthesis of hydroxylated 4-thiaflavanes used in this study. (**B**) Structure of 4-thiaflavanes tested in this study. (**C**) Biomimetism of 4-thiaflavanes with 2-arylchromane (flavane) and flavone skeletons.

Dihydrobenzo[1,4]oxathines, and in particular compounds possessing a polyhydroxylated 4-thiaflavane skeleton, have received great attention in chemical, medicinal, and pharmaceutical research. During the last decades, their syntheses as well as their abilities as antioxidants, hypertensive agents, estrogen receptor modulators, adrenoreceptor antagonists, and artificial sweeteners have been reported in papers and patents [27]. As shown in Figure 1, such 4-thiaflavane derivatives are structural mimickers of several natural polyphenols. In particular, compounds **4** and **5** (see Figure 1) are thia-substituted biomimetic examples of catechin derivatives showing a free catechol residue on the B ring. In this work, we investigated whether selected hydroxylated 4-thiaflavane derivatives may act as an inhibitor of amyloid aggregation for the widely used and accepted HEWL amyloid model.

2. Materials and Methods

2.1. Materials

Reagents and chemicals, unless otherwise specified, were purchased from Sigma-Aldrich (St. Louis, MO, USA) and used without further purification, including lysozyme (HEWL, code L6876).

2.2. Preparation of 4-Thiaflavane Derivatives

Hydroxylated 4-thiflavanes were prepared as previously reported by inverse electron demand hetero Diels–Alder reaction of transient *ortho*-thioquinones with properly substituted styrenes. This allowed for the direct synthesis of compounds bearing hydroxy or methoxy groups on the selected position of the A and/or B ring of the thiaflavane skeleton (Figure 1) [28–30].

Lyophilized thiaflavane powders were weighted and dissolved in pure dimethyl sulfoxide (DMSO) at a concentration of 100 mM. For assays, they were freshly diluted at appropriate concentrations in assay buffers. The chemical structures of the thiaflavanes used in this study are depicted in Figure 1 and their absorbance spectra shown in Figure 2A.

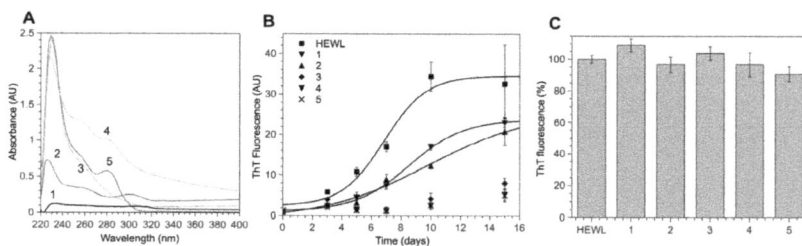

Figure 2. 4-Thiaflavane derivatives inhibit hen egg-white lysozyme (HEWL) amyloid aggregation. (**A**) Absorption spectra of 4-thiaflavanes derivatives (compound **1–5**) used in this study. (**B**) Aggregation kinetic of HEWL in the presence of 4-thiaflavane compounds as followed by a thioflavin T (ThT) assay for 15 days. (**C**) The absence of competition of 4-thiaflavane compounds with ThT on preformed HEWL fibrils. Values in (**B,C**) are expressed as the mean ± standard deviation of at least three independent measurements. AU: Arbitrary units.

2.3. Preparation of Hen-Egg White Lysozyme for Aggregation

HEWL solutions were freshly prepared before each assay. HEWL powder was weighted and dissolved in 10 mM HCl (pH 2, HEWL buffer) via vortexing at room temperature at a concentration of 1 mM (14.7 mg/mL). The solution was then filtered using 0.2 μm syringe filter discs (EMD Millipore, Milan, Italy).

2.4. Aggregation Conditions

HEWL aggregation was achieved by incubating the pH 2 solution at 65 °C for up to 15 days [11,31]. 4-Thiaflavanes were mixed to HEWL at a 1:1 molar ratio at a 1 mM concentration, immediately before incubation at 65 °C. Control HEWL aggregation was performed in the presence of 1% DMSO. All aggregation experiments were repeated at least three times.

2.5. Thioflavin T Assay

For the thioflavin T (ThT) assay [32], 25 µM ThT was freshly prepared from a 2.5 mM stock solution in a 25 mM sodium phosphate buffer, pH 6 (ThT buffer). For each measurement, 245 µL of the diluted ThT solution was added to a sample volume of 5 µL on a polystyrene multiwell plate. ThT fluorescence was measured at 25 °C with a Fluoroskan Ascent FL multiwell plate reader (Thermo Scientific, Waltham, MA, USA) using 440 nm and 485 nm as excitation and emission wavelengths, respectively. All measurements were performed in triplicate. Kinetic traces were analyzed via non-linear fitting to the sigmoidal function $F = F_f + (F_i - F_f)/(1 + \exp((x - t_0)/d_x))$, F being the time-dependent fluorescence intensity, F_i the fluorescence at the beginning of the aggregation, F_f the fluorescence at the end of the aggregation process, t_0 the time at which 50% of the total variation in fluorescence is reached, and d_x the time constant. The apparent rate constant (kf) for the growth of fibrils is given by $1/d_x$, the lag time is calculated as $t_0 - 2d_x$, and the fluorescence amplitude is given by $Ff - Fi$. All these analyses were performed with QtiPlot v0.9.8.0 software (http://www.qtiplot.com).

2.6. Atomic Force Microscopy

For atomic force microscopy (AFM) analysis, a drop of aggregating solutions (HEWL with or without 4-thiaflavanes) were vortexed and laid onto a freshly cleaved mica disc (Ted Pella Inc., Redding, CA) for about 2 min. Excess of sample was removed by washing twice with 1 mL of bidistilled water, the preparation was then dried with a soft nitrogen flow. AFM experiments were performed in air, in non-contact mode, using a PicoSPM microscope equipped with an AAC-Mode controller (Molecular Imaging, Phoenix, AZ, USA). The probes were non-contact Silicon cantilevers (model NSG-01, NT-MDT Co., Moscow, Russia) with a 150 KHz typical resonance frequency. Scanner calibration was periodically checked by means of a reference grid (TGZ02 by MikroMash, Tallin, Estonia) with a known pitch of 3 µm and a step height of 100 nm. Scan size ranged from 450 × 450 nm to 30 × 30 µm. Images were processed and analyzed with Gwyddion software v2.34 (http://gwyddion.net). For the analysis, the pre-processing involved (i) levelling the map by mean plane subtraction, (ii) correcting lines by matching height median, (iii) correcting horizontal artefacts (scars), (iv) applying a Gaussian smoothing filter of 2 px, and (v) shifting minimum data value to zero.

2.7. Cell Growth and Citotoxicity Assay

Human SH-SY5Y neuroblastoma cells (American Type Culture Collection, Manassas, VA, USA) were cultured in Dulbecco's modified Eagle's medium DMEM F-12 Ham with 25 mM HEPES (N-2-hydroxyethylpiperazine-N-2-ethane sulfonic acid) and NaHCO3 (1:1) supplemented with 10% fetal bovine serum (FBS, Sigma-Aldrich), 1 mM glutamine, and antibiotics.

The cytotoxicity of the aggregates was assessed by an MTT (3-(4,5-dimethylthiazol-2-yl)-2,5-diphenyltetrazolium bromide) reduction inhibition assay [33]. Briefly, SHSY-5Y cells in exponential growth were incubated for 48 h in the presence of HEWL aggregates matured alone or in the presence of 4-thiaflavanes. The growth medium was removed, and the plates were incubated for 2 h in a 5% CO_2-humidified atmosphere at 37 °C in the presence of a medium solution containing 0.5 mg/mL of the MTT reagent. After 2 h, the solution was removed and replaced with a lysis buffer containing 20% sodium dodecyl sulfate (SDS) and 50% dimethylformamide (DMF, pH 4.7), and further incubated for 1 h. The absorbance of blue formazan was measured at 570 nm with an iMark^{TM} microplate reader (BioRad, Hercules, CA, USA).

3. Results

3.1. 4-Thiaflavane Derivatives Obstacle/Impair Amyloid Aggregation Kinetics

HEWL aggregation is primed by a fragmentation process that may be induced by heating a concentrated HEWL solution in acidic conditions. We tested the anti-aggregation properties of hydroxylated 4-thiaflavanes (Figure 1) on HEWL by incubating them at a 1:1 molar ratio prior to heating, thus allowing the fragmentation to occur in the presence of 4-thiaflavanes under study. The aggregation kinetic of HEWL in the presence of 4-thiaflavanes was followed by ThT assay (a universally accepted fluorogenic probe for cross β-aggregates). HEWL aggregation proved to be deeply altered by three out of the five 4-thiaflavanes tested, namely **4**, **5**, both containing a catechol residue, and **3** (Figure 2B). While compounds **3**, **4** and **5** were able to almost completely inhibit ThT signals, suggesting a deep impact on amyloid aggregate maturation, compounds **1** and **2** proved to have little effect on the estimated lag phase of the aggregation process (3 ± 2 and 5 ± 0.5 days for **1** and **2**, respectively, compared to 4 ± 1 days for HEWL) and a moderate effect on the plateau phase (about 30%). This analysis suggested that aggregation is not completely abolished and possibly stabilized in a different final conformation, with lower affinity for the ThT dye or with a lower concentration of amyloid aggregates.

3.2. 4-Thiaflavane Derivatives Do Not Compete with Thioflavin T

It has been recently demonstrated that ThT signals can be strongly affected by compounds with phenolic moieties, due to the optical or physical competition with the ThT binding site on aggregates [34], leading to misinterpretation of amyloid inhibition by exogenous compounds. In order to exclude optical interferences, we measured the absorption spectra of 4-thiaflavane derivatives dissolved at a 1 mM concentration in a ThT buffer using 1% DMSO as a blank. Figure 2A shows that none of the substances tested showed significant absorption in the 400–500 nm region, where ThT is excited and emitted when cross-β structures are present. Secondly, we tested whether 4-thiaflavanes could compete with ThT by incubating them with ThT on preformed HEWL amyloid fibrils. The rationale behind this assay was based on the assay time: amyloid aggregates are considered extremely stable and resistant to disaggregation, being solubilized only by strong denaturing agents such as DMSO or hexafluoroisopropanol (HFIP). It therefore appears unrealistic that disaggregation could occur in a few minutes of assay time in very mild conditions such as those required for the ThT assay. As shown in Figure 2C, none of the 4-thiaflavanes proved to compete significantly with ThT, since fluorescence signals recorded in the presence of preformed fibrils were not reduced by the addition of 4-thiaflavanes at concentrations ten times higher than that of ThT. We therefore concluded that competition or interference could not have been the reason for the fluorescence loss in our kinetic experiments.

3.3. Polyhydroxylated 4-Thiaflavanes Inhibit the Formation of Amyloid Fibrils

In order to study the morphology of the HEWL amyloid aggregates after 10 days (the time required to HEWL to convert into mature fibrils), we deposited the aggregating solutions on freshly cleaved mica and its surface was scanned by AFM. As shown in Figure 3, HEWL fibrils with a height of about 4 nm were highly abundant in the absence of 4-thiaflavane derivatives. Their burden was found to be highly reduced in the presence of all 4-thiaflavane derivatives.

By combining morphology maps, height distribution analysis, and z-profile analysis, we studied the morphology of aggregates formed at this time point. Compound **1** showed a drastic reduction in the number of fibrils formed and the accumulation of round particles of about 2–4 nm in height, a size similar to the diameter of the residual fibrils. A similar behavior was observed for compound **2** with an apparent increase in the ability of reducing fibrillar structures. Both compounds were only in part effective in reducing ThT signals, suggesting that round particles partly maintain the cross-β amyloid structure targeted by such dye, resembling the structure of early aggregates. Compound **3** proved to completely abolish the formation of fibrils, resulting in the accumulation of round particles again with an apparent size of 2–4 nm but with a drastically reduced affinity for ThT. The aggregation of HEWL in the presence of compound **4** exhibited a completely different morphological pattern, showing a mixture of round particles of different diameters and small fibrils. The fibrils observed in this case were almost invariantly larger than that observed with other compounds or untreated HEWL, with apparent diameters above 10 nm, and unable to give rise to an increase in ThT signals. In addition, and peculiar to this compound, particles with heights surpassing 100 nm were detected. A complete absence of fibrils and a drastic reduction in globular particles of 3–5 nm were eventually observed for compound **5**. Details of the results of the z-profile measurements for all samples are further available as accompanying supplementary materials.

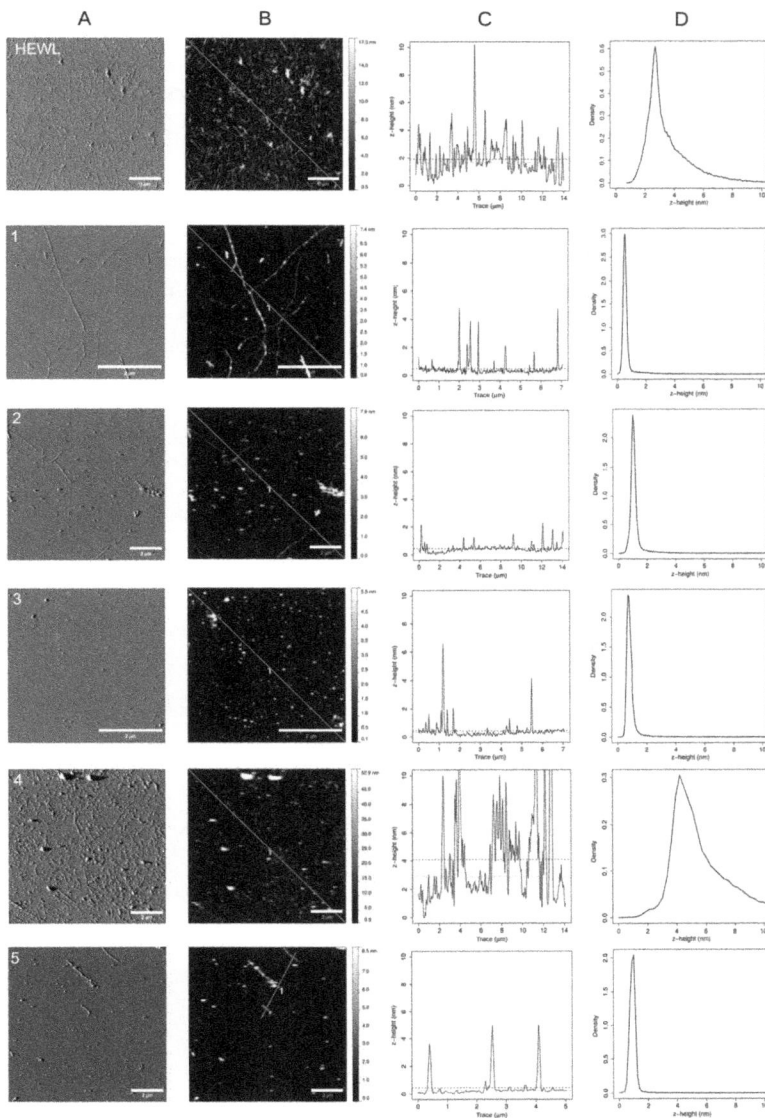

Figure 3. Atomic force microscopy (AFM) maps of HEWL incubated in aggregation conditions for 10 days in the presence or absence of the 4-thiaflavane derivatives (compounds **1–5**): (**A**) amplitude maps; (**B**) morphology maps and the corresponding (**C**) profile traces, together with the mean and median z-high values plotted as horizontal dashed or dotted lines, respectively; (**D**) height distribution of the whole map. Scale bar: 2 μm.

3.4. 4-Thiaflavane Derivatives Do Not Prevent Toxicity Induced by Early Amyloid Aggregates

For assessing whether the drastic reduction of aggregate load evidenced in both ThT assay and AFM was accompanied by a reduction of aggregate-induced cytotoxicity, we incubated actively growing SHSY-5Y neuroblastoma cells with HEWL aggregates that had matured for 3 days both alone or in the presence of thiaflavanes. In fact, in a preliminary cytotoxicity kinetic analysis, the 3-day incubation time proved to be the most toxic phase of HEWL aggregates formed in our conditions (data not shown). 4-Thiaflavane derivatives alone proved to be safe for cells at concentrations used in the assay (Figure 4A). On the contrary, we found that 4-thiaflavanes were able to show little or no protective effects on cells treated with 3-day aggregates (Figure 4B), indicating that premature toxic aggregates were present in the solutions and that their formation was not prevented by the presence, during the aggregation process, of such compounds. This is in clear contrast to the impairment of the polymerization process evidenced by ThT and by AFM. In order to exclude possible artefacts due to the heat-induced conversion into toxic compounds of 4-thiaflavanes tested, we also heated compounds alone. We verified that 4-thiaflavanes were also safe for cells after heating, resulting in cells that showed no signs of toxicity. We also verified the opposite condition, i.e., that heating could have induced a loss of protective effect that 4-thiaflavane derivatives could have exerted per se (e.g., as an antioxidant, or with some other property independent of the effect of amyloid material). To do this, we added freshly prepared 4-thiaflavanes to cells, concomitantly with untreated 3-day aggregates. Additionally, in this case, we did not find particular signs of protection. When the cells were treated with aggregates that had matured for 10 days, a completely different picture was observed. HEWL alone, in the form of mature fibrillar aggregates (see Figure 3), showed a markedly reduced toxicity with respect to the untreated control. On the contrary, thiaflavane-treated HEWL showed a pronounced toxicity, similar to that observed for the 3-day maturation especially for compounds **4** and **5** (Figure 4B).

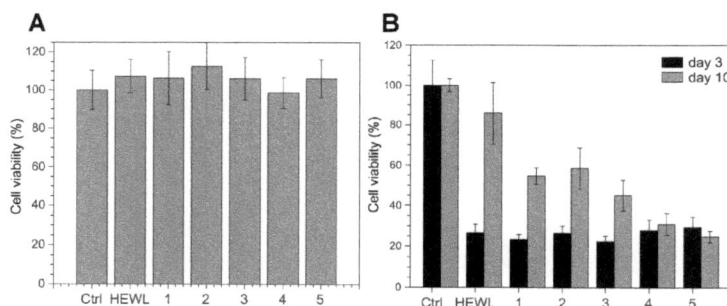

Figure 4. Cytotoxicity assay on SHSY5Y cells. (**A**) Absence of cell toxicity after 48 h incubation with monomeric HEWL or 4-thiaflavanes (compounds **1–5**). (**B**) Cell toxicity after 48 h incubation with 3 and 10 days old aggregates of HEWL incubated with or without 4-thiaflavanes (compounds **1–5**). Bars represent the mean ± standard deviation of at least three independent measurements. Ctrl: untreated cells.

4. Discussion

The possibility of blocking or reverting amyloid aggregation with small molecules may have a great impact on worldwide health. Amyloidosis, a group of over 20 different and heterogeneous diseases, are directly linked to the accumulation of amyloid matter into organs and are severely affecting human population. Amyloidosis, including neurodegenerative diseases such as Alzheimer's disease or Parkinson's disease as well as systemic diseases such as reactive systemic amyloidosis, TTR, and light chain amyloidosis are increasingly recognized as important death factors for public health systems.

Many studies have so far demonstrated that molecules with peculiar structural features are able to impair the formation or the elongation of amyloid fibrils, among which are a number of synthetic or natural compounds with polyphenolic rings [23]. Several activity–structure studies have shown that, though the effect of the addition of such exogenous compounds could be beneficial in some cases in terms of the prevention of amyloid formation or the reduction of amyloid load, the activity spectrum is bound to a defined experimental set and confined to certain amyloid related proteins and peptides, such as Aβ peptides, α-synuclein, and TTR.

In this study, we used HEWL as a model of amyloid aggregation. The reason for this choice was based on the need to decouple the effect of metal ions on aggregation to the documented chelating properties of 4-thiaflavanes [35]. In fact, many amyloid models have been found to be deeply affected by the presence of copper or iron ions, and in traces, leading to alterations of the aggregation kinetics, of the lag phase, or of the morphology of the resulting fibrils. Such effects have never been reported for HEWL, so we considered it a good model for our class of compounds, allowing us to establish a direct link between the structure of the molecules and the effect on aggregation inhibition. In fact, we found that the poor solubility of compounds such as 1 and 2 is not of benefit in reducing amyloid signals, which is contrary to what has been found for other compounds, such as curcumin, which has been declared to be extremely active in aggregation in vitro despite its absolute insolubility in aqueous buffers. In our conditions, the most active 4-thiaflavanes were molecules with a higher degree of hydroxylation, leading to increased solubility and a reasonably higher capability of interacting with nascent aggregates, blocking their elongation into higher order assemblies such as mature fibrils. Moreover, despite the lack of a direct relationship between antioxidant activity and the inhibition of amyloid aggregation, it is worth mentioning that compounds 4 and 5, containing a catechol moiety, showed a much higher antioxidant activity with respect to compounds 1, 2, and 3 [36].

A relevant body of literature exists trying to shed light on the true nature of the cytotoxicity induced by amyloid material. Initially, it was strongly believed that toxicity was due to early aggregates only (a stage at which large portions of hydrophobic protein regions are exposed, waiting to gain stabilization, with polymerization and burial of these regions inside highly ordered structures). Recently, it has been shown that this cannot be considered a general rule because of the existence of proteins for which mature fibrils largely surpass the toxicity of early aggregates [37]. HEWL has been previously shown to behave in the "classic" way, losing toxicity as fibrils grow and as early aggregates are progressively sequestered from the medium to the fibril. We verified this behavior in our experimental conditions, selecting day 3 as the one displaying the maximal toxicity that was progressively lost at longer times, until day 10, when fibrils were completely mature, as shown by AFM. At this time point, we measured extremely reduced ThT signals for compounds 3, 4, and 5, but we found a toxicity comparable to that of untreated HEWL. When the incubation time was increased to allow HEWL controls to develop into non-toxic mature fibrils, the HEWL samples treated with active compounds were found to maintain a toxicity comparable to samples tested at early aggregation stages. Nevertheless, the ThT signal proved to be, in most cases, highly reduced. Our results seem to suggest that the effect of 4-thiaflavane derivatives was to impair (in part or almost completely, depending on the compound) the elongation of HEWL fibrils, leading to the formation of round particles with diameters similar to that of amyloid species and that are highly toxic to cells, stable over time, and poorly responsive to ThT binding.

According to a recent finding, molecules containing catecholic and (less efficiently) hydroquinone moieties may be able to drive the formation of quinoproteins, i.e., to covalently modify proteins in a hot acidic environment [26] and hamper HEWL fibril formation. Compounds 4 and 5 bear in the B ring a catechol-like structure that, given aggregation conditions (acidic pH and high temperature), may induce protein derivatization and form *ortho*-quinonic adducts. Similarly, compounds 1, 2, and 3 present a phenol ring that, in the same conditions, can undergo acid hydrolysis and further oxidation, leading to the formation of p-quinoic adducts. Our results on hydroxylated 4-thiaflavanes seem to suggest that the abovementioned reaction mechanisms may drive the observed reduction in amyloid formation without a loss of cell toxicity. The ability of compound **3** to drastically reduce the load in fibrils (despite its structural similarity with the less effective compounds **1** and **2**) partly contrasts this vision and suggests that additional research is required to definitively validate the above hypothesis.

Although our findings strongly discourage the usage of the investigated 4-thiaflavanes as potential drugs for amyloidosis, the opposite route seems interesting. Since these compounds have been shown to be safe for cells at the tested doses, their contribution to cell toxicity following incubation with toxic aggregates is minimal, allowing a coherent assay of aggregate toxicity. In fact, one of the most challenging aspects of amyloid studies is that intermediate, toxic structures are transient in nature and their effect is confined in a restricted, highly variable temporal frame.

Further biophysical studies will be needed to elucidate the structures and the features of the toxic aggregates stabilized by the 4-thiaflavane derivatives tested in this work and their mechanism of action. Of particular interest is the observation that such particles give reduced or impaired ThT signals, a fact that is counterintuitive given their toxicity, which is similar to 3-day HEWL early aggregates, which proved to give rise to aggregates that efficiently bind ThT. Furthermore, it cannot be excluded that the structural features of such aggregates may be different in response to different compounds, although the results collected so far seem to suggest as the major player a population of small, globular particles with a quite uniform size.

If confirmed on other amyloid systems, these results suggest that hydroxylated 4-thiaflavanes are promising stabilizing agents for toxic aggregates that are useful for studying, for example, the effect of single or multiple mutations on a uniform and homogeneous population of toxic species.

Acknowledgments: This work was supported by grants from Fondi di Ateneo 2015, University of Florence.

Author Contributions: M.R. conceived the experiments, performed amyloid assays, and drafted the manuscript together with D.D.; P.P. performed cell-based assays; B.T. performed AFM experiments; C.V. and S.M. designed and prepared the 4-thiaflavane molecules. All authors contributed to manuscript revision and approved its final version.

Conflicts of Interest: The authors declare no conflict of interest.

References

1. Sideras, K.; Gertz, M.A. Amyloidosis. *Adv. Clin. Chem.* **2009**, *47*, 1–44. [PubMed]
2. Chiti, F.; Dobson, C.M. Protein misfolding, functional amyloid, and human disease. *Annu. Rev. Biochem.* **2006**, *75*, 333–366. [CrossRef] [PubMed]
3. Gertz, M.A.; Benson, M.D.; Dyck, P.J.; Grogan, M.; Coelho, T.; Cruz, M.; Berk, J.L.; Plante-Bordeneuve, V.; Schmidt, H.H.; Merlini, G. Diagnosis, prognosis, and therapy of transthyretin amyloidosis. *J. Am. Coll. Cardiol.* **2015**, *66*, 2451–2466. [CrossRef] [PubMed]
4. Bemporad, F.; Calloni, G.; Campioni, S.; Plakoutsi, G.; Taddei, N.; Chiti, F. Sequence and structural determinants of amyloid fibril formation. *Acc. Chem. Res.* **2006**, *39*, 620–627. [CrossRef] [PubMed]
5. Kelly, J.W. Amyloid fibril formation and protein misassembly: A structural quest for insights into amyloid and prion diseases. *Structure* **1997**, *5*, 595–600. [CrossRef]
6. Stefani, M.; Dobson, C.M. Protein aggregation and aggregate toxicity: New insights into protein folding, misfolding diseases and biological evolution. *J. Mol. Med. (Berl)* **2003**, *81*, 678–699. [CrossRef] [PubMed]
7. Dobson, C.M. Principles of protein folding, misfolding and aggregation. *Semin. Cell Dev. Biol.* **2004**, *15*, 3–16. [CrossRef] [PubMed]
8. Merlini, G.; Bellotti, V. Lysozyme: A paradigmatic molecule for the investigation of protein structure, function and misfolding. *Clin. Chim. Acta* **2005**, *357*, 168–172. [CrossRef] [PubMed]
9. Dumoulin, M.; Kumita, J.R.; Dobson, C.M. Normal and aberrant biological self-assembly: Insights from studies of human lysozyme and its amyloidogenic variants. *Acc. Chem. Res.* **2006**, *39*, 603–610. [CrossRef] [PubMed]
10. Malisauskas, M.; Ostman, J.; Darinskas, A.; Zamotin, V.; Liutkevicius, E.; Lundgren, E.; Morozova-Roche, L.A. Does the cytotoxic effect of transient amyloid oligomers from common equine lysozyme in vitro imply innate amyloid toxicity? *J. Biol. Chem.* **2004**, *280*, 6269–6275. [CrossRef] [PubMed]

11. Frare, E.; Polverino De Laureto, P.; Zurdo, J.; Dobson, C.M.; Fontana, A. A highly amyloidogenic region of hen lysozyme. *J. Mol. Biol.* **2004**, *340*, 1153–1165. [CrossRef] [PubMed]
12. Arnaudov, L.N.; de Vries, R. Thermally induced fibrillar aggregation of hen egg white lysozyme. *Biophys. J.* **2004**, *88*, 515–526. [CrossRef] [PubMed]
13. Sethuraman, A.; Belfort, G. Protein structural perturbation and aggregation on homogeneous surfaces. *Biophys. J.* **2004**, *88*, 1322–1333. [CrossRef] [PubMed]
14. Vieira, M.N.; Forny-Germano, L.; Saraiva, L.M.; Sebollela, A.; Martinez, A.M.; Houzel, J.C.; De Felice, F.G.; Ferreira, S.T. Soluble oligomers from a non-disease related protein mimic Aβ-induced tau hyperphosphorylation and neurodegeneration. *J. Neurochem.* **2007**, *103*, 736–748. [CrossRef] [PubMed]
15. Walzem, R.L. Wine and health: State of proofs and research needs. *Inflammopharmacology* **2008**, *16*, 265–271. [CrossRef] [PubMed]
16. Higdon, J.V.; Frei, B. Tea catechins and polyphenols: Health effects, metabolism, and antioxidant functions. *Crit. Rev. Food Sci. Nutr.* **2003**, *43*, 89–143. [CrossRef] [PubMed]
17. Virgili, F.; Marino, M. Regulation of cellular signals from nutritional molecules: A specific role for phytochemicals, beyond antioxidant activity. *Free Radic. Biol. Med.* **2008**, *45*, 1205–1216. [CrossRef] [PubMed]
18. Kim, H.; Park, B.S.; Lee, K.G.; Choi, C.Y.; Jang, S.S.; Kim, Y.H.; Lee, S.E. Effects of naturally occurring compounds on fibril formation and oxidative stress of beta-amyloid. *J. Agric. Food Chem.* **2005**, *53*, 8537–8541. [CrossRef] [PubMed]
19. Manach, C.; Williamson, G.; Morand, C.; Scalbert, A.; Rémésy, C. Bioavailability and bioefficacy of polyphenols in humans. I. Review of 97 bioavailability studies. *Am. J. Clin. Nutr.* **2005**, *81*, 230S–242S. [PubMed]
20. Manach, C.; Scalbert, A.; Morand, C.; Rémésy, C.; Jiménez, L. Polyphenols: Food sources and bioavailability. *Am. J. Clin. Nutr.* **2004**, *79*, 727–747. [PubMed]
21. Leri, M.; Nosi, D.; Natalello, A.; Porcari, R.; Ramazzotti, M.; Chiti, F.; Bellotti, V.; Doglia, S.M.; Stefani, M.; Bucciantini, M. The polyphenol Oleuropein aglycone hinders the growth of toxic transthyretin amyloid assemblies. *J. Nutr. Biochem.* **2016**, *30*, 153–166. [CrossRef] [PubMed]
22. Leri, M.; Bemporad, F.; Oropesa-Nuñez, R.; Canale, C.; Calamai, M.; Nosi, D.; Ramazzotti, M.; Giorgetti, S.; Pavone, F.S.; Bellotti, V.; et al. Molecular insights into cell toxicity of a novel familial amyloidogenic variant of β2-microglobulin. *J. Cell. Mol. Med.* **2016**, *20*, 1443–1456. [CrossRef] [PubMed]
23. Porat, Y.; Abramowitz, A.; Gazit, E. Inhibition of amyloid fibril formation by polyphenols: Structural similarity and aromatic interactions as a common inhibition mechanism. *Chem. Biol. Drug Des.* **2006**, *67*, 27–37. [CrossRef] [PubMed]
24. Ono, K.; Yamada, M. Antioxidant compounds have potent anti-fibrillogenic and fibril-destabilizing effects for α-synuclein fibrils in vitro. *J. Neurochem.* **2006**, *97*, 105–115. [CrossRef] [PubMed]
25. Ramazzotti, M.; Melani, F.; Marchi, L.; Mulinacci, N.; Gestri, S.; Tiribilli, B.; Degl'Innocenti, D. Mechanisms for the inhibition of amyloid aggregation by small ligands. *Biosci. Rep.* **2016**, *36*, e00385. [CrossRef] [PubMed]
26. Feng, S.; Song, X.H.; Zeng, C.M. Inhibition of amyloid fibrillation of lysozyme by phenolic compounds involves quinoprotein formation. *FEBS Lett.* **2012**, *586*, 3951–3955. [CrossRef] [PubMed]
27. Viglianisi, C.; Menichetti, S. Dihydrobenzo[1,4]oxathiine: A multi-potent pharmacophoric heterocyclic nucleus. *Curr. Med. Chem.* **2010**, *17*, 915–928. [CrossRef] [PubMed]
28. Menichetti, S.; Aversa, M.C.; Cimino, F.; Contini, A.; Viglianisi, C.; Tomaino, A. Synthesis and "double-faced" antioxidant activity of polyhydroxylated 4-thiaflavans. *Org. Biomol. Chem.* **2005**, *3*, 3066–3072. [CrossRef] [PubMed]
29. Menichetti, S.; Amorati, R.; Bartolozzi, M.G.; Pedulli, G.F.; Salvini, A.; Viglianisi, C. A straightforward hetero-Diels–Alder approach to (2-ambo,4′R,8′R)-α/β/γ/δ-4-thiatocopherol. *Eur. J. Org. Chem.* **2010**, *10*, 2218–2225. [CrossRef]
30. Viglianisi, C.; Bartolozzi, M.G.; Pedulli, G.F.; Amorati, R.; Menichetti, S. Optimization of the antioxidant activity of hydroxy-substituted 4-thiaflavanes: A proof-of-concept study. *Chemistry* **2011**, *17*, 12396–12404. [CrossRef] [PubMed]
31. Krebs, M.R.; Wilkins, D.K.; Chung, E.W.; Pitkeathly, M.C.; Chamberlain, A.K.; Zurdo, J.; Robinson, C.V.; Dobson, C.M. Formation and seeding of amyloid fibrils from wild-type hen lysozyme and a peptide fragment from the β-domain. *J. Mol. Biol.* **2000**, *300*, 541–549. [CrossRef] [PubMed]

32. LeVine, H. Thioflavine T interaction with synthetic Alzheimer's disease β-amyloid peptides: Detection of amyloid aggregation in solution. *Protein Sci.* **1993**, *2*, 404–410. [CrossRef] [PubMed]

33. Mosmann, T. Rapid colorimetric assay for cellular growth and survival: Application to proliferation and cytotoxicity assays. *J. Immunol. Methods* **1983**, *65*, 55–63. [CrossRef]

34. Hudson, S.A.; Ecroyd, H.; Kee, T.W.; Carver, J.A. The thioflavin T fluorescence assay for amyloid fibril detection can be biased by the presence of exogenous compounds. *FEBS J.* **2009**, *276*, 5960–5972. [CrossRef] [PubMed]

35. Lodovici, M.; Menichetti, S.; Viglianisi, C.; Caldini, S.; Giuliani, E. Polyhydroxylated 4-thiaflavans as multipotent antioxidants: Protective effect on oxidative DNA damage in vitro. *Bioorg. Med. Chem. Lett.* **2006**, *16*, 1957–1960. [CrossRef] [PubMed]

36. Amorati, R.; Fumo, M.G.; Pedulli, G.F.; Menichetti, S.; Pagliuca, C.; Viglianisi, C. Antioxidant and antiradical activity of hydroxy-substituted 4-thiaflavanes. *Helv. Chim. Acta* **2006**, *89*, 2462–2472. [CrossRef]

37. Stefani, M. Biochemical and biophysical features of both oligomer/fibril and cell membrane in amyloid cytotoxicity. *FEBS J.* **2010**, *277*, 4602–4613. [CrossRef] [PubMed]

Chapter 3:
Catechol Applications in Materials Science

biomimetics

MDPI

Review

Composite Materials and Films Based on Melanins, Polydopamine, and Other Catecholamine-Based Materials

Vincent Ball [1,2]

[1] Faculté de Chirurgie Dentaire, Université de Strasbourg, 8 rue Sainte Elisabeth, 67000 Strasbourg, France; vball@unistra.fr; Tel.: +33-03-6885-3384

[2] Unité Mixte de Recherche 1121, Institut National de la Santé et de la Recherche Médicale, 11 rue Humann, 67085 Strasbourg Cedex, France

Academic Editors: Marco d'Ischia and Daniel Ruiz-Molina
Received: 8 May 2017; Accepted: 26 June 2017; Published: 6 July 2017

Abstract: Polydopamine (PDA) is related to eumelanins in its composition and structure. These pigments allow the design, inspired by natural materials, of composite nanoparticles and films for applications in the field of energy conversion and the design of biomaterials. This short review summarizes the main advances in the design of PDA-based composites with inorganic and organic materials.

Keywords: polydopamine; melanin; composite films; core-shell nanoparticles

1. Introduction

The strong underwater adhesion of mussels [1], and the exceptional mechanical properties of squid beaks [2] and see worm jaws [3], which are achieved without high inorganic content contrarily to bones and teeth where the inorganic material content is high, raised the question how such exceptional properties are reached. Analysis of the composition and structure of the mussel byssus, squid beaks, and marine hydroid perisarc [4] showed that those materials are essentially composites between proteins, polysaccharides, and melanin-related materials. The chemical principles implied in the formation of those composites rely on the rich chemistry of catechol and catecholamines, the molecular building blocks of melanins: pH-dependent redox chemistry, electrostatic interactions, and complexation with metallic cations are implied in the formation of melanins, their interaction with proteins and other materials [5,6]. The need for simplified design principles of adhesives mimicking the properties of the mussel foot proteins (mfps) led to the use of dopamine as the precursor of films adhering to the surface of almost all known materials [7]. Indeed, dopamine contains a catechol moiety (like the L-3,4-dihydroxyphenylalanine (L-DOPA) residue in mfps) and a primary amine function (like L-lysine in mfps). The films obtained during the oxidation of dopamine using either O_2 dissolved in water (at basic pH values) or exogeneous oxidants ($NaIO_4$, sodium peroxodisulfate) [8] offer fascinating opportunities not only to coat all kinds of materials (metals, oxides, polymers), but also to post-functionalize them with either inorganic or organic compounds. During all synthesis routes of polydopamine (PDA) and related materials, the catecholamine solubilized in an appropriate buffer solution has simply to be mixed with an excess of oxidant to yield to the formation of particles/precipitates in solution and to a homogeneous film at interfaces after an initial island growth regime. Note that the growth rate of the PDA films and their composition depends markedly on the used buffer and the used oxidant [8–10].

The strong adhesion afforded by those PDA, and related catechol- and catecholamine-based coatings (norepinephrine [9], L-DOPA), as well as their broad reactivity with cations and nucleophiles (thiols and amines), offers a broad range of applications, allowing the design of new composite materials, particles, and films [10–12]. One of the driving principles in the manipulation of PDA is to consider its close compositional and structural analogy with eumelanins [13,14], the brown-black dye affording photoprotection to the skin [15].

It is the aim of this short review to summarize the main recent advances in the design of PDA-, melanin-, and catechol-based composite materials, particles, and films. The main principles in the design of those composites will be explained without trying to be exhaustive, but rather illustrative. We will, hence, classify the composites obtained with inorganic and organic materials rather than focusing on their possible applications. The rationale behind this classification is that the composites with inorganic and organic materials rely on the different interaction modes afforded by PDA, melanins, or catechols. This mini-review will end with a section devoted to highlighting perspectives in this field.

2. Composites of Polydopamine and Inorganic Materials

2.1. Carbon-Based Composites

Graphene oxide can be reduced in graphene in the presence of norepinephrine which, in turn, is oxidized and polymerized in poly(norepinephrine). The resulting material covers graphene and affords colloidal stability to the composite [16]. When dopamine is modified with an azide function, the PDA-covered graphene can be further functionalized with molecules carrying alkynes through 1,3-dipolar cycloadditions (i.e., via "click chemistry") [17].

Graphene oxide functionalized with PDA is an efficient support to fix S and CS_2 in the design of cathodes for Li–S batteries. In this case the volume expansion of sulphur during discharge, and the leaching of sulphur-containing species is strongly limited and allows to increase the number of charge–discharge cycles of the batteries [18].

Carbon nanotubes coated with PDA can be modified with an initiator of atom transfer radical polymerization to coat the nanotubes with polydimethylamino-ethyl methacrylate brushes, which can be subsequently quaternized with CH_3I. Finally, Pd nanoparticles can be deposited on that composite structure which displayed an excellent electrochemical behavior when deposited on indium tin oxide electrodes [19].

The pyrolysis of PDA at temperatures above 500 °C yields a nitrogen-doped graphitic material because O is lost during the pyrolysis [20]. This concept was used to improve the electrochemical performance of spray-dried Si/graphite (Si/G) composite anodes in Li batteries [21]. The Si and carbon nanopowders were blended and reacted with dopamine (at pH = 8.5 during 24 h using O_2 as the oxidant) resulting in a PDA-coated blend which was subsequently pyrolyzed to yield a composite displaying an electronic conductivity of 1.44 S m^{-1} compared to 1.2 S m^{-1} for the Si/G blend. More interesting, the capacity retention is of 82% after 100 lithiation–deliathiation cycles for the Si/G-PDA composite, whereas it falls to 6.8% for the uncoated Si/G blend [21].

Multi-walled carbon nanotubes (MWNT) decorated with PDA are excellent supports for Pt nanoparticles allowing for an optimal Pt utilization (6051 mW mg^{-1} of catalyst) and a high-power density in polymer electrolyte membrane fuel cells (Figure 1). In such a membrane design, PDA plays the same role as Nafion® (DuPont) in other cells to afford a high protonic conductivity, between the anode (oxidation of H_2) and the cathode (reduction of O_2). In addition, PDA was shown to protect the electronic conductor, the MWNTs, against corrosion in the hard operation condition of the fuel cell [22].

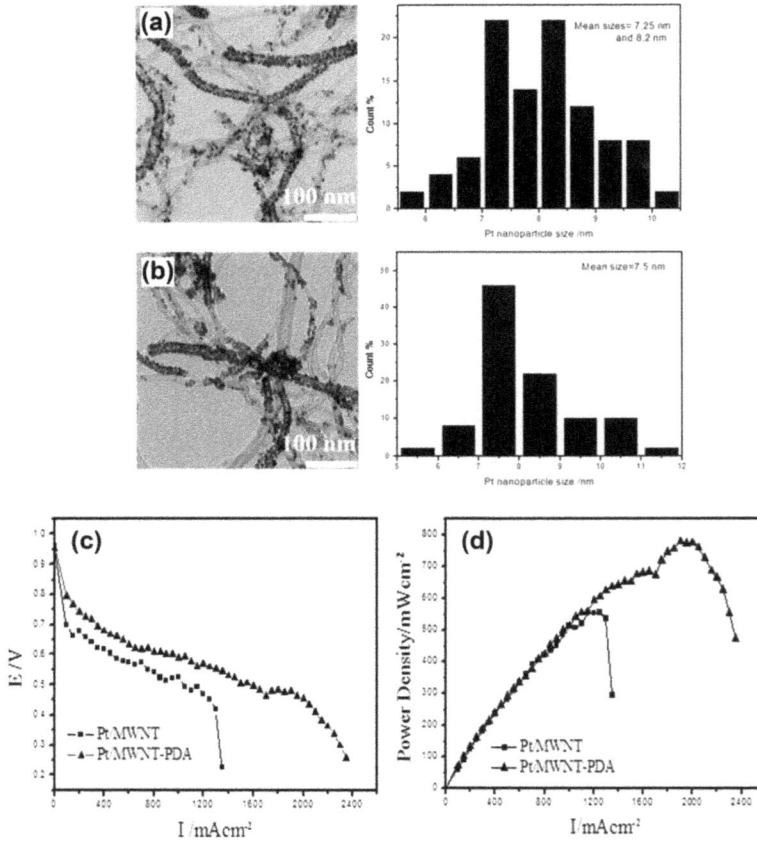

Figure 1. Transmission electron microscopy (TEM) images and corresponding particle size distribution histograms of (**a**) Pt-decorated multi-walled carbon nanotubes (Pt/MWNTs) and (**b**) Pt/MWNTs with polydopamine (Pt/MWNTs-PDA) catalysts used in the design of polyelectrolyte membrane-based fuel cells. (**c**) Polarization and (**d**) power density curves of fuel cell (Pt/MWNTs)$_{50}$ and (Pt/MWNTs-PDA)$_{50}$. The composite membranes were produced by 50 spray cycles on a hot substrate. Reprinted from [22], Copyright (2016), with permission from Elsevier.

2.2. Composites with Ions and Nanoparticles

The strong interactions between melanins and Na$^+$ cations has been used to evaluate melanin–Na$^+$ composites as anodes in fuel cells with a λ-MnO$_2$ cathode [23].

PDA films deposited on planar surfaces contain enough catechol groups to reduce metallic cations into nanoparticles [7]. In particular cotton coated with PDA was exposed to a solution containing Ag$^+$ cations to deposit silver nanoparticles affording excellent antibacterial activity to the fabric [24]. X ray photoelectron spectroscopy has allowed to show that the reduction of Ag$^+$ cations into Ag is accompanied by an oxidation of catechol groups into quinones but without affecting the concentration of free radicals in the PDA film [25].

Coating of superparamagnetic Fe$_3$O$_4$ nanoparticles of different core size with PDA allowed to increase the biocompatibility of those core-shell nanoparticles [26]. The deposition of melanin–Fe$_3$O$_4$

composite films on Au(111) electrodes allows to catalyse the electroreduction of hydrogen peroxide in alkaline and neutral solutions [27].

Citrate-capped gold nanoparticles are useful as localized surface plasmon resonance (SPR) elements and are hence useful for label-free biosensing. Indeed, a binding event causes a change in the refractive index at the surface of the particle and a concomitant red shift in the plasmonic absorbtion bands. When used in solution, separation–redispersion steps are required to separate the particles from the unbound ligands. These separation–redispersion steps induce some particle loss and aggregation. In the latter case, the sensitivity of the aggregates to further changes in the refractive index are lost, as well as their sensing ability. To overcome these problems, and to ensure multiple uses of these sensing elements, their immobilization on transparent surfaces is helpful, provided the particles do not desorb or move on the substrate to finish in a cluster. Such an aggregation process would result in a loss of active surface area. The deposition of a thin PDA film on the surface of citrate-capped Au nanoparticles was found not to hinder the responsiveness of the nanoparticles and to obtain a stable SPR spectrum after several washing–drying cycles which are relevant for real world sensing applications [28].

Core-shell magnetite–polystyrene particles 90 nm in diameter can be aligned to form chains in a magnetic field. When dopamine is oxidized in the presence of this magnetic field the nanoparticles remain aligned even after removing the external magnet. Stable nanochains of up to 20 μm in length can be obtained this way. They can subsequently be aligned in an external magnetic field and used as nanostirrers. The PDA shell can be used to reduce metal cations into metallic nanoparticles affording some catalytic activity to the nanostirrers. These catalysts can be easily separated from the reaction mixture and allow for self-mixing. For instance, Au-decorated PDA-capped nanochains allow the catalysis of the transformation of 4-nitrophenol into 4-aminophenol in the presence of $NaBH_4$. When self-stirred in the presence of an external magnetic field, the rate constant of the reaction is of 0.208 min^{-1} in comparison to 0.132 min^{-1} in the absence of a magnetic field. Finally, the PDA-capped nanochains can be easily functionalized with thiol-modified poly(ethylene glycol) (PEG) or with thiol-modified DNA aptamers [29].

TiO_2 nanoparticles were coated with a catechol-based bifunctional initiator of poly(methyl methacrylate) to synthetize inorganic core-organic shell stable particles [30].

2.3. Composites with Clays and Zeolites

When L-DOPA is oxidized in the presence of laponite, its polymerization is considerably accelerated with respect to homogeneous L-DOPA solutions, the clay is delaminated and embedded in a gel [31]. Later on, the influence of another clay, saponite, on the formation of a eumelanin-like composite material was investigated [32].

3-(3,4-Dihydroxyphenyl)-DL-alanine (DL-DOPA) was reacted with $V_2O_5 \cdot nH_2O$ gels to form a dark blue metallic-colored film after solvent casting of the DL-DOPA + $V_2O_5 \cdot nH_2O$ mixture. The lamellar structure of V_2O_5 is preserved upon the intercalation of the melanin like material [33], but with an increase in the interlayer spacing. This allowed an increase in the reproducibility of the Li^+ insertion–deinsertion process in the V_2O_5-based gel. In addition, electron paramagnetic resonance spectroscopy showed that the incorporation of the melanin-like material in V_2O_5 was accompanied by a partial reduction of V^{+5} cations into V^{+4} cations. Finally, the room temperature electrical conductivity of the composite was increased from 1.1×10^{-4} S cm^{-1} to 5.2×10^{-3} S cm^{-1} upon incorporation of about 2% (*w/w*) of melanin-like material [33].

When 5,6-dihydroxyindole (DHI) and its *N*-methyl derivative are oxidized in the nanopores of zeolite L at a ratio of 10 mg indole derivative to 300 mg porous material, the composite is red. Its dissolution with HF (to remove the inorganic core) and the analysis of the organic content reveals the presence of indole dimers. Hence, the self-assembly process of DHI is hindered in the nanoporous environment of the zeolite. However, when the same oxidation process is performed in the larger pores of SBA-15, a black eumelanin derivative is obtained [34].

3. Composites of Polydopamine with Organic Materials

3.1. Interactions between Melanins and Porphyrins

Melanins are known to interact strongly with cationic porphyrins to change the optical properties of the dye, in particular its delayed luminescence is substantially reduced in the presence of melanin [35].

These strong interactions (of an electrostatic nature and completed by π–π interactions) between porphyrins and phtalocyanines and melanin-like materials have been exploited to incorporate cationic Cu(II) phtalocyanines in PDA films by dissolving Alcyan Blue (AB) up to 0.2 mM in a 10.6 mM dopamine solution at pH 8.5. Dopamine was oxidized by dissolved O_2 to yield PDA [7]. The obtained films incorporate AB as manifested by a marked blue color and the presence of Cu in the X-ray photoelectron spectra (XPS). This incorporation of dye is apparent after short reaction times, but the spectral signature of AB and the detection of Cu is not possible anymore after longer reaction times, typically after about 10 h (Figure 2) [36]. This decrease of the characteristic optical and compositional signature of AB does not originate from leaching out from the film but from a quantitative incorporation in the composite PDA–AB film: all the available AB (with a solubility in water limited to about 0.3 mM) is incorporated in the film after a short reaction time and when the deposition is continued for longer deposition times, only small oligomers of oxidized dopamine are incorporated in the film to allow for the deposition of almost "pure" PDA. Then the upper parts of the film, close to the film solution/interface reflect the composition of PDA and not of PDA–AB. If homogenous PDA–AB films have to be deposited, the substrate should be exposed regularly to fresh dopamine–AB mixtures [36].

3.2. Layer-by-Layer Deposition of Polydopamine-Based Materials

Negatively-charged PDA aggregates and polycations can be assembled into composite thin films using the layer-by-layer deposition method [37–39]. The obtained films are more transparent and more permeable for a redox probe like hexacyanoferrate anions, than PDA films directly deposited on the same substrate from a dopamine solution [40].

The PDA particles obtained by adding poly(allylamine hydrochloride) (PAH) in the reaction mixture are of a controlled size (depending on the dopamine/PAH ratio) and positively-charged. They can be assembled with polyoxometalate anions to yield electroactive films [41].

An alternative strategy to incorporate PDA in films deposited according to the layer-by-layer deposition method is to use those films as templates for the formation of PDA from dopamine solutions in the presence of O_2 as an oxidant [42,43].

When dopamine is oxidized in the presence of a polyelectrolyte multilayer film made from the alternated deposition of poly(L-lysine) (PLL) and hyaluronic acid (HA), the film becomes rougher, and homogeneously filled over a thickness of about 1 μm with PDA. The mobility of PLL labeled with fluorescein as a fluorescent probe is reduced (its diffusion coefficient decreases from 5.8 to 3.2×10^{-2} μm^2 s^{-1} after 1 h of dopamine oxidation) and the mechanical properties of the composite film are affected: a (PLL–HA)$_{24}$ film spontaneously dissolves when put in ultrapure water but after only 4 h of contact with dopamine in oxidizing conditions, it can be detached from its quartz support as a free-standing membrane after short immersion in a 0.1 M HCl solution [42]. The same concept has been adapted for films made from the alternate adsorption of poly(allylamine) and clay platelets (montmorillonite, MMT). The obtained composite films kept the layered structure of the original (PAH–MMT)$_n$ films (Figure 3a,b), became impermeable to hexacyanoferrate anions (Figure 3c), smoother and more homogeneous in their elasticity even if the average Young's modulus (1 GPa) decreased with respect to the unmodified multilayered film (4.5 GPa) (Figure 3d) [43].

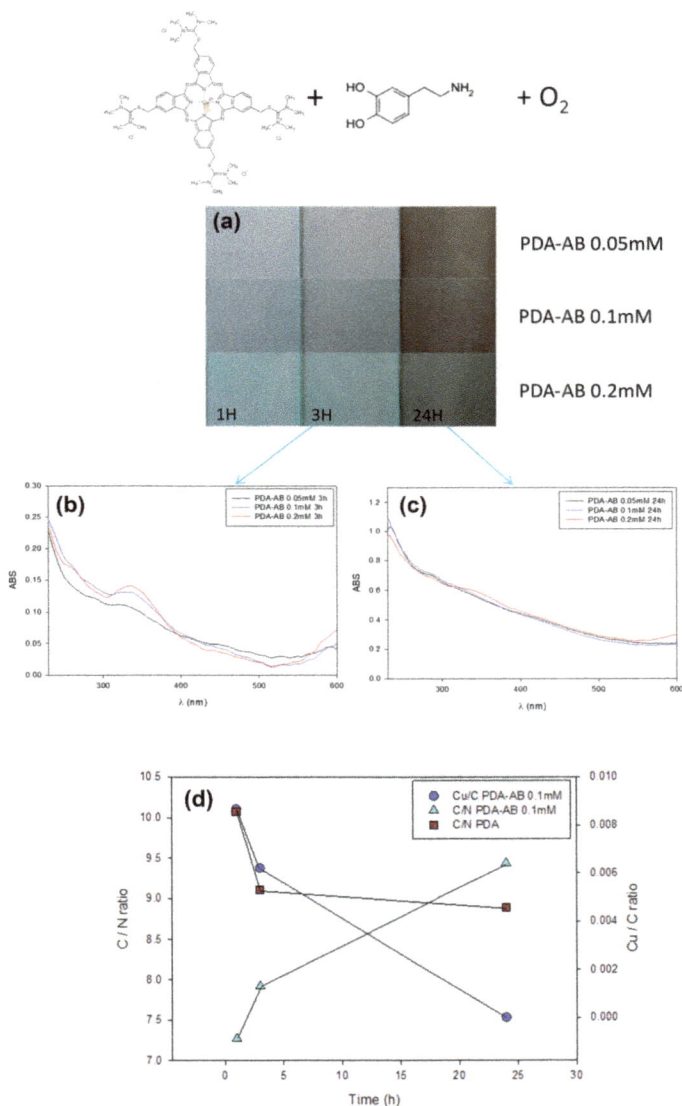

Figure 2. Composites obtained by adding Alcyan Blue (AB) in an oxygenated dopamine solution as shown at the top. (**a**) Digital pictures of glass slides after being put in dopamine–AB blends with AB concentrations of 0.05, 0.1, and 0.2 mM after 1, 3, and 24 h of reaction. The dopamine concentration was the same, 10.6 mM, in all experiments and the films were deposited from 50 mM Tris buffer at pH 8.5 using dissolved O_2 as the oxidant. (**b**) Ultraviolet-visible (UV-Vis) spectra taken on quartz slides put in dopamine–AB mixtures after 3 h of reaction and increasing the AB concentration from 0.05 to 0.2 mM as indicated in the inset. (**c**) UV-Vis spectra taken on quartz slides put in dopamine–AB mixtures after 24 h of reaction and increasing the AB concentration from 0.05 to 0.2 mM as indicated in the inset. (**d**) Evolution with time of the Cu/C (•) and the C/N (▲) ratios for the polydopamine (PDA)–AB (0.1 mM) films, and the C/N (■) ratio for the PDA films. These atomic ratios are obtained from X-ray photoelectron spectroscopy (XPS). Reprinted from [36], Copyright (2015), with permission from Elsevier.

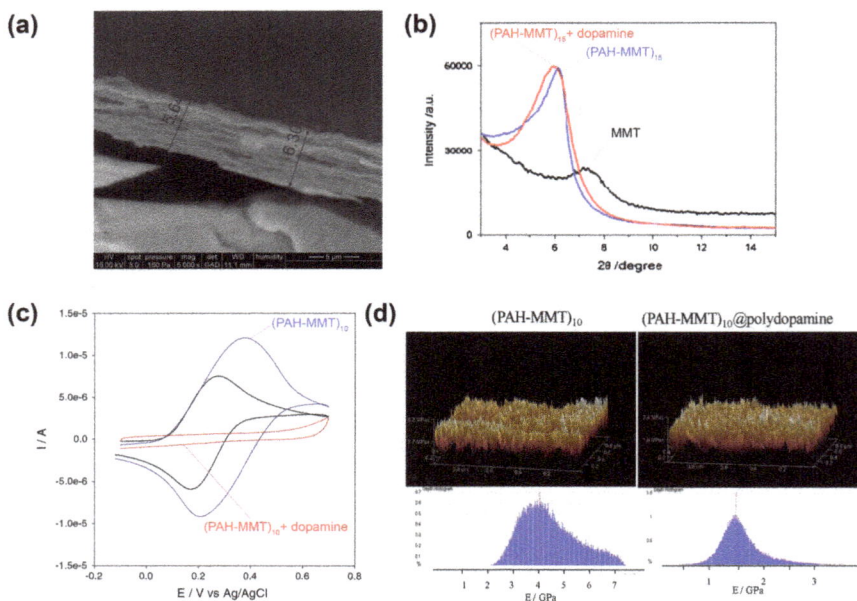

Figure 3. Layer-by-layer deposition of clay and poly(allylamine hydrochloride) (PAH) followed by post-modification with polydopamine. (**a**) Cross-sectional scanning electron microscope (SEM) image of a poly(allylamine hydrochloride)–montmorillonite (PAH-MMT)$_{15}$@polydopamine-coated film obtained after 14 h of contact with dopamine (2 mg mL^{-1}) in oxidizing conditions (pH = 8.5, O$_2$ as the oxidant). (**b**) X-ray diffractogram of the MMT powder (——) of (PAH–MMT)$_{15}$ film before (——) and after (——) 14 h of contact with dopamine in oxidizing conditions. (**c**) Cyclic voltammetry (at a potential scan rate of 100 mV s^{-1}) measured on a pristine amorphous carbon electrode (——), on the same electrode covered with a (PAH-MMT)$_{10}$–PAH film (——) and after 14 h of dopamine oxidation (——). The redox probe was 1 mM K$_4$Fe(CN)$_6$ dissolved in 50 mM Tris buffer + 150 mM NaCl (pH = 8.5). (**d**) Surface topography over (1 μm × 1 μm) of (PAH–MMT)$_{10}$ and (PAH–MMT)$_{10}$@polydopamine films and distribution of their corresponding elastic moduli. Reproduced with permission from [43].

A multi-layered film made of (3-aminopropyl)trietoxysilane (as an anchoring layer), PDA (as a covalent attachment platform) and stearoyl chloride (as a final lubricating layer) was deposited on glass to produce a robust lubricating composite with a two-fold decreased friction coefficient with respect to glass [44].

3.3. Dopamine Grafted on Polymers and Gels

Many researchers have shown the possibility to graft dopamine onto polymers [45,46] and to use these materials to design gels for biomedical applications. Of the highest interest was the demonstration that vanadyl cations (VO^{2+}) can gelify chitosan modified with catechol groups at a low vanadyl/catechol ratio, whereas Fe^{3+} cations allow the production of more rigid gels, only at higher cation concentrations. This will allow the production of polymer–catechol composite hydrogels based on metal coordination with a lower cytotoxicity as evaluated with NIH3T3 cells [47].

A mixture of alginate modified with catechol groups and pluronic bis-SH (comporting end-chain functional groups) allows the production of physical gels above the lower critical solution temperature of pluronic bis-SH. These gels can subsequently be cross-linked via catechol and thiol moieties in the presence of a strong oxidant, NaIO$_4$. The obtained gels not only display excellent antibacterial activities against *Porphyromonas gingivalis*, but also excellent biocompatibility towards a human line of gingival fibroblasts [48].

3.4. Polydopamine—Protein Composites

Inspired by the strong binding of proteins on natural melanin grains [49], and the fact that PDA is an excellent support for protein binding [50] in an active state [51], it was believed that PDA synthesis in solution would be influenced by the presence of proteins. Indeed, it was found that human serum albumin (HSA) resulted in a decrease in the size of PDA aggregates in solution after 24 h of dopamine oxidation in a protein/dopamine ratio-dependent manner (Figure 4). Simultaneously the deposition of PDA on the solid–liquid interface was reduced when the concentration of dissolved HSA increased [52].

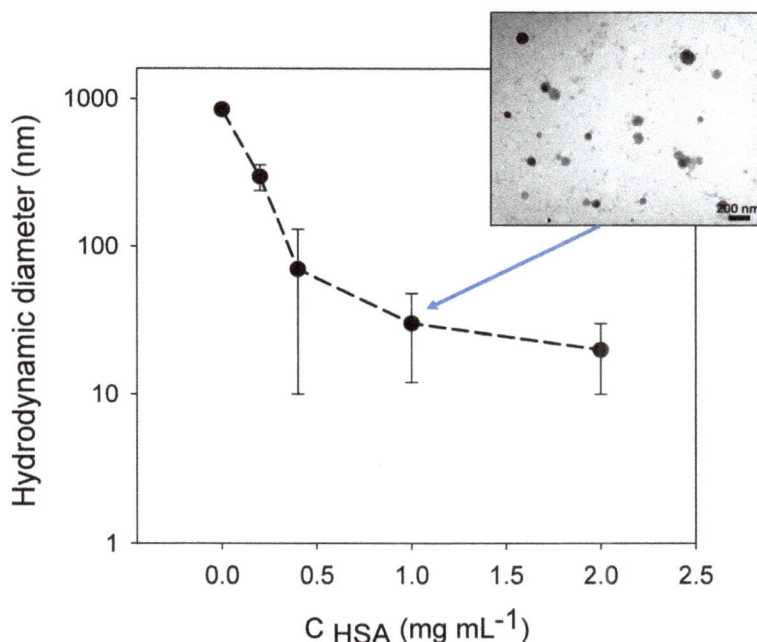

Figure 4. Variation of the hydrodynamic diameter of polydopamine (PDA) particles, as measured by dynamic light scattering; with the concentration of human serum albumin (HSA) added in the dopamine solution (10.6 mM in the presence of 50 mM Tris buffer at pH = 8.5). Dopamine has been oxidized with dissolved O_2 (open vessel) for 24 h at room temperature before the measurement. Reprinted from [52], Copyright (2014), with permission from Elsevier. The inset (personal data from the author) represents a transmission electron micrograph of the PDA suspension obtained in the presence of HSA at 1 mg mL^{-1}.

Similar results have been obtained when dopamine was oxidized in the presence of proteins from chicken egg white [53] suggesting that a large, but yet unknown, set of proteins allows the interaction and interference with PDA during its formation.

3.5. Polydopamine–Polymer Composites

When dopamine is oxidized in the presence of poly(vinyl alcohol) (PVA) the obtained polydopamine particles are of controlled size [54] in a manner similar to what has been described in the case of HSA and the proteins from chicken egg white.

PVA is also incorporated in PDA coatings when this polymer is added in the dopamine solution. Surprisingly, when poly(N-vinylpyrrolidone) is added to the dopamine solution, no PDA film deposition was found [55]. When poly(N-isopropylacrylamide) (PNIPAM) is added to the dopamine solution, the obtained PDA@PNIPAM coatings display temperature-dependant interactions with proteins and cells owing to the presence of the temperature-sensitive polymer in the composite coating [56].

Polymer membranes can be easily functionalized with PDA which allows to confer higher selectivity in filtration processes as reviewed recently [57]. Such composite membranes of high hydrophilicity or hydrophobicity allow for the separation of oil from water [58]

PDA not only forms films at the solid–liquid and liquid–liquid interfaces, but also at the water–air interface. However, the obtained films are apparent only in the absence of agitation because the formed PDA films are too brittle to withstand strong shear stresses [59]. Nevertheless, these materials can be useful because they are transferable to the surface of solids, for instance, poly(tetrafluoroethylene), by means of a horizontal Langmuir–Schaeffer transfer. This offers the interesting opportunity to coat only one face of a planar substrate. However, for many applications, a robust and flexible PDA-based film may be useful. To that aim, it has been shown that the addition of branched poly(ethylene imine) (PEI) to the dopamine solution allows the fabrication of robust free-standing membranes that can be handled with tweezers and which display an anisotropic composition: they are PDA-rich close to the film–air interface, and PEI-rich close to the water–film interface [60]. Later, this concept has been extended with alginate–catechol in the dopamine solution to yield Janus-like free-standing films that change their shape in response to changes in the relative humidity (Figure 5) [61]. The adhesion strength of those membranes are humidity- and side-dependent.

3.6. Polydopamine and Conductive Polymers

The electrical conductivity of melanins and PDA films is low, humidity-dependent, and lies typically in the 10^{-13}–10^{-5} S cm^{-1} range [62]. This conductivity is most probably of protonic nature, but also implies some redox process due to the composition of melanins [63,64]. The simultaneous presence of hydroquinones, semiquinones (radical species), quinones, and quinone imine forms allow for many kinds of charge storage, hence, allowing interesting applications of melanins and PDA films as supercapacitors [65].

In addition, PDA is highly biocompatible, which is not the case for many conductive polymers. The emerging concept is then to try to blend both kinds of materials, PDA and conductive polymers, to find a good compromise between acceptable biocompatibility and improved electrical conductivity. This would allow the design of new flexible electrodes for stimulation of cells, neurons, etc.

PDA has been blended with polyaniline (PANI) in a reactive layer-by-layer manner. PDA was deposited from dopamine solutions at pH 8.5 using O_2 as the oxidant whereas PANI was deposited from cold acidic solutions using formic acid as the dopant and sodium peroxodisulfate as the oxidant [66]. The obtained layered films displayed a conductivity close to 10^{-2}–10^{-1} S cm^{-1} after immersion in a 1.5 M HCl solution. The conductivity of the films decreased when the mass fraction of PDA increased. Nevertheless, this deposition method is cumbersome and of multistep in nature.

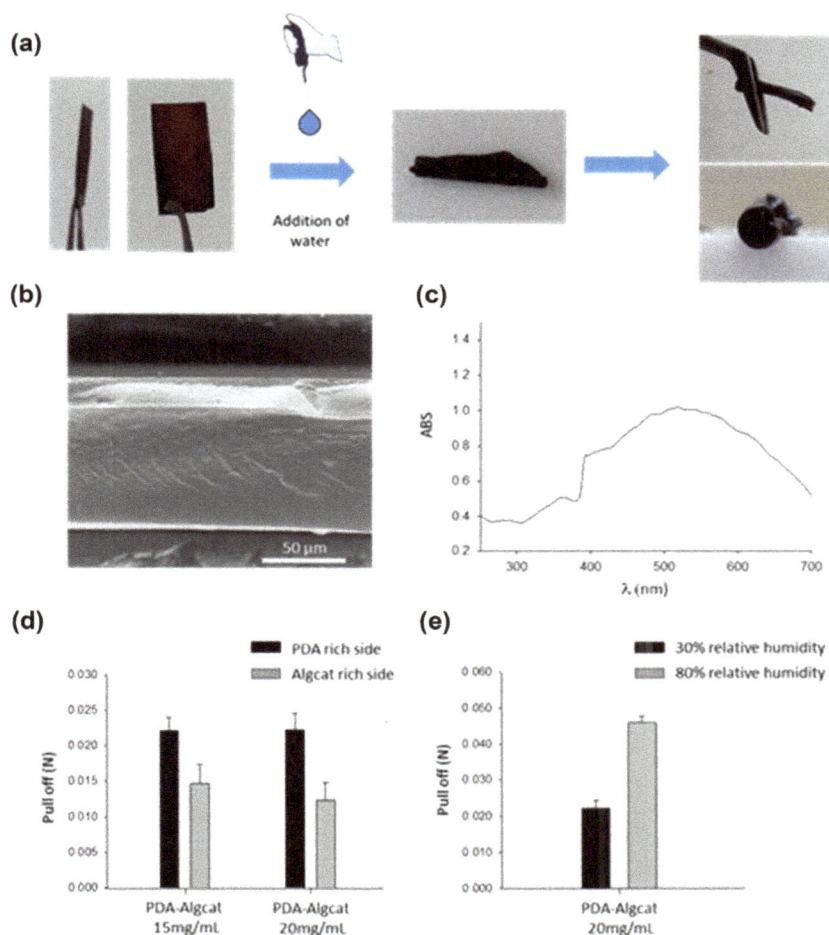

Figure 5. Composite alginate–catechol@polydopamine membranes. (**a**) Images showing the responsivity of PDA-alginate–catechol (PDA-Algcat) membranes to the addition of water. (**b**) Scanning electron microscopy (SEM) images of the dry membrane obtained from an Algcat solution at 20 mg mL^{-1} and (**c**) its UV-Vis spectrum. ABS: Absorbance. (**d**) Pull-off data of a PDA-Algcat membrane at different concentrations of Algcat depending on the side of the membrane (30% relative humidity), and (**e**) pull-off data as a function of the relative humidity. Reprinted from [61]. Copyright (2017) American Chemical Society.

The possibility to deposit PDA films from slightly acidic solutions (pH = 5, 50 mM sodium acetate buffer) using strong oxidants (like sodium periodate) [67] will perhaps offer the opportunity to deposit composite PDA–PANI films from solutions containing a mixture of dopamine and aniline. However, it is not excluded that aniline could react with 5,6-dihydroxyindole (the final oxidation state of dopamine) to yield a new compound and a film with different properties. Indeed, when mixtures of dopamine and pyrrole are oxidized with ammonium persulfate, the obtained particles are of smaller size than those obtained in the presence of pyrrole only (Figure 6a), they adhere much more strongly on the surface of solid substrates (Figure 6b), and the electrical conductivity passes through a maximum (of about 1 S cm^{-1}) at dopamine/pyrrole mole fraction of about 0.06 [68]. The electrical conductivity of films and pellets of the obtained particles decreased when the synthesis was performed at a higher mole fraction in dopamine because the fraction of insulating PDA increased (Figure 6c). The marked increase in conductivity observed at low mole fraction in dopamine was attributed to better cohesion and adhesion between neighboring particles. Similar results were obtained with L-DOPA or 1,2-dihydroxybenzene added to pyrrole in the polymerization batch. All these results were interpreted on the basis of the formation of a copolymer in which the catechol units are incorporated in the polypyrrole chains.

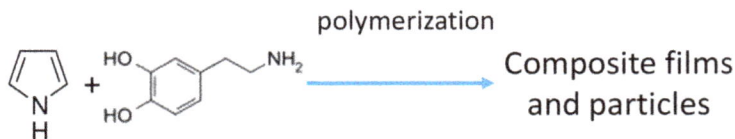

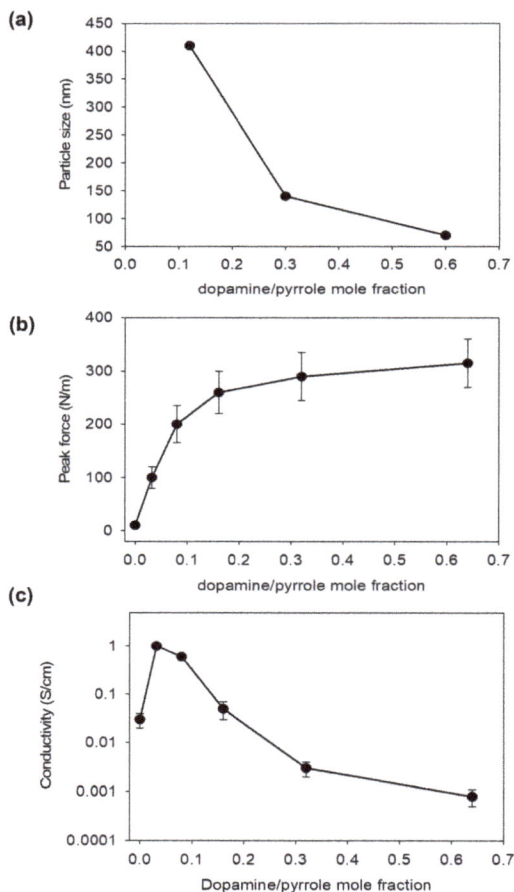

Figure 6. Influence of the presence of dopamine in pyrrole solutions on the properties of the obtained particles. (**a**) Particle size as determined by dynamic light scattering and films; (**b**) Peak force in peeling tests; (**c**) Electrical conductivity after polymerization in the presence of ammonium persulfate. Pyrrole and dopamine structures are shown at the top. Adapted with permission from [68].

3.7. Polydopamine-Based Composites for Improved Sensing

PDA was deposited in the internal lumen of a 0.5 mm polyetheretherketone (PEEK) tube, a dialdehyde starch porous composite was then fixed on this adhesion coating. The composite coating allowed for the detection of hexanal and 2-butanone derivatized with 2,4-dinitrophenyl hydrazine. The released conjugates where then quantified by means of high-performance liquid chromatography (HPLC). This eco-friendly and cheap conjugate could be used up to 30 times without a notable decrease in performance. The limit of detection for hexanal and 2-butanone as specific markers for liver cancer, was of 1.4 and 1.6 nmol L^{-1}, respectively, comparable with other solid phase extraction methods. Statistical analysis showed that this method allowed to distinguish patients with liver cancer from healthy patients [69].

4. Future Perspectives of Polydopamine-Based Composites

Even if PDA has shown to be an interesting material to build up composites to yield either core shell nanoparticles, composites with carbon-based nanomaterials, multilayered films, intercalated composites, and stable PDA–protein bioinspired nanoparticles, no doubt that the design principles of such composites will be improved if we gain a better understanding of the structure of PDA. Moreover, the question of a polymeric structure or more probably a self-assembled aggregate of small oligomers of 5,6-dihydroxyindole is still open [70]. If PDA is, indeed, a self-assembled structure with local graphitic order, as suggested by high-resolution transmission electron micrographs [71,72], one may think about processing methods to increase the size of ordered domains in PDA to produce composite materials having a higher electrical conductivity. In this context, defining synthesis conditions allowing to produce a conductive material from a mixture of aniline (yielding to polyaniline in oxidative conditions) and dopamine (or other catecholamines) is challenging. Indeed, PDA is most often obtained in basic conditions, whereas doped polyaniline is ideally obtained in strongly acidic conditions.

The design of PDA-based composites for the design of biomaterials with antibacterial properties and composites allowing for improved adhesion on two different materials will certainly benefit from a better understanding of the structure of PDA and melanin. The ageing process of such composites in real usage conditions has to be investigated in a more detailed manner even if the ageing of PDA in the harsh conditions of proton exchange membranes in fuel cells appears promising [22].

The de-excitation mode of photoexcited melanins and PDA into phonons [73] to produce heat has almost not been exploited in the design of eumelanin- and PDA-based composites.

Finally, the most important recent finding is the possibility to produce stable PDA-based suspensions, in a biomimetic manner, in the presence of proteins in the reaction medium. Such nanoparticles could be processed in two- and three-dimensional printing processes to yield new scaffolds for the design of new biomaterials and materials aimed for energy conversion processes.

Conflicts of Interest: The author declares no conflict of interest.

References

1. Lee, B.P.; Messersmith, P.B.; Israelachvili, J.N.; Waite, J.H. Mussel-inspired adhesives and coatings. *Ann. Rev. Mater. Sci.* **2011**, *41*, 99–132. [CrossRef] [PubMed]
2. Miserez, A.; Schneberk, T.; Sun, C.; Zok, F.W. Role of melanin in mechanical properties of *Glycera* jaws. *Science* **2008**, *319*, 1816–1819. [CrossRef] [PubMed]
3. Lichtenegger, H.C.; Schoberl, T.; Bartl, M.H.; Waite, J.H.; Stucky, G.D. High abrasion resistance with sparse mineralization: Copper biomineral in worm jaws. *Science* **2002**, *298*, 389–392. [CrossRef] [PubMed]
4. Hwang, D.S.; Masic, A.; Prajatelistia, E.; Iordachescu, M.; Waite, J.H. Marine hydroid perisarc: A chitin-and melanin-reinforced composite with DOPA-iron(II) complexes. *Acta Biomater.* **2013**, *9*, 8110–8117. [CrossRef] [PubMed]
5. Krogsgaard, M.; Nue, V.; Birkedal, H. Mussel-inspired materials: Self-healing through coordination chemistry. *Chem. Eur. J.* **2016**, *22*, 844–857. [CrossRef] [PubMed]

6. Sedo, J.; Saiz-Poseu, J.; Busque, F.; Ruiz-Molina, D. Catechol-based biomimetic functional materials. *Adv. Mater.* **2013**, *25*, 653–701. [CrossRef] [PubMed]

7. Lee, H.; Dellatore, S.M.; Miller, W.M.; Messersmith, P.B. Mussel-inspired surface chemistry for multifunctional coatings. *Science* **2007**, *318*, 426–430. [CrossRef] [PubMed]

8. Wei, Q.; Zhang, F.; Li, J.; Li, B.; Zhao, C. Oxidant-induced dopamine polymerization for multifunctional coatings. *Polym. Chem.* **2010**, *1*, 1430–1433. [CrossRef]

9. Hong, S.; Kim, J.; Na, Y.S.; Park, J.; Kim, S.; Singha, K.; Im, G.-I.; Han, D.-K.; Kim, W.J.; Lee, H. Poly(norepinephrine): Ultrasmooth material-independent surface chemistry and nanodepot for nitric oxide. *Angew. Chem. Int. Ed.* **2013**, *52*, 9187–9191. [CrossRef] [PubMed]

10. Liu, Y.; Ai, K.; Lu, L. Polydopamine and its derivative materials: Synthesis and promising applications in energy, Eenvironmental, and biomedical fields. *Chem. Rev.* **2014**, *114*, 5057–5115. [CrossRef] [PubMed]

11. Ball, V.; Del Frari, D.; Michel, M.; Buehler, M.J.; Toniazzo, V.; Singh, M.K.; Gracio, J.; Ruch, D. Deposition mechanism and properties of thin polydopamine films for high added value applications in surface science at the nanoscale. *Bionanoscience* **2012**, *2*, 16–34. [CrossRef]

12. Lynge, M.E.; van der Westen, R.; Postma, A.; Stadler, B. Polydopamine—A nature-inspired polymer coating for biomedical science. *Nanoscale* **2012**, *3*, 4916–4928. [CrossRef] [PubMed]

13. D'Ischia, M.; Napolitano, A.; Ball, V.; Chen, C.-T.; Buehler, M.-J. Polydopamine and eumelanin: From structure–property relationships to a unified tailoring strategy. *Acc. Chem. Res.* **2014**, *47*, 3541–3550. [CrossRef] [PubMed]

14. D'Ischia, M.; Wakamatsu, K.; Napolitano, A.; Briganti, S.; Garcia-Borron, J.-C.; Kovacs, D.; Meredith, P.; Pezzella, A.; Picardo, M.; et al. Melanins and melanogenesis: Methods, standards, protocols. *Pig. Cell Melanoma Res.* **2013**, *26*, 616–633. [CrossRef] [PubMed]

15. Simon, J.D.; Peles, D.N. The red and the black. *Acc. Chem. Res.* **2010**, *43*, 1452–1460. [CrossRef] [PubMed]

16. Kang, S.M.; Park, S.; Kim, D.; Park, S.S.; Ruoff, R.S.; Lee, H. Simultaneous reduction and surface functionalization of graphene oxide by mussel-inspired chemistry. *Adv. Funct. Mater.* **2011**, *21*, 108–112. [CrossRef]

17. Kaminska, I.; Das, M.R.; Coffinier, Y.; Niedziolka-Jonsson, J.; Sobczak, J.; Woisel, P.; Lyskawa, J.; Opallo, M.; Boukherroub, R.; Szunerits, S. Reduction and functionalization of graphene oxide sheets using biomimetic dopamine derivatives in one step. *ACS Mater. Interfaces* **2012**, *4*, 1016–1020. [CrossRef] [PubMed]

18. Zhou, L.; Zong, Y.; Liu, Z.; Yu, A. A polydopamine coating ultralight graphene matrix as a highly effective polysulfide adsorbent for high energy Li–S batteries. *Renew. Energy* **2016**, *96*, 333–340. [CrossRef]

19. Hu, H.; Yu, B.; Ye, Q.; Gu, Y.; Zhou, F. Modification of carbon nanotubes with a nanothin polydopamine layer and polydimethylamino-ethyl mathacrylate brushes. *Carbon* **2010**, *48*, 2347–2353. [CrossRef]

20. Li, R.; Parvez, K.; Hinkel, F.; Feng, X.; Müllen, K. Bioinspired wafer-scale production of highly stretchable carbon films for transparent conductive electrodes. *Angew. Chem. Int. Ed.* **2013**, *52*, 5535–5538. [CrossRef] [PubMed]

21. Zhou, R.; Guo, H.; Yang, Y.; Wang, Z.; Li, X.; Zhou, Y. N-doped carbon layer derived from polydopamine to improve the electrochemical performance of spray-dried Si/graphite composite anode material for Li ion batteries. *J. Alloys Compd.* **2016**, *689*, 130–137. [CrossRef]

22. Long, H.; Del Frari, D.; Martin, A.; Didierjean, J.; Ball, V.; Michel, M.; Ibn El Ahrach, H. Polydopamine as a promising candidate for the design of high performance and corrosion tolerant polymer electrolyte fuel cell electrodes. *J. Power Sources* **2016**, *307*, 569–577. [CrossRef]

23. Kim, Y.J.; Wu, W.; Chun, S.-E.; Whitacre, J.F.; Bettinger, C.J. Biologically derived melanin electrodes in aqueous sodium-ion energy storage devices. *Proc. Nat. Acad. Sci. USA* **2013**, *110*, 20912–20917. [CrossRef] [PubMed]

24. Xu, H.; Shi, X.; Ma, H.; Lv, Y.; Zhang, L.; Mao, Z. The preparation and antibacterial effects of dopa-cotton/AgNPs. *Appl. Surf. Sci.* **2011**, *257*, 6799–6803. [CrossRef]

25. Ball, V.; Nguyen, I.; Haupt, M.; Oehr, C.; Arnoult, C.; Toniazzo, V.; Ruch, D. The reduction of Ag$^+$ in metallic silver on pseudomelanin films allows for antibacterial activity but does not imply unpaired electrons. *J. Colloid Interface Sci.* **2011**, *364*, 359–365. [CrossRef] [PubMed]

26. Si, J.; Yang, H. Preparation and characterization of bio-compatible Fe$_3$O$_4$@polydopamine spheres with core/shell nanostructure. *Mater. Chem. Phys.* **2011**, *128*, 519–524. [CrossRef]

27. González Orive, A.; Dip, P.; Gimeno, Y.; Díaz, P.; Carro, P.; Hernández Creuz, A.; Benítez, G.; Schilardi, P.L.; Andrini, L.; Requejo, F.; et al. Electroctalytic and magnetic properties of ultrathin nanostructured iron-melanin films on Au(111). *Chem. Eur. J.* **2007**, *13*, 473–482. [CrossRef] [PubMed]

28. Chen, H.; Zhao, L.; Chen, D.; Hu, W. Stabilization of gold nanoparticles on glass surface with polydopamine thin film for reliable LSPR sensing. *J. Colloid Interface Sci.* **2015**, *460*, 258–263. [CrossRef] [PubMed]

29. Zhou, J.; Wang, C.; Wang, P.; Messersmith, P.B.; Duan, H. Multifunctional magnetic nanochains: Exploiting self-polymerization and versatile reactivity of mussel-inspired polydopamine. *Chem. Mater.* **2015**, *27*, 3071–3076. [CrossRef]

30. Fan, X.; Lin, L.; Messersmith, P.B. Surface-initiated polymerization from TiO_2 nanoparticle surfaces through a biomimetic initiator: A new route toward polymer-matrix nanocomposites. *Compos. Sci. Technol.* **2006**, *66*, 1198–1204. [CrossRef]

31. Jaber, M.; Lambert, J.-F. A new nanocomposite: L-DOPA/Laponite. *J. Phys. Chem. Lett.* **2010**, *1*, 85–88. [CrossRef]

32. Jaber, M.; Bouchoucha, M.; Delmotte, L.; Méthivier, C.; Lambert, J.-F. Fate of L-DOPA in the presence of inorganic matrices: Vectorization or composite material formation? *J. Phys. Chem. C* **2011**, *115*, 19216–19225. [CrossRef]

33. Oliveira, H.P.; Graeff, C.F.O.; Zanta, C.L.P.S.; Galina, A.C.; Gonçalves, P.J. Synthesis, characterization, and properties of a melanin-like vanadium pentoxide hybrid compound. *J. Mater. Chem.* **2000**, *10*, 371–375. [CrossRef]

34. Prasetyanto, E.K.; Manini, P.; Napolitano, A.; Crescenzi, O.; D'Ischia, M.; De Cola, L. Towards eumelanin@zeolite hybrids: Pore-size-controlled 5,6-dihydroxyindole polymerization. *Chem. Eur. J.* **2014**, *20*, 1597–1601. [CrossRef] [PubMed]

35. Wróbel, D.; Planner, A.; Haniz, I.; Sarna, T. Melanin-porphyrin interaction monitored by delayed luminescence and photoacoustics. *J. Photochem. Photobiol. B Biol.* **1997**, *41*, 45–52. [CrossRef]

36. Ponzio, F.; Bour, J.; Ball, V. Composite films of polydopamine-Alcian Blue for colored coating with new physical properties. *J. Colloid Interface Sci.* **2015**, *459*, 29–35. [CrossRef] [PubMed]

37. Decher, G. Fuzzy Nanoassemblies: Toward layered polymeric multicomposites. *Science* **1997**, *277*, 1232–1237. [CrossRef]

38. Lavalle, P.; Voegel, J.-C.; Vautier, D.; Senger, B.; Schaaf, P.; Ball, V. Dynamic aspects of films prepared by a sequential deposition of species: Perspectives for smart and responsive materials. *Adv. Mater.* **2011**, *23*, 1191–1221. [CrossRef] [PubMed]

39. Richardson, J.J.; Cui, J.; Björnmalm, M.; Braunger, J.A.; Ejima, H.; Caruso, F. Innovation in layer-by-layer assembly. *Chem. Rev.* **2016**, *116*, 14828–14867. [CrossRef] [PubMed]

40. Bernsmann, F.; Ersen, O.; Voegel, J.-C.; Jan, E.; Kotov, N.A. Melanin-containing films: Growth from dopamine solutions versus layer-by-layer deposition. *ChemPhysChem* **2010**, *11*, 3299–3305. [CrossRef] [PubMed]

41. Ball, V.; Haider, A.; Kortz, U. Composite films of poly(allylamine)-capped polydopamine nanoparticles and P8W48 polyoxometalates with electroactive properties. *J. Colloid Interface Sci.* **2016**, *481*, 125–130. [CrossRef] [PubMed]

42. Bernsmann, F.; Richert, L.; Senger, B.; Lavalle, P.; Voegel, J.-C.; Schaaf, P.; Ball, V. Use of dopamine polymerization to produce free-standing membranes from (PLL-HA)$_n$ exponentially growing multilayer films. *Soft Matter* **2008**, *4*, 1621–1624. [CrossRef]

43. Ball, V.; Apaydin, K.; Laachachi, A.; Toniazzo, V.; Ruch, D. Changes in permeability and in mechanical properties of layer-by-layer films made from poly(allylamine) and montmorillonite postmodified upon reaction with dopamine. *Biointerphases* **2012**, *7*, 59. [CrossRef] [PubMed]

44. Ou, J.; Wang, J.; Liu, S.; Zhou, J.; Yang, S. Self-assembly and tribological property of a novel 3-layer organic film on silicon wafer with polydopamine coating as the interlayer. *J. Phys. Chem. C* **2009**, *113*, 20429–20434. [CrossRef]

45. Ryu, J.H.; Lee, Y.; Kong, W.H.; Kim, T.G.; Park, T.G.; Lee, H. Catechol-functionalized chitosan/Pluronic hydrogels for tissue adhesives and hemostatic materials. *Biomacromolecules* **2011**, *12*, 2653–2659. [CrossRef] [PubMed]

46. Kim, K.; Ryu, J.H.; Lee, D.H.; Lee, H. Bio-inspired catechol conjugation converts water-insoluble chitosan into a highly water-soluble, adhesive chitosan derivative for hydrogels and LbL assembly. *Biomater. Sci.* **2013**, *1*, 783–790. [CrossRef]

47. Park, J.P.; Song, I.T.; Lee, J.; Ryu, J.H.; Lee, Y.; Lee, H. Vanadyl–catecholamine hydrogels inspired by ascidians and mussels. *Chem. Mater.* **2015**, *27*, 105–111. [CrossRef]
48. Mateescu, M.; Baixe, S.; Garnier, T.; Jierry, L.; Ball, V.; Haikel, Y.; Metz-Boutigue, M.-H.; Nardin, M.; Schaaf, P.; Etienne, O.; et al. Antibacterial peptide-based gel for prevention of medical implanted-device infection. *PLoS ONE* **2015**, *10*, e0145143. [CrossRef] [PubMed]
49. Kushimoto, T.; Basrur, V.; Valencia, J.; Matsunaga, J.; Vieira, W.D.; Ferrans, V.J.; Muller, J.; Appella, E.; Hearing, V.J. A model for melanosome biogenesis based on the purification analysis of early melanosomes. *Proc. Natl. Acad. Sci. USA* **2001**, *98*, 10698–10703. [CrossRef] [PubMed]
50. Lee, H.; Rho, J.; Messersmith, P.B. Facile conjugation of biomolecules onto surfaces via mussel adhesive protein inspired coatings. *Adv. Mater.* **2009**, *21*, 431–435. [CrossRef] [PubMed]
51. Ball, V. Activity of alkaline phosphatase adsorbed and grafted on "polydopamine" films. *J. Colloid Interface Sci.* **2014**, *429*, 1–7. [CrossRef] [PubMed]
52. Chassepot, A.; Ball, V. Human serum albumin and other proteins as templating agents for the synthesis of nanosized dopamine-eumelanin. *J. Colloid Interface Sci.* **2014**, *414*, 97–102. [CrossRef] [PubMed]
53. Della Vecchia, N.F.; Cerruti, P.; Gentile, G.; Errico, M.E.; Ambrogi, V.; D'Errico, G.; Longobardi, S.; Napolitano, A.; Paduano, L.; Carfagna, C.; et al. Artificial biomelanin: Highly light-absorbing nano-sized eumelanin by biomimetic synthesis in chicken egg white. *Biomacromolecules* **2014**, *15*, 3811–3816. [CrossRef] [PubMed]
54. Arzillo, M.; Mangiapia, G.; Pezzella, A.; Heenan, R.K.; Radulescu, A.; Paduano, L.; d'Ischia, M. Eumelanin buildup on the nanoscale: Aggregate growth/assembly and visible absorption development in biomimetic 5,6-dihydroxyindole polymerization. *Biomacromolecules* **2012**, *13*, 2379–2390. [CrossRef] [PubMed]
55. Zhang, Y.; Thingholm, B.; Goldie, K.N.; Ogaki, R.; Städler, B. Assembly of poly(dopamine) films mixed with a nonionic polymer. *Langmuir* **2012**, *28*, 17585–17592. [CrossRef] [PubMed]
56. Zhang, Y.; Panneerselvam, K.; Ogaki, R.; Hosta-Rigau, L.; van der Westen, R.; Jensen, B.E.B.; Teo, B.M.; Zhu, M.; Städler, B. Assembly of poly(dopamine)/poly(*N*-isopropylacrylamide) mixed films and their temperature-dependent interaction with proteins, liposomes, and cells. *Langmuir* **2013**, *29*, 10213–10222. [CrossRef] [PubMed]
57. Yang, H.-C.; Luo, J.; Lv, Y.; Shen, P.; Xu, Z.-K. Surface engineering of polymer membranes via mussel-inspired chemistry. *J. Memb. Sci.* **2015**, *483*, 42–59. [CrossRef]
58. Garcia, B.; Saiz-Poseu, J.; Gras-Charles, R.; Hernando, J.; Alibés, R.; Novio, F.; Sedó, J.; Busqué, F.; Ruiz-Molina, D. Mussel-inspired hydrophobic coatings for water-repellent textiles and oil removal. *ACS Appl. Mater. Interfaces* **2014**, *6*, 17616–17625. [CrossRef] [PubMed]
59. Ponzio, F.; Payamyar, P.; Schneider, A.; Winterhalter, M.; Bour, J.; Addiégo, F.; Krafft, M.-P.; Hemmerlé, J.; Ball, V. Polydopamine films from the forgotten air/water Interface. *J. Phys. Chem. Lett.* **2014**, *5*, 3436–3440. [CrossRef] [PubMed]
60. Hong, S.; Schaber, C.F.; Dening, K.; Appel, E.; Gorb, S.N.; Lee, H. Air/water interfacial formation of freestanding, stimuli-responsive, self-healing catecholamine Janus-faced microfilms. *Adv. Mater.* **2014**, *26*, 7581–7587. [CrossRef] [PubMed]
61. Ponzio, F.; Le Houerou, V.; Zafeiratos, S.; Gauthier, C.; Garnier, T.; Jierry, L.; Ball, V. Robust alginate-catechol@polydopamine free standing membranes obtained from the water/air interface. *Langmuir* **2017**, *33*, 2420–2426. [CrossRef] [PubMed]
62. Meredith, P.; Sarna, T. The physical and chemical properties of eumelanin. *Pig. Cell Res.* **2006**, *19*, 572–594. [CrossRef] [PubMed]
63. Wünsche, J.; Cicoira, F.; Graeff, C.F.O.; Santato, C. Eumelanin thin films: Solution-processing, growth, and charge transport properties. *J. Mater. Chem. B* **2013**, *1*, 3836–3842. [CrossRef]
64. Wünsche, J.; Deng, Y.; Kumar, P.; Di Mauro, E.; Josberger, E.; Sayago, J.; Pezzella, A.; Soavi, F.; Cicoira, F.; Rolandi, M.; et al. Protonic and electronic transport in hydrated thin films of the pigment eumelanin. *Chem. Mater.* **2015**, *27*, 436–442.
65. Kumar, P.; Di Mauro, E.; Zhang, S.; Pezzella, A.; Soavi, F.; Santato, C.; Cicoira, F. Melanin-based flexible supercapacitors. *J. Mater. Chem.* **2016**, *4*, 9516–9525. [CrossRef]
66. Mihai, I.; Addiégo, F.; Del Frari, D.; Bour, J.; Ball, V. Associating oriented polyaniline and eumelanin in a reactive layer-by-layer manner: Composites with high electrical conductivity. *Colloids Surfaces A Physicochem. Eng. Asp.* **2013**, *434*, 118–125. [CrossRef]

67. Ponzio, F.; Barthès, J.; Bour, J.; Michel, M.; Bertani, P.; Hemmerlé, J.; d'Ischia, M.; Ball, V. Oxidant control of polydopamine surface chemistry in acids: A mechanism-based entry to superhydrophilic-superoleophobic coatings. *Chem. Mater.* **2016**, *28*, 4697–4705. [CrossRef]
68. Zhang, W.; Pan, Z.; Yang, F.K.; Zhao, B. A facile in situ approach to polypyrrole functionalization through bioinspired catechols. *Adv. Funct. Mater.* **2015**, *25*, 1588–1597. [CrossRef]
69. Wu, S.; Cai, C.; Cheng, J.; Cheng, M.; Zhou, H.; Deng, J. Polydopamine/dialdehyde starch/chitosan composite coating for in-tube solid-phase microextraction and in-situ derivation to analysis of two liver cancer biomarkers in human blood. *Anal. Chim Acta* **2016**, *935*, 113–120. [CrossRef] [PubMed]
70. Dreyer, D.R.; Miller, D.J.; Freeman, B.D.; Paul, D.R.; Bielawski, C.W. Perspectives on poly(dopamine). *Chem. Sci.* **2013**, *4*, 3796–3802. [CrossRef]
71. Watt, A.A.R.; Bothma, J.P.; Meredith, P. The supramolecular structure of melanin. *Soft Matter* **2009**, *5*, 3754–3760. [CrossRef]
72. Chen, C.-T.; Ball, V.; de Almeida Gracio, J.; Singh, M.K.; Toniazzo, V.; Ruch, D.; Buehler, M.J. Self-assembly of tetramers of 5,6-dihydroxyindole explains the primary physical properties of eumelanin: Experiment, simulation, and design. *ACS Nano* **2013**, *7*, 1524–1532. [CrossRef] [PubMed]
73. Meredith, P.; Powell, B.J.; Riesz, J.; Nighswander-Rempel, S.P.; Pederson, M.R.; Moore, E.G. Towards structure–property–function relationships for eumelanin. *Soft Matter* **2006**, *2*, 37–44. [CrossRef]

biomimetics

MDPI

Article

Copolymerization of a Catechol and a Diamine as a Versatile Polydopamine-Like Platform for Surface Functionalization: The Case of a Hydrophobic Coating

Salvio Suárez-García , Josep Sedó, Javier Saiz-Poseu * and Daniel Ruiz-Molina *

Catalan Institute of Nanoscience and Nanotechnology (ICN2), CSIC and BIST, Campus UAB, Bellaterra, 08193 Barcelona, Spain; salvio.suarez@icn2.cat (S.S.-G.); josep.sedo@icn2.cat (J.S.)
* Correspondence: javier.saiz@icn2.cat (J.S.-P.); dani.ruiz@icn2.cat (D.R.-M.); Tel.: +34-093-737-2638 (J.S.-P.); +34-093-737-3614 (D.R.-M.)

Academic Editor: Josep Samitier
Received: 6 October 2017; Accepted: 7 November 2017; Published: 13 November 2017

Abstract: The covalent functionalization of surfaces with molecules capable of providing new properties to the treated substrate, such as hydrophobicity or bioactivity, has been attracting a lot of interest in the last decades. For achieving this goal, the generation of a universally functionalizable primer coating in one-pot reaction and under relatively mild conditions is especially attractive due to its potential versatility and ease of application. The aim of the present work is to obtain such a functionalizable coating by a cross-linking reaction between pyrocatechol and hexamethylenediamine (HDMA) under oxidizing conditions. For demonstrating the efficacy of this approach, different substrates (glass, gold, silicon, and fabric) have been coated and later functionalized with two different alkylated species (1-hexadecanamine and stearoyl chloride). The success of their attachment has been demonstrated by evaluating the hydrophobicity conferred to the surface by contact angle measurements. Interestingly, these results, together with its chemical characterization by means of X-ray photoelectron spectroscopy (XPS) and Fourier-transform infrared spectroscopy (FT-IR), have proven that the reactivity of the primer coating towards the functionalizing agent can be tuned in function of its generation time.

Keywords: catechol; coating; surface modification; hydrophobicity; cross-linking; polydopamine; surface functionalization

1. Introduction

Macro- and micro/nano-surface functionalization by means of thin films with thicknesses measuring a few nanometres has become a topic of considerable interest in the past years due to the possibility of tuning their physico-chemical properties without significantly affecting bulk features or particle sizes. In this sense, the generation of molecular self-assembled monolayers (SAMs) represents a relatively simple and efficient way for surface modification [1]. Nevertheless, this approach usually requires chemical specificity between substrate and anchoring group. For example, thiols are useful for noble metals such as gold [2–6], platinum [7–10] and palladium [7,11], phosphates and phosphonates for metal oxides [12–15] and silanes for hydroxyl-bearing materials such as cellulose [16,17] and silicon oxides [18–22], zinc oxides [23,24], iron oxides [25,26], boron [27], and alumina [6] among others. In order to overcome this limitation, several research groups have drawn inspiration from mussel adhesive proteins, since it is known that they are able to strongly stick onto virtually any kind of surface, thus being usable as models for universal anchoring elements [28–30]. Research on these proteins suggests that the presence of considerable amounts of the non-essential catecholic amino acid L-3,4-dihydroxyphenylalanine (L-DOPA) is responsible, to a large extent, for this gluing effect, and particularly to the presence of the catechol moiety, as has been demonstrated in single molecule experiments [31]. Thus, catechol-based molecular SAMs could be good candidates for the functionalization of almost any kind of surface. Nonetheless, stable SAMs can only be obtained on certain substrates where the interaction between catechol and surface is strong enough for achieving a proper stabilization of the molecular monolayer. This is the case of some metal oxides, where a coordination bond is established [32–34]. If only weak interactions come into play, such as hydrogen bonding, electrostatic forces or π–π stacking, cooperativity between catechol moieties is essential for generating a stable thin film on the surface. This cooperativity can be achieved by polymerizing a catechol bearing a certain functional group whose properties are to be transferred to the treated surface, such as hydrophobicity [35], or making the catechol part of a more complex polymeric backbone [36–39]. However, this strategy usually entails following more or less complex synthetic pathways for obtaining the final catechol-based coating material. An alternative approach is the self-polymerization of catecholic molecules under oxidative conditions for obtaining coatings with thicknesses of few nanometres, which can be later functionalized. This is the case of polydopamine, which is obtained by oxidative polymerization of dopamine [40]. This first coating, which has been reported to adhere to a wide range of materials [40], can be reacted a posteriori with an alkanethiol, for example, for transferring its hydrophobicity to a given substrate [41]. Based on this same strategy, several authors have been searching for alternative approaches to polydopamine using other catechols as anchoring elements, like caffeic acid [42] and gallic acid [43,44], which are copolymerized along with diamines to favor their cross-linking. It has been demonstrated that the successfully generated copolymer films can be later tailored with different kinds of molecules thanks to the presence of reactive groups like amines and carbonyls.

In the present work, we report the copolymerization, under mild basic conditions, of pyrocatechol with different nitrogen-based cross-linkers. As a proof of concept, the obtained polymeric primer is functionalized ex situ by reaction with different alkyl-bearing molecules with the objective of transferring their hydrophobicity to the treated substrates. The chemical and morphological characterization of the coatings at different stages of the process demonstrate that the copolymerization of the simplest catecholic molecule with hexamethylenediamine (HMDA) is enough for successfully generating a coating that can be later used as a chemically functionalizable platform onto a wide range of materials (Figure 1).

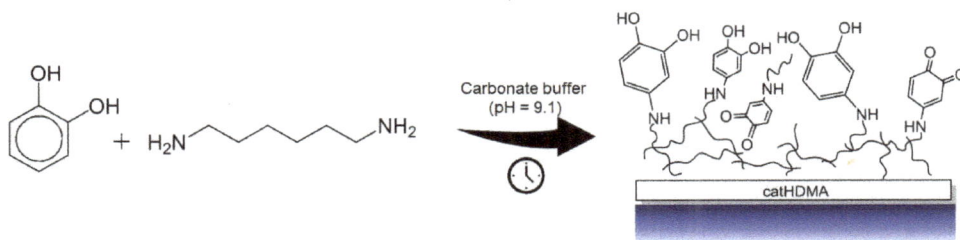

Figure 1. Schematic representation of the copolymerization between pyrocatechol and hexamethylenediamine (catHMDA).

2. Materials and Methods

2.1. Materials

All reagents were purchased and used without further purification from Sigma-Aldrich (Madrid, Spain). Solvents were used as received without additional drying or degasification. Aqueous carbonate buffer (pH = 9.1) was prepared by dissolving 265 mg of Na_2CO_3 and 1.89 g of $NaHCO_3$ in 250 mL of Milli-Q® water (Merck Chemicals & Life Science S.A., Madrid, Spain). Glass (cut from microscope borosilicate glass slides (Labbox, Madrid, Spain)); and silicon substrates (CNM, Barcelona, Spain) were cleaned in an ultrasonic bath for 10 min, first in acetone and then in ethanol, rinsed with Milli-Q® water and finally dried under a nitrogen flow. Gold-coated glass substrates (Labbox) for Fourier-transform infrared spectroscopy (FT-IR) analysis were obtained by deposition with a high vacuum sputter coater (Leica, Wetzlar, Germany), a 100 nm gold layer on glass substrates previously cut with a diamond tip. Polyester fabric substrates (Servei Estació, Barcelona, Spain) were cut and blown with nitrogen before use.

2.2. Preparation of the Primer Coatings for Their Physicochemical Characterization

Primer coatings were synthesized in 15 mL glass vials by dissolving benzene-1,2-diol (10 mmol) and hexamethylenediamine (HMDA, 10 mmol) in carbonate buffer (6 mL, pH = 9.1). The substrates (glass, gold or silicon) were placed in the vials together with the reagents. Then, the carbonate buffer was added under homogenous magnetic stirring. The vials were covered with pierced Parafilm® (Labbox) in order to allow the entrance of oxygen to the reaction mixture. Finally, the samples were washed with Milli-Q® water, dried with air flow and stored in a desiccator under vacuum.

2.3. Functionalization of the Primer Coatings with Hydrophobizing Agents

Primer-coated glass substrates were placed in 15 mL glass vials together with 6 mg of the corresponding hydrophobizing agent (1-hexadecanamine or stearoyl chloride). Then, 6 mL of hexane were added, the vial closed and kept under magnetic stirring for 24 h. Finally, the substrates were washed with hexane, dried with air flow and stored in a desiccator under vacuum conditions.

Variations of the contact angle values due to the presence of adsorbed solvent molecules were discarded by preparing blank samples of the coatings, following the above protocol but without addition of hydrophobizing agents.

2.4. Material Coating and Functionalization under Optimized Conditions

The substrates (fabric, silicon and gold) were placed horizontally in glass petri dishes for the deposition and formation of the primer coatings following the process described in Section 2.2. The resulting primer-coated substrates were then incubated for 24 h in glass petri dishes containing 1 mg/mL of hydrophobizing agent (1-hexadecanamine or stearoyl chloride) in hexane. Finally, the substrates were washed with hexane, dried with air flow and stored in a desiccator under vacuum conditions.

2.5. Contact Angle

Static contact angle measurements were performed with an EasyDrop contact angle meter (KRÜSS GmbH, Hamburg, Germany) using 15 µL water droplets. Each substrate was measured at three different points for obtaining an average of the whole surface. The measurements were performed approximately one minute after the droplet deposition.

2.6. Chemical Characterization

Ultraviolet–visible (UV–Vis) spectra have been acquired on glass substrates using a Cary 4000 Spectrometer (Agilent Technologies, Santa Clara, CA, USA). Surface FT-IR experiments have been performed with the Hyperion 2000 FT-IR microscope (Bruker Optik GmbH, Ettlingen, Germany) in reflection mode equipped with a nitrogen cooled mercury–cadmium–telluride (MCT) detector (InfraRed Associates, Inc., Stuart, FL, USA) using a 15× reflection objective, a gold mirror as reference and scanning for 30 min with a resolution of 4 cm^{-1}. X-ray photoelectron spectroscopy (XPS) measurements were performed with a Phoibos 150 analyser (SPECS EAS10P GmbH, Berlin, Germany) in ultra-high vacuum conditions (based pressure 10^{-10} mbar, residual pressure around 10^{-7} mbar).

Biomimetics **2017**, *2*, 22

Monochromatic Al Kα line was used as X-ray source (1486.6 eV and 300 W). The electron energy analyser was operated with pass energy of 50 eV. The hemispherical analyser was located perpendicular to the sample surface. The data was collected every eV with a dwell time of 0.5 s. A flood gun of electrons, with energy lower than 20 eV, was used to compensate the charge. The primer coatings were deposited on silicon substrates during three different times: 12, 24 and 48 h. All the data was treated with CasaXPS version 2.3.17PR1.1 (Casa Software LTD, Teignmouth, UK) and OriginPro version 8.0988 (OriginLab Corporation, Northampton, MA, USA) software.

2.7. Morphological Characterization of the Coatings

Measurements of the coating thickness were obtained from a Stylus Profilometer (D-500, KLA Tencor, Milpitas, CA, USA), with a vertical range of 1200 μm; 0.38 Å and 100 nm as vertical and lateral resolution, respectively. The measurements were performed with stylus force in the range of 1–5 mg. The primer coatings were deposited on glass substrates with a mask in the middle for the formation of a step suitable for the thickness measurement.

Images of treated fabric were obtained from scanning electron microscopy (SEM) (FEI Quanta 650 FEG, Thermo Fisher Scientific, Eindhoven, The Netherlands) in secondary electron mode with a beam voltage between 15 and 20 kV. Samples were coated with 15 nm of platinum by sputter coater (Leica).

2.8. Statistical Analysis

The statistical analysis of the contact angle measurements was carried out by performing three replicates per sample. The obtained results per sample were used for calculating the standard deviation using the $n-1$ method with Excel version 2010 (Microsoft Office, Redmond, WA, USA) software.

3. Results and Discussion

3.1. Generation of the Primer Coating

In a first attempt to generate a functionalizable primer coating by cross-linking of pyrocatechol with a nitrogenated species, three different reagents were tested for copolymerization in basic media: NH_3 aqueous solution (25% w/w), *p*-phenylenediamine and HMDA. Reactions using NH_3 or *p*-phenylenediamine yielded no coating on glass substrates. Despite the fact that an evident darkening of the solutions took place with time during these reactions, no material remained on the glass substrates after cleaning with water, as could be clearly observed with the naked eye, but also with UV–Vis spectroscopy. Finally, contact angle measurements were also performed on these samples, but values did not significantly vary from those of the untreated glass. Thus, the pyrocatechol polymerization products obtained with NH_3 or *p*-phenylenediamine did not seem to be able to generate a stable coating on this substrate. In the case of HMDA, a brownish coating remained on glass after thoroughly cleaning with water. Ultraviolet–visible spectroscopy showed a band around 345 nm, which was assigned to the polymerization product between pyrocatechol and HMDA (catHMDA). This technique was also used for following coating growth with time (Figure 2A). Surprisingly, no maximum was reached even after four days of reaction, and the deposition of material seemed to follow a linear trend, without appreciable shift of the position of the band (inset in Figure 2A).

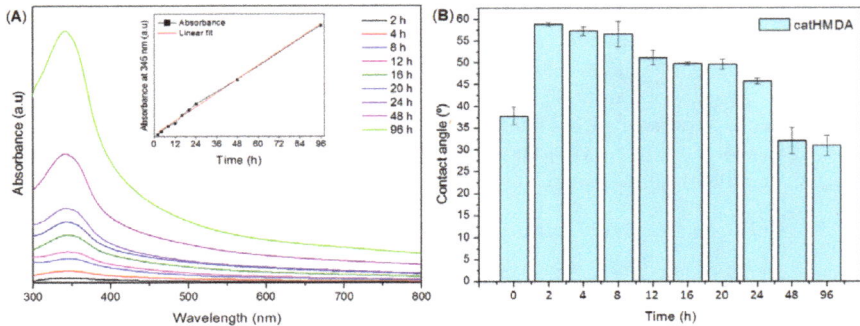

Figure 2. Study of the catHMDA growth with time. (**A**) Ultraviolet–visible (UV–Vis) spectra of catHMDA on glass vs. reaction time. Inset shows the linear trend of the maximum of absorbance at 345 nm as a function of the reaction time. (**B**) Static contact angle measurements of catHMDA-coated glass as a function of the reaction time. Data is shown as mean ± standard deviation. a.u.: Arbitrary units.

This observation is also supported by profilometry measurements, showing that the measured thickness of samples at 12, 24, 48 and 96 h also follows a nearly linear trend, being of 55 ± 8, 95 ± 5, 160 ± 10 and 386 ± 7 nm, respectively. Thus, it seems that under the reaction conditions used in these experiments, catHMDA grows on the glass substrates at an essentially constant rate during at least the first four days. A feasible explanation for this phenomenon is that, since the cross-linking reaction needs the prior oxidation of the catechol to the corresponding quinone and, in turn, this step needs the presence of oxygen, the rationed entrance of this last to the reaction mixture makes the process to run gradually. However, static contact angle measurements of the same samples showed that a significant decrease of this value takes place with time, parallel to the growth of the material (Figure 2B). A first quick increase of the contact angle in relation to the naked surface can be observed after 2 h of reaction. From here, the surface becomes more hydrophilic as the reaction time increases. These results suggest that the outer surface of the coating exposes more hydrophilic moieties as the reaction progresses. Nevertheless, and unlike layer thickness, no linearity was observed for this parameter. In fact, three different stages could be clearly differentiated regarding the variation of contact angle with time: between 2 and 8 h, when the highest contact angles around 57° are obtained; from 12 to 24 h, where a slight decrease down to around 50° may be appreciated; and between 48 and 96 h, when a dramatic drop of this parameter down to around 30° takes place, being even lower than that of the blank surface. It is worth mentioning that these values were reproducible after several weeks, demonstrating the robustness of the coating. Overall, results suggest that although the generation of the coating seems to take place gradually as mentioned above, its chemical composition should be varying with time, considering the availability of reactants in the medium (starting ratio of catechol:HMDA 1:1.5). As the catechol monomer is depleted, excess amine might favor the formation of more hydrophilic layers with free amine chain ends, which would explain the observed drop in contact angle values. In order to shed some light on this issue, XPS was used for the analysis of atomic chemical bonding of the primer coatings on silicon substrates at three different times: 12, 24 and 48 h. In all the cases, the C1s, N1s and O1s peaks yielded a very similar spectrum as expected and appeared in a closely position with a concentration of 75.83 ± 0.27%, 14.34 ± 0.13% and 7.81 ± 0.08%, respectively (Figure 3).

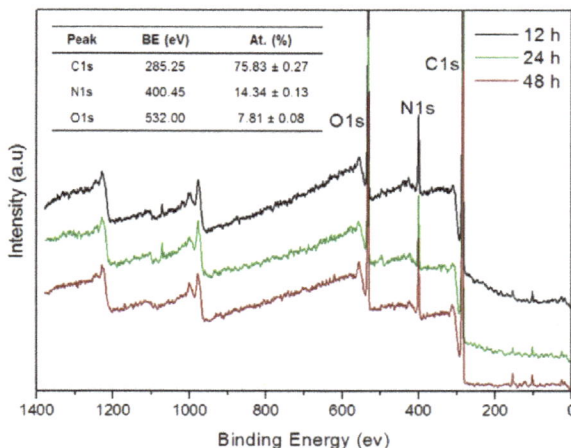

Peak	BE (eV)	At. (%)
C1s	285.25	75.83 ± 0.27
N1s	400.45	14.34 ± 0.13
O1s	532.00	7.81 ± 0.08

Figure 3. X-ray photoelectron spectroscopy (XPS) spectra of the primer coatings deposited on silicon substrate at different reaction times. The inset table shows the binding energy (BE) and the atomic concentration (At.) of C, N and O. a.u.: Arbitrary units.

To study in detail the bonding environment of the coating deposited on SiO$_2$, high-resolution XPS curve-fitting was performed. Figure 4 shows the XPS spectra with the fitting results for each peak of the primer coatings at the studied times. Table 1 shows the corresponding assignments for the resulting components. The energies were in agreement with previously reported values for similar systems [43–47].

Figure 4A shows the C1s spectra for each time. Five peaks have been properly fitted, which would correspond to five different chemical environments in the primer coating. They confirm the presence of catechol coexisting with its oxidized quinone state as can be noted by the C–OH signals at 286.45, 286.64 and 286.83 eV, and the C=O signals at 288.31, 288.42 and 288.43 eV for 12, 24 and 48 h reaction times, respectively. The O1s spectra (Figure 4C) also support the coexistence of these two species. In addition, it can be observed that the peaks corresponding to the reduced (catechol) state (both at C1s and O1s spectra) decrease with time, whereas those assigned to quinonic species increase, which would indicate an over-oxidation of catechol moieties at progressively longer reaction times. Regarding N1s spectra (Figure 4B), an increase of both aliphatic and aromatic amine-related components can be observed. C–N aromatic contributions could be assigned to the amino groups that are directly bonded to the catechol rings and are supposed to be involved in the cross-linking of the material, whereas C–NH aliphatic would be indicating the presence of unreacted amine tail ends. Although both signals rise with time, it can be noted that the one corresponding to the unreacted (aliphatic) amine tails undergoes a more significant increase.

This would be in agreement with our previous statement about the drop in the contact angle values for long reaction times, since free amine-rich top layers would confer hydrophilicity to the primer coating surface.

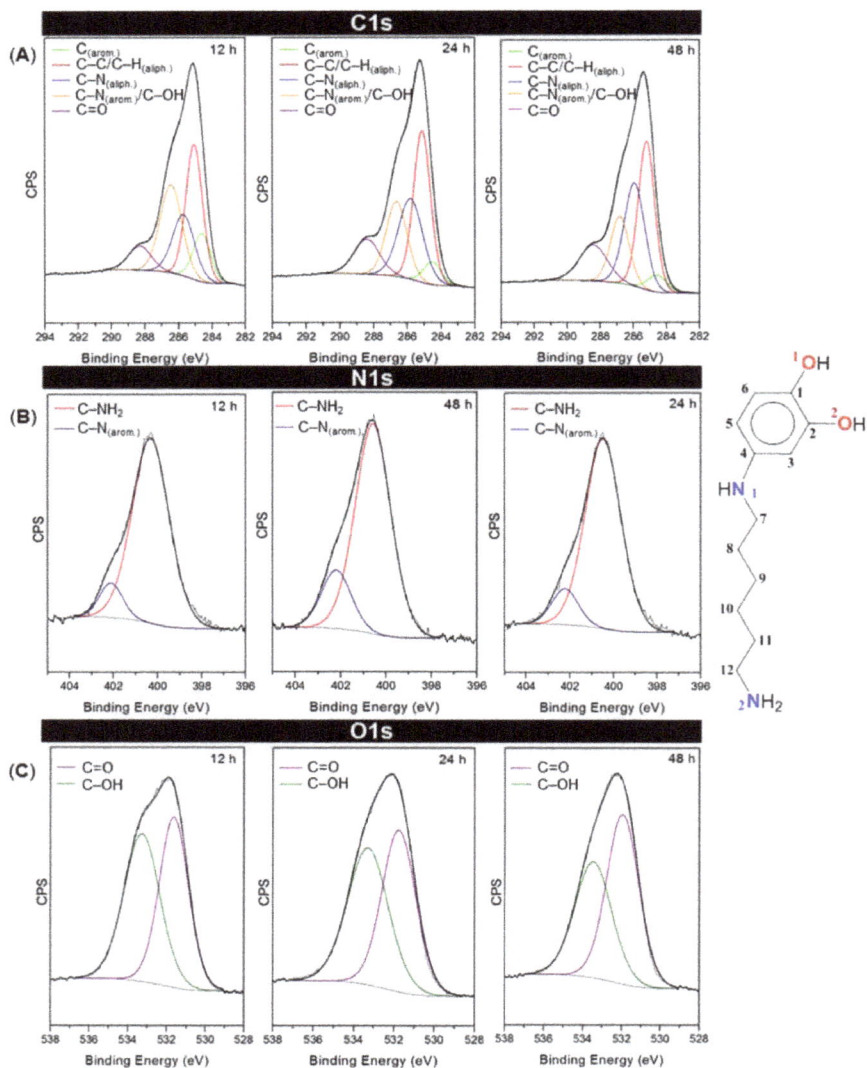

Figure 4. Curve-fitting results for C1s, N1s and O1s high-resolution XPS spectra at 12, 24 and 48 h. CPS: Counts per second. The scheme on the right represents the different kinds of chemical bonds in catHMDA. The atoms are arbitrary numbered for the XPS peak assignment.

Table 1. Curve-fitting assignments for C1s, N1s and O1s peaks at 12, 24 and 48 h.

Peak	Time [1] (h)	Atom	Atomic Bond	BE (eV)	At. (%)
C1s	12	C3,5,6	$C_{(arom.)}$	284.61	12.06
		C8,9,10,11	$C–C/C–H_{(aliph.)}$	285.07	32.52
		C7,12	$C–N_{(aliph.)}$	285.71	19.48
		C4/C1,2	$C–N_{(arom.)}/C–OH$	286.45	27.87
		C1,2	$C=O$	288.31	8.07
	24	C3,5,6	$C_{(arom.)}$	284.53	5.34
		C8,9,10,11	$C–C/C–H_{(aliph.)}$	285.13	34.64
		C7,12	$C–N_{(aliph.)}$	285.81	26.61
		C4/C1,2	$C–N_{(arom.)}/C–OH$	286.64	21.04
		C1,2	$C=O$	288.42	12.37
	48	C3,5,6	$C_{(arom.)}$	284.57	4.52
		C8,9,10,11	$C–C/C–H_{(aliph.)}$	285.22	36.52
		C7,12	$C–N_{(aliph.)}$	285.94	29.64
		C4/C1,2	$C–N_{(arom.)}/C–OH$	286.83	15.27
		C1,2	$C=O$	288.43	14.05
N1s	12	N1	$C–NH_2$	399.87	81.26
		N2	$C–N_{(arom.)}$	401.11	10.51
	24	N1	$C–NH_2$	399.08	82.14
		N2	$C–N_{(arom.)}$	401.65	11.26
	48	N1	$C–NH_2$	399.87	88.49
		N2	$C–N_{(arom.)}$	401.78	11.51
O1s	12	O1,2	$C=O$	531.63	48.39
		O1,2	$C–OH$	533.27	51.61
	24	O1,2	$C=O$	531.75	49.41
		O1,2	$C–OH$	533.29	50.59
	48	O1,2	$C=O$	531.95	56.08
		O1,2	$C–OH$	533.45	43.92

[1] catHMDA generation time. aliph.: Aliphatic; arom.: Aromatic; At.: Atomic concentration; BE: Binding energy.

Characterization of the primer coatings obtained at 12, 24 and 48 h by means of FT-IR also shows, on the one hand, an increase in the thickness of the coating with time, as can be noted by the increase in absorbance of the corresponding spectra (Figure 5).

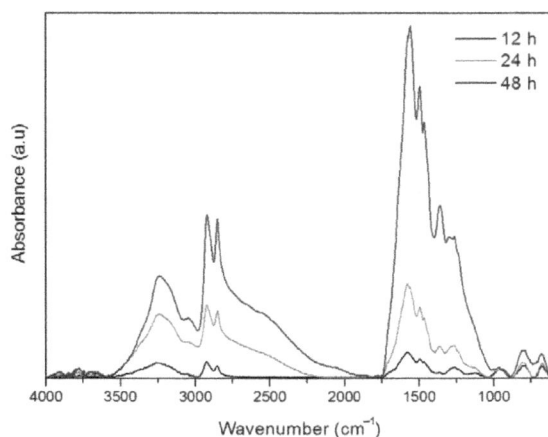

Figure 5. Fourier-transform infrared spectroscopy (FT-IR) spectra of the primer coatings after 12, 24 and 48 h of reaction. a.u.: Arbitrary units.

On the other hand, FT-IR clearly shows that both HMDA and catecholic/quinonic species are present in the coating. The broad band around 3240 cm^{-1} and its shoulder at higher wavenumbers around 3400 cm^{-1} may be assigned to the NH$_2$ and O–H vibrations from catechol and amine moieties, respectively. The bands observed in all the spectra at 3050 cm^{-1}, and between 1560 and 1580 cm^{-1} can be respectively assigned to C=C–H and C=C vibrations from the catecholic/quinonic rings, whereas peaks around 2853 and 2924 cm^{-1} would correspond to the asymmetric and symmetric stretching vibrations of the methylene C–H bonds from the alkyl chain of the HMDA. All spectra present shoulders of the intense bands around 1575 cm^{-1}, and between 1620 and 1710 cm^{-1}. Although they are not perfectly resolved, it is feasible to assign them to C=O quinonic groups, that otherwise have also been observed, as described above, by XPS. Finally, the peak at 1264 cm^{-1} observed in all the spectra could be assigned to a secondary amine bridging an alkyl and an aromatic ring. Since these measurements are a read of the whole material (unlike XPS, as it is not a superficial technique) and, in addition, the resolution of the bands is not optimal mainly for shorter reaction times, it is complicated by only means of FT-IR to establish an evolution of the functional groups with time. Nevertheless, these findings along with the XPS, UV–Vis and profilometry data would suggest that the primer coating consists in a cross-linkage of catechol/quinone rings through HDMA that gradually increases its thickness with time in an almost linear trend. The longer the time of reaction, the higher the concentration of quinones, and unreacted tail-end amines can be found in the surface of the coating. Thus, its reactivity towards a given functionalizing agent should be considered in function of the time of its generation.

3.2. Functionalization of the Primer Coating

Considering the presence of quinones and primary amines in the surface of the catHMDA coating, its reactivity towards a nucleophilic and an electrophilic attack as functionalization strategies was explored and evaluated. For this, an amine and an acyl chloride, both bearing alkyl chains, were used as functionalization agents, since their successful attachment may be easily tested by measuring the resulting contact angle. Thus, catHMDA coatings were generated at different times and later incubated with the functionalizing agents (Figure 6). As mentioned above, the reactivity of the primer coating would be expected to change with reaction time, concomitantly with changes in its chemical composition.

In Figure 6, different trends for the functionalizing agents can be observed. Although both maxima achieve similar values around 85°, the amine nucleophilic attack seems to be more successful on the catHMDA obtained after 16 h (catHMDA-16 h), whereas the electrophilic attack of the acyl chloride seems to be optimum for the catHMDA coating generated for 48 h (catHMDA-48 h).

This trend would support the primer coating characterization results: on one hand, the longer time for the generation of catHMDA, the larger number of primary amines can be found in the surface, thus increasing the probability of being attacked by an electrophilic reagent during the subsequent functionalization incubation. On the other hand, it has also been observed that the amount of quinones, liable to later reacting with nucleophiles, increases with the generation time of catHMDA. Nonetheless, HMDA can react with these quinones during the generation of the primer coating, thus blocking reactive positions towards other nucleophiles in the later functionalization process. In fact, this could be the reason why a larger number of tail-end amines can be found as an overlayer on the coating for longer reaction times by means of XPS (Table 1, N1s). Considering these observations, it seems feasible that at a certain point the reactivity of catHMDA towards nucleophiles would decrease, with a concomitant increase in reactivity towards electrophiles. This point of inflection can be observed in Figure 6 approximately after 16 h. The extremely low contact angles obtained up to 20 h for the incubation with stearoyl chloride, being even lower than the uncoated glass substrate, could be due to a chemical modification of the coating surface during this process. An increase in the number of primary amines on the surface for longer generation times, besides increasing the reactivity towards an acyl chloride, could be buffering such side effects. Nevertheless, further studies would be needed to shed light on this issue.

In order to confirm the attachment of the functionalizing agents, as well as to determine the mechanisms involved in the corresponding processes, the functionalized coatings showing optimal contact angles (catHMDA-16 h/amine and catHMDA-48 h/stearoyl) were analysed by XPS. Figure 7 shows the wide scan XPS spectra and the compositional analysis, where an increase of the atomic percentage of carbon can be noted, reaching values higher than 80%, which would be compatible with the attachment of the bulky alkyl chains to the surface. Additionally, the presence of N and O was also confirmed.

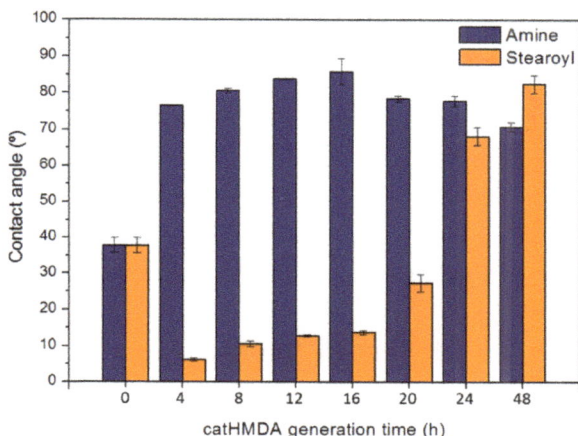

Figure 6. Contact angles of catHMDA generated at different times (4–48 h), after incubation with 1-hexadecanamine (blue) and stearoyl chloride (orange). Both incubations are carried out for 24 h in hexane. Data is shown as mean ± standard deviation.

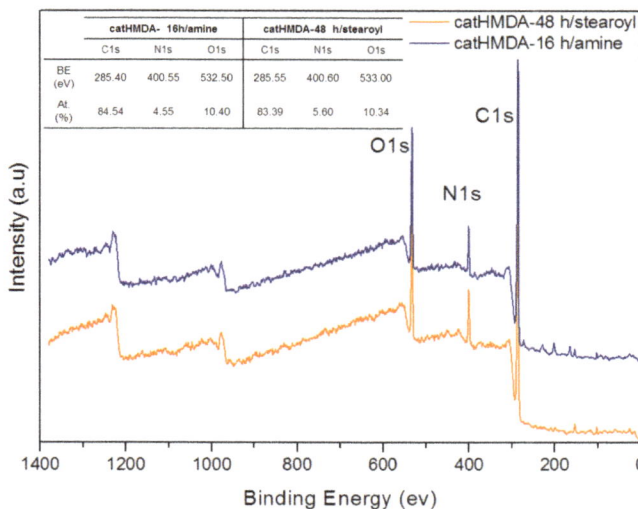

	catHMDA- 16h/amine			catHMDA-48 h/stearoyl		
	C1s	N1s	O1s	C1s	N1s	O1s
BE (eV)	285.40	400.55	532.50	285.55	400.60	533.00
At. (%)	84.54	4.55	10.40	83.39	5.60	10.34

Figure 7. XPS spectra of the functionalized primer coatings deposited on a silicon substrate. The inset table shows their chemical composition. a.u.: Arbitrary units.

The core-level spectra C1s, N1s and O1s for both functionalized coatings are presented in Figure 8, and relevant peak data and assignments listed in Table 2. The energies are in agreement with reported values for similar chemical bonding [43,48,49]. In both systems, besides the signals coming from the underlying catHMDA primer coating, new signals which could be assigned to the new established bonds after the incubation process can be observed. In the case of catHMDA-16 h/amine, the C1s spectrum (Figure 8A) shows a component at 286.23 eV that can be assigned to the formation of an imine bond (C=N), also confirmed in the N1s spectrum with a signal at 399.73 eV.

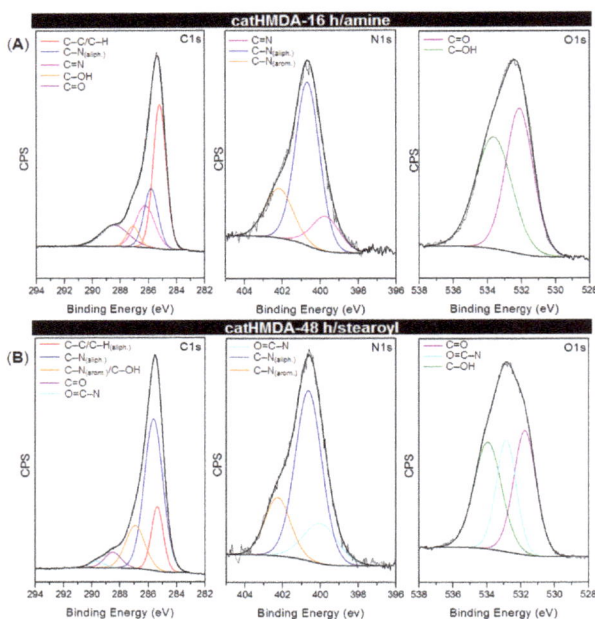

Figure 8. Curve-fitting results corresponding to C1s, N1s and O1s high-resolution XPS spectra for (**A**) catHMDA-16 h/amine and (**B**) catHMDA-48 h/stearoyl.

This would hint at covalent bonds being formed by nucleophilic attacks of amines on the carbonyl groups of the quinone rings (Figure 9B), which seem to be present in large amounts for intermediate reaction times (Figure 9A). With regard to Michael-type additions, they cannot be completely ruled out, since signals at 400.67 and 402.17 eV, corresponding to secondary amines bridging aromatic and aliphatic carbons, are also observed in this spectrum. Although cross-linking of catechol rings through HMDA is the main origin of such signals, some contribution coming from the attachment of 1-hexadecanamine to the aromatic ring by means of a 1,4-addition is also feasible in this context. In the case of catHMDA-48 h/stearoyl (Figure 8B), one of the fitted component in C1s peak with highest binding energy (289.67 eV) can be attributed to an amidic carbon atom (N–C=O), thus confirming the condensation between free amines in the catHMDA and stearoyl chloride (Figure 9C). The presence of the amide bond is also confirmed by the 532.86 eV peak in the O1s high-resolution spectrum.

Chemical characterization by means of FT-IR has also been performed after the functionalization processes, but no significant peaks coming from these overlayers have been detected due to the considerable larger thickness of the underlying catHMDA coating, which absorbs most of the FT-IR signal. Nevertheless, contact angle and XPS measurements have been enough for confirming the successful functionalization of the catHMDA primer coating with 1-hexadecanamine and stearoyl chloride.

Table 2. Curve-fitting assignments corresponding to C1s, N1s and O1s peaks for catHMDA-16 h/amine and catHMDA-48 h/stearoyl.

catHMDA [1]	Peak	Atomic Bond	BE (eV)	At. (%)
Amine	C1s	C–C/C–H$_{(aliph.)}$	285.20	42.92
		C–N$_{(aliph.)}$	285.81	18.93
		C=N	286.23	18.51
		C–OH	287.11	5.73
		C=O	288.43	13.91
	N1s	C=N	399.73	15.62
		C–N$_{(aliph.)}$	400.67	62.72
		C–N$_{(arom.)}$	402.17	21.66
	O1s	C=O	532.12	49.84
		C–OH	533.64	50.16
Stearoyl	C1s	C–C/C–H$_{(aliph.)}$	285.37	16.29
		C–N$_{(aliph.)}$	285.63	58.94
		C–N$_{(arom.)}$/C–OH	286.93	17.20
		C=O	288.51	5.56
		O=C–N	289.67	2.01
	N1s	O=C–N	399.98	18.51
		C–N$_{(aliph.)}$	400.63	61.30
		C–N$_{(arom.)}$	402.26	20.19
	O1s	C=O	531.76	35.53
		O=C–N	532.86	27.88
		C–OH	533.95	36.59

[1] Optimized catHMDA for each functionalization. aliph.: Aliphatic; arom.: Aromatic; At.: Atomic concentration; BE: Binding energy.

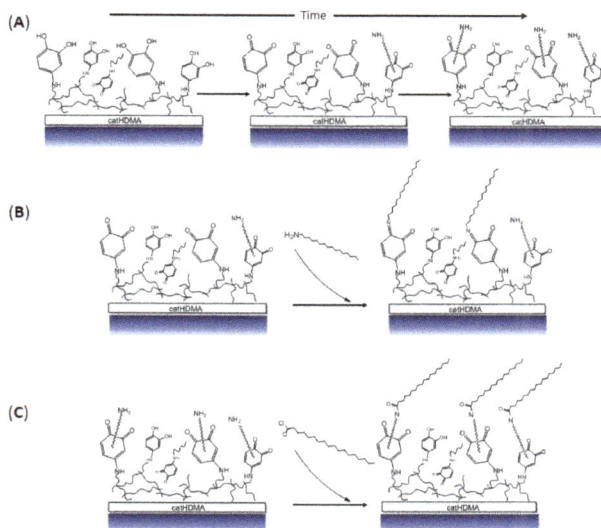

Figure 9. Chemical structure of the catHMDA coating and its functionalization. (**A**) Tentative schematic representation of the primer coating (catHDMA) and its evolution in function of the reaction time, where an increase in the amount of quinones and non-reacted amine tail ends can be observed. (**B**) Proposed schematic mechanism of the attachment of 1-hexadecanamine to catHDMA-16 h (the optimal generation time of the primer coating for its functionalization with this reagent is 16 h) (**C**) Proposed schematic mechanism of the attachment of stearoyl chloride to catHDMA-48 h (the optimal generation time of the primer coating for its functionalization with this reagent is 48 h).

Finally, three different substrates (gold, silicon and fabric) were rendered hydrophobic in the optimal conditions for the generation of the primer coatings and their further functionalization. Contact angle values at the different stages of this treatment are shown in Figure 10.

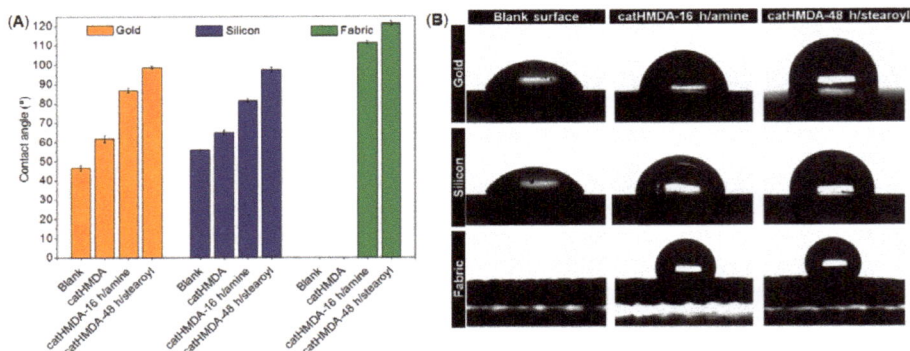

Figure 10. Wettability study after functionalization of catHDMA. (**A**) Contact angles of water on gold, silicon and fabric substrates (blank/uncoated, coated with catHMDA, and coated + functionalized with 1-hexadecanamine and stearoyl chloride). Data is shown as mean ± standard deviation. (**B**) Images of water droplets on the three substrates before and after functionalization with catHMDA-16 h/amine and catHMDA-48 h/stearoyl.

As can be seen in Figure 10A, in all the cases catHMDA-48 h/stearoyl leads to the highest contact angle values, approximately 10° above those obtained with catHMDA-16 h/amine. Interestingly, it can be noted that for substrates having similar roughness, like gold and silicon, the contact angle values are almost equal for the same treatment. This could be evidence for the universality of this coating, which would provide the same functionalization capacities independently of the chemical nature of the treated surface. The greater roughness of the fabric substrate causes a significant increase in the final contact angles, due to the structuration of the surface. The morphology of this last sample coated with catHMDA-16 h/amine has been analyzed by SEM. As shown in Figure 11, almost no differences in appearance can be seen between the untreated (Figure 11A,B) and the treated fabric (Figure 11C,D), where the presence of a coating is hinted at by small cracks across its surface, as observed at higher magnification (Figure 11D and zoom inset). Thus, even after 16 h of generation of the coating, the submicron thickness of the coating does not significantly modify the morphology of the treated textile, which is important for preserving its mechanical and breathability properties.

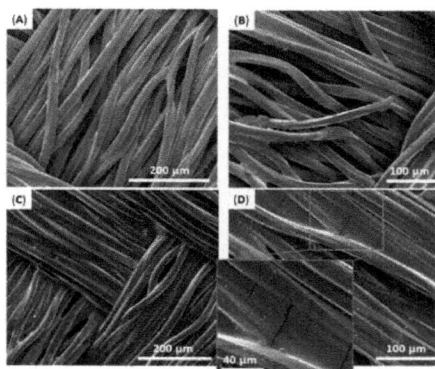

Figure 11. Scanning electron microscopy (SEM) images of fabric samples (**A**,**B**) before and (**C**,**D**) after being coated with catHMDA-16 h/amine. Inset in (**D**) shows zoom area of the coated polyester fabric fibers, where small cracks can be observed on its surface.

4. Conclusions

A functionalizable polydopamine-like coating (catHMDA) was successfully obtained by cross-linking polymerization of pyrocatechol with HMDA under oxidizing conditions. Regarding the growth of the coating, UV–Vis spectroscopy and profilometry measurements show that the thickness of catHMDA on flat surfaces follows an almost linear trend with time under the reaction conditions. The XPS measurements suggest that its superficial chemical composition also changes with time, thus opening the possibility of tuning its reactivity in function of this parameter. As a proof-of-concept, different substrates were coated with catHMDA and later functionalized with 1-hexadecanamine and stearoyl chloride, demonstrating the versatility of this platform for its functionalization following both nucleophilic and electrophilic attack approaches. The bulky alkyl chain in both species successfully conferred robust hydrophobicity to four different surfaces (glass, gold, silicon and fabric). These results demonstrate the potential of catHDMA as a universal coating that can act as a flexible platform for further functionalization with, for example, biomolecules of interest in biomaterials science, amongst others. However, future work should be directed in order to explore the degradability of these coatings. In addition, the use of more superficial techniques could be of interest for shedding more light on the chemical composition of the top atomic layers of these systems.

Acknowledgments: This work was supported by grant MAT 2015-70615-R from the Spanish Government funds and by the European Regional Development Fund (ERDF). Funded by the CERCA Program/Generalitat de Catalunya. ICN2 is supported by the Severo Ochoa program from Spanish Ministry of Economy, Industry and Competitiveness (MINECO) (Grant No. SEV-2013-0295).

Author Contributions: J.S.-P. conceived and designed the experiments; S.S.-G. performed the experiments; S.S.-G. and J.S.-P. analyzed the data; S.S.-G, J.S., D.R.-M. and J.S.-P. wrote the paper.

Conflicts of Interest: The authors declare no conflict of interest.

References

1. Ye, Q.; Zhou, F.; Liu, W. Bioinspired catecholic chemistry for surface modification. *Chem. Soc. Rev.* **2011**, *40*, 4244–4258. [CrossRef] [PubMed]
2. Grumelli, D.; Cristina, L.J.; Lobo Maza, F.; Carro, P.; Ferrón, J.; Kern, K.; Salvarezza, R.C. Thiol adsorption on the Au(100)-hex and Au(100)-(1 × 1) surfaces. *J. Phys. Chem. C* **2015**, *119*, 14248–14254. [CrossRef]
3. Park, C.S.; Lee, H.J.; Jamison, A.C.; Lee, T.R. Robust thick polymer brushes grafted from gold surfaces using bidentate thiol-based atom-transfer radical polymerization initiators. *ACS Appl. Mater. Interfaces* **2016**, *8*, 5586–5594. [CrossRef] [PubMed]

4. Bartucci, M.A.; Florián, J.; Ciszek, J.W. Spectroscopic evidence of work function alterations due to photoswitchable monolayers on gold surfaces. *J. Phys. Chem. C* **2013**, *117*, 19471–19476. [CrossRef]

5. Lebec, V.; Landoulsi, J.; Boujday, S.; Poleunis, C.; Pradier, C.-M.; Delcorte, A. Probing the orientation of β-lactoglobulin on gold surfaces modified by alkyl thiol self-assembled monolayers. *J. Phys. Chem. C* **2013**, *117*, 11569–11577. [CrossRef]

6. Escobar, C.A.; Zulkifli, A.R.; Faulkner, C.J.; Trzeciak, A.; Jennings, G.K. Composite fluorocarbon membranes by surface-initiated polymerization from nanoporous gold-coated alumina. *ACS Appl. Mater. Interfaces* **2012**, *4*, 906–915. [CrossRef] [PubMed]

7. Kahsar, K.R.; Schwartz, D.K.; Medlin, J.W. Liquid- and vapor-phase hydrogenation of 1-epoxy-3-butene using self-assembled monolayer coated palladium and platinum catalysts. *Appl. Catal. A Gen.* **2012**, *445–446*, 102–106. [CrossRef]

8. Kim, Y.R.; Kim, H.J.; Kim, J.S.; Kim, H. Rhodamine-based "turn-on" fluorescent chemodosimeter for Cu(II) on ultrathin platinum films as molecular switches. *Adv. Mater.* **2008**, *20*, 4428–4432. [CrossRef]

9. Makosch, M.; Lin, W.I.; Bumbalek, V.; Sa, J.; Medlin, J.W.; Hungerbuhler, K.; van Bokhoven, J.A.; Bokhoven, J.A. Organic thiol modified Pt/TiO$_2$ catalysts to control chemoselective hydrogenation of substituted nitroarenes. *ACS Catal.* **2012**, *2*, 2079–2081. [CrossRef]

10. Silien, C.; Dreesen, L.; Cecchet, F.; Thiry, P.A.; Peremans, A. Orientation and order of self-assembled *p*-benzenedimethanethiol films on Pt(111) obtained by direct adsorption and via alkanethiol displacement. *J. Phys. Chem. C* **2007**, *111*, 6357–6364. [CrossRef]

11. Marshall, S.T.; O'Brien, M.; Oetter, B.; Corpuz, A.; Richards, R.M.; Schwartz, D.K.; Medlin, J.W. Controlled selectivity for palladium catalysts using self-assembled monolayers. *Nat. Mater.* **2010**, *9*, 853–858. [CrossRef] [PubMed]

12. Liu, H.B.; Venkataraman, N.; Spencer, N.D.; Textor, M.; Xiao, S.J. Structural evolution of self-assembled alkanephosphate monolayers on TiO$_2$. *ChemPhysChem* **2008**, *9*, 1979–1981. [CrossRef] [PubMed]

13. Spori, D.M.; Venkataraman, N.V.; Tosatti, S.G.P.; Durmaz, F.; Spencer, N.D.; Zürcher, S. Influence of alkyl chain length on phosphate self-assembled monolayers. *Langmuir* **2007**, *23*, 8053–8060. [CrossRef] [PubMed]

14. Hoque, E.; DeRose, J.A.; Kulik, G.; Hoffmann, P.; Mathieu, H.J.; Bhushan, B. Alkylphosphonate modified aluminum oxide surfaces. *J. Phys. Chem. B* **2006**, *110*, 10855–10861. [CrossRef] [PubMed]

15. Pawsey, S.; Yach, K.; Reven, L. Self-assembly of carboxyalkylphosphonic acids on metal oxide powders. *Langmuir* **2002**, *18*, 5205–5212. [CrossRef]

16. Hassanpour, A.; Asghari, S.; Lakouraj, M.M. Synthesis, characterization and antibacterial evaluation of nanofibrillated cellulose grafted by a novel quinolinium silane salt. *RCS Adv.* **2017**, *7*, 23907–23916. [CrossRef]

17. Achoundong, C.S.K.; Bhuwania, N.; Burgess, S.K.; Karvan, O.; Johnson, J.R.; Koros, W.J. Silane modification of cellulose acetate dense films as materials for acid gas removal. *Macromolecules* **2013**, *46*, 5584–5594. [CrossRef]

18. Watté, J.; Van Gompel, W.; Lommens, P.; De Buysser, K.; VanDriessche, I. Titania nanocrystal surface functionalization through silane chemistry for low temperature deposition on polymers. *ACS Appl. Mater. Interfaces* **2016**, *8*, 29759–29769. [CrossRef] [PubMed]

19. Rodriguez, C.; Laplace, P.; Gallach-Perez, D.; Pellacania, P.; Martin-Palma, R.J.; Torres-Costa, V.; Ceccone, G.; Silvan, M.M. Hydrophobic perfluoro-silane functionalization of porous silicon photoluminescent films and particles. *App. Surf. Sci.* **2016**, *380*, 243–248. [CrossRef]

20. Lee, S.; Kim, J.Y.; Cheon, S.; Kim, S.; Kim, D.; Ryu, H. Stimuli-responsive magneto-/electro-chromatic color-tunable hydrophobic surface modified Fe$_3$O$_4$@SiO$_2$ core-shell nanoparticles for reflective display approaches. *RSC Adv.* **2017**, *7*, 6988–6993. [CrossRef]

21. Garcia-Uriostegui, L.; Melendez-Ortiz, H.I.; Toriz, G.; Delgado, E. Post-grafting and characterization of mesoporous silica MCM-41 with a thermoresponsive polymer TEVS/NIPAAm/β-cyclodextrin. *Mater. Lett.* **2017**, *196*, 26–29. [CrossRef]

22. Kelleher, S.M.; Nooney, R.I.; Flynn, S.P.; Clancy, E.; Burke, M.; Daly, S.; Smith, T.J.; Daniels, S.; McDonagh, C. Multivalent linkers for improved covalent binding of oligonucleotides to dye-doped silica nanoparticles. *Nanotechnology* **2015**, *26*, 365703–365715. [CrossRef] [PubMed]

23. Matthews, R.; Glasser, E.; Sprawls, S.C.; French, R.H.; Peshek, T.J.; Pentzer, E.; Martin, I.T. Facile synthesis of unique hexagonal nanoplates of Zn/Co hydroxy sulfate for efficient electrocatalytic oxygen evolution reaction. *ACS Appl. Mater. Interfaces* **2017**, *9*, 17621–17629. [CrossRef]

24. Rohe, B.; Veeman, W.S.; Tausch, M. Synthesis and photocatalytic activity of silane-coated and UV-modified nanoscale zinc oxide. *Nanotechnology* **2006**, *17*, 277–282. [CrossRef]
25. Cano, M.; Nunez-Lozano, R.; Lumbreras, R.; Gonzalez-Rodriguez, V.; Delgado-Garcia, A.; Jimenez-Hoyuela, J.M.; de la Cueva-Mendez, G. Partial PEGylation of superparamagnetic iron oxide nanoparticles thinly coated with amine-silane as a source of ultrastable tunable nanosystems for biomedical applications. *Nanoscale* **2017**, *9*, 812–822. [CrossRef] [PubMed]
26. Dincer, C.A.; Yildiz, N.; Aydogan, N.; Calimli, A. A comparative study of Fe$_3$O$_4$ nanoparticles modified with different silane compounds. *App. Surf. Sci.* **2014**, *318*, 297–304. [CrossRef]
27. Du, M.J.; Li, G. Preparation of silane-capped boron nanoparticles with enhanced dispersibility in hydrocarbon fuels. *Fuel* **2017**, *194*, 75–82. [CrossRef]
28. Waite, J.H. Mussel adhesion—Essential footwork. *J. Exp. Biol.* **2017**, *220*, 517–530. [CrossRef] [PubMed]
29. Petrone, L.; Kumar, A.; Sutanto, C.N.; Patil, N.J.; Kannan, S.; Palaniappan, A.; Amini, S.; Zappone, B.; Verma, C.; Miserez, A. Mussel adhesion is dictated by time-regulated secretion and molecular conformation of mussel adhesive proteins. *Nat. Commun.* **2015**, *6*, 8737–8748. [CrossRef] [PubMed]
30. Sedó-Vegara, J.; Saiz-Poseu, J.; Busqué, F.; Ruiz-Molina, D. Catechol-based biomimetic functional materials. *Adv. Mater.* **2013**, *25*, 653–701. [CrossRef] [PubMed]
31. Lee, H.; Scherer, N.F.; Messersmith, P.B. Single-molecule mechanics of mussel adhesion. *Proc. Natl. Acad. Sci. USA* **2006**, *103*, 12999–13003. [CrossRef] [PubMed]
32. Chouirfa, H.; Evans, M.D.M.; Castner, D.G.; Bean, P.; Mercier, D.; Galtayries, A.; Falentin-Daudre, C.; Migonney, V. Grafting of architecture controlled poly(styrene sodium sulfonate) onto titanium surfaces using bio-adhesive molecules: Surface characterization and biological properties. *Biointerphases* **2017**, *12*. [CrossRef] [PubMed]
33. Wang, D.H.; Guo, S.T.; Zhang, Q.; Wilson, P.; Haddleton, D.M. Mussel-inspired thermoresponsive polymers with a tunable LCST by Cu(0)-LRP for the construction of smart TiO$_2$ nanocomposites. *Polym. Chem.* **2017**, *8*, 3679–3688. [CrossRef]
34. Chao, C.G.; Kumar, M.P.; Riaz, N.; Khanoyan, R.T.; Madrahimov, S.T.; Bergbreiter, D.E. Polyisobutylene oligomers as tools for iron oxide nanoparticle solubilization. *Macromolecules* **2017**, *50*, 1494–1502. [CrossRef]
35. Saiz-Poseu, J.; Sedó, J.; García, B.; Benaiges, C.; Parella, T.; Alibés, R.; Hernando, J.; Busqué, F.; Ruiz-Molina, D. Versatile nanostructured materials via direct reaction of functionalized catechols. *Adv. Mater.* **2013**, *25*, 2066–2070. [CrossRef] [PubMed]
36. Yamamoto, S.; Uchiyama, S.; Miyashita, T.; Mitsuishi, M. Multimodal underwater adsorption of oxide nanoparticles on catechol-based polymer nanosheets. *Nanoscale* **2016**, *8*, 5912–5919. [CrossRef] [PubMed]
37. Kim, S.M.; In, I.; Park, S.Y. Study of photo-induced hydrophilicity and self-cleaning property of glass surfaces immobilized with TiO$_2$ nanoparticles using catechol chemistry. *Surf. Coat. Technol.* **2016**, *294*, 75–82. [CrossRef]
38. Zhong, J.; Ji, H.; Duan, J.; Tu, H.Y.; Zhang, A.D. Coating morphology and surface composition of acrylic terpolymers with pendant catechol, OEG and perfluoroalkyl groups in varying ratio and the effect on protein adsorption. *Colloids Surf. B Biointerfaces* **2016**, *140*, 254–261. [CrossRef] [PubMed]
39. Faure, E.; Falentin-Daudré, C.; Jérôme, C.; Lyskawa, J.; Fournier, D.; Woisel, P.; Detrembleur, C. Catechols as versatile platforms in polymer chemistry. *Prog. Polym. Sci.* **2013**, *38*, 236–270. [CrossRef]
40. Liu, Y.L.; Ai, K.L.; Lu, L.H. Polydopamine and its derivative materials: Synthesis and promising applications in energy, environmental, and biomedical fields. *Chem. Rev.* **2014**, *114*, 5057–5115. [CrossRef] [PubMed]
41. Haeshin, L.; Dellatore, S.M.; Miller, W.M.; Messersmith, P.B. Mussel-inspired surface chemistry for multifunctional coatings. *Science* **2007**, *318*, 426–430. [CrossRef]
42. Iacomino, M.; Paez, J.I.; Avolio, R.; Carpentieri, A.; Panzella, L.; Falco, G.; Pizzo, E.; Errico, M.E.; Napolitano, A.; del Campo, A.; et al. Multifunctional thin films and coatings from caffeic acid and a cross-linking diamine. *Langmuir* **2017**, *33*, 2096–2102. [CrossRef] [PubMed]
43. Chen, S.; Zhang, J.; Chen, Y.Q.; Zhao, S.; Chen, M.Y.; Li, X.; Maitz, M.F.; Wang, J.; Huang, N. Application of phenol/amine copolymerized film modified magnesium alloys: Anticorrosion and surface biofunctionalization. *ACS Appl. Mater. Interfaces* **2015**, *7*, 24510–24522. [CrossRef] [PubMed]
44. Chen, S.; Li, X.; Yang, Z.L.; Zhou, S.; Luo, R.F.; Maitz, M.F.; Zhao, Y.C.; Wang, J.; Xiong, K.Q.; Huang, N. A simple one-step modification of various materials for introducing effective multi-functional groups. *Colloids Surf. B Biointerfaces* **2014**, *113*, 125–133. [CrossRef] [PubMed]

45. Sen, R.; Gahtory, D.; Carvalho, R.R.; Albada, B.; van Delft, F.L.; Zuilhof, H. Ultrathin covalently bound organic layers on mica: Formation of atomically flat biofunctionalizable surfaces. *Angew. Chem. Int. Ed.* **2017**, *56*, 4130–4134. [CrossRef] [PubMed]
46. Jackman, M.J.; Syres, K.L.; Cant, D.J.H.; Hardman, S.J.O.; Thomas, A.G. Adsorption of dopamine on rutile TiO$_2$ (110): A photoemission and near-edge X-ray absorption fine structure study. *Langmuir* **2014**, *30*, 8761–8769. [CrossRef] [PubMed]
47. Syres, K.L.; Thomas, A.G.; Flavell, W.R.; Spencer, B.F.; Bondino, F.; Malvestuto, M.; Preobrajenski, A.; Grätzel, M. Adsorbate-induced modification of surface electronic structure: Pyrocatechol adsorption on the anatase TiO$_2$ (101) and rutile TiO$_2$ (110) surfaces. *J. Phys. Chem. C* **2012**, *116*, 23515–23525. [CrossRef]
48. Das, S.K.; Dickinson, C.; Lafir, F.; Brougham, D.F; Marsili, E. Synthesis, characterization and catalytic activity of gold nanoparticles biosynthesized with *Rhizopus oryzae* protein extract. *Green Chem.* **2012**, *14*, 1322–1334. [CrossRef]
49. Zorn, G.; Liu, L.-H.; Árnadóttir, L.; Wang, H.; Gamble, L.J.; Castner, D.G.; Yan, M. X-ray photoelectron spectroscopy investigation of the nitrogen species in photoactive perfluorophenylazide-modified surfaces. *J. Phys Chem. C* **2014**, *118*, 376–383. [CrossRef] [PubMed]

MDPI

Article

Mechanically Reinforced Catechol-Containing Hydrogels with Improved Tissue Gluing Performance

Jun Feng [1,3,†], Xuan-Anh Ton [2,†], Shifang Zhao [1,3], Julieta I. Paez [1] and Aránzazu del Campo [1,3,*]

1   INM – Leibniz Institute for New Materials, Campus D2 2, 66123 Saarbrücken, Germany;
    Jun.Feng@leibniz-inm.de (J.F.); shifang.zhao@leibniz-inm.de (S.Z.); julieta.paez@leibniz-inm.de (J.I.P.)
2   Max-Planck-Institut für Polymerforschung, Ackermannweg 10, 55128 Mainz, Germany;
    xuananh.ton@gmail.com
3   Chemistry Department, Saarland University, 66123 Saarbrücken, Germany
*   Correspondence: aranzazu.delcampo@leibniz-inm.de; Tel.: +49-(0)681-9300-510; Fax: +49-(0)681-9300-223
†   These authors contributed equally to this work.

Academic Editors: Marco d'Ischia and Daniel Ruiz-Molina
Received: 25 August 2017; Accepted: 2 November 2017; Published: 13 November 2017

Abstract: In situ forming hydrogels with catechol groups as tissue reactive functionalities are interesting bioinspired materials for tissue adhesion. Poly(ethylene glycol) (PEG)–catechol tissue glues have been intensively investigated for this purpose. Different cross-linking mechanisms (oxidative or metal complexation) and cross-linking conditions (pH, oxidant concentration, etc.) have been studied in order to optimize the curing kinetics and final cross-linking degree of the system. However, reported systems still show limited mechanical stability, as expected from a PEG network, and this fact limits their potential application to load bearing tissues. Here, we describe mechanically reinforced PEG–catechol adhesives showing excellent and tunable cohesive properties and adhesive performance to tissue in the presence of blood. We used collagen/PEG mixtures, eventually filled with hydroxyapatite nanoparticles. The composite hydrogels show far better mechanical performance than the individual components. It is noteworthy that the adhesion strength measured on skin covered with blood was >40 kPa, largely surpassing (>6 fold) the performance of cyanoacrylate, fibrin, and PEG–catechol systems. Moreover, the mechanical and interfacial properties could be easily tuned by slight changes in the composition of the glue to adapt them to the particular properties of the tissue. The reported adhesive compositions can tune and improve cohesive and adhesive properties of PEG–catechol-based tissue glues for load-bearing surgery applications.

Keywords: tissue glues; reinforced hydrogels; bioinspired adhesives; catechol-functionalized polymers; nanocomposite

1. Introduction

Tissue adhesives are promising biomaterials for wound closure, fixation of implants, and medical devices. Compared with sutures, tissue glues are less traumatic, and easier and faster to use, especially on soft tissues such as lung, liver, or kidney. Commercially available tissue glues can be classified into two categories: natural-based tissue adhesives including fibrin (Tisseel® and Artiss®, Baxter; Evarrest®, Ethicon), albumin (BioGlue®, CryoLife), gelatin (FloSeal®, Baxter) or chitosan (Celox®, MedTrade); and synthetic tissue adhesives based on cyanoacrylate (Dermabond®, Ethicon; Leukosan®, BSN Medical; Histoacryl®, B. Braun), poly(ethylene glycol) (PEG) (CoSeal®, Baxter), or isocyanate (TissuGlu®, B. Braun), among others [1]. Several properties are desirable for a good tissue adhesive: strong adhesion in the presence of body fluids, biocompatibility, biodegradability, fast polymerization, adaptability to the mechanical properties of the tissue, and low cost. Cyanoacrylates and fibrin glues are among the most extensively used tissue adhesives. Despite this, none of them meets all requirements listed above [2]. Cyanoacrylate glues show high adhesion strength, rapid curing, and low cost. However, their degradation products are cytotoxic, making them unsuitable for internal use. Fibrin glues are biocompatible and biodegradable, but they exhibit poor mechanical strength. Considerable effort has been expended to develop in situ cross-linking hydrogels compatible with physiological conditions as alternative tissue gluing systems [3–11]. The main focus has been reinforcing interfacial properties by exploiting chemistries to improve reactivity with tissue [3]. Relevant outcomes in terms of interfacial strength in the presence of blood have been reported. However, all systems come up short for load-bearing applications in the clinic because of poor mechanical strength under physiological conditions [12,13].

Over the last few years, reinforced hydrogels with improved mechanical performance have been developed. Reinforcement can be achieved by mixing with micelles or micro- or nanoparticles that strongly interact with the polymer chains [14,15], with fibrillar nanostructures [16–18], or with other polymers [19,20]. Examples of PEG-based nanocomposite hydrogels include PEG diacrylate hydrogel containing hydroxyapatite [21] or Laponite [22] nanoparticles, or dopamine-functionalized 4-arm PEG hydrogel containing Laponite [23] or Fe_3O_4 [15] nanoparticles. Alternatively, double or interpenetrating networks combining two types of polymers have demonstrated impressive mechanical performance [14,24]. Examples of double networks include alginate/collagen-I mixtures for skin wound healing [25] or PEG diacrylate–collagen interpenetrated networks as cell scaffolds [26] and for vascular tissue engineering [27].

Catecholamine-functionalized polymers have attracted considerable attention as tissue adhesives in the last few years [3,10]. The idea of the catechol moiety as adhesive unit is inspired by the high content of dihydroxyphenylalanine (DOPA) amino acid of the adhesive mussel foot proteins (mfps) used by the mussel to attach to rocks in the sea. In tissue applications, oxidized catecholamines can undergo self-reaction (polymerization) or can bind to nucleophiles (such as $-NH_2$ or $-SH$ in the protein components of the tissue) under physiological conditions. This leads to rapid and effective cross-linking and tissue-binding of the catechol-derived material. A broad range of catecholamine-functionalized hydrogels have been reported, such as alginates [28,29], hyaluronan [30–32], heparin [33], gelatin [11], polyvinylpyrrolidone [34], PEG [8,35–40], and chitosan [10]. Reported applications of these materials include wound closure [35], hemostasis [10], adhesion in blood vessels [28], cell therapy [31,32], cell culture [33], biological glue [11], islet transplantation in the liver [8], or fetal membrane repair [37]. The catechol-based tissue glues show good interfacial strength in in vivo applications. However, as hydrogel-based materials, they present modest mechanical performance in wet environments, and their use in load-bearing applications is severely limited.

Cohesive and adhesive properties need always to be considered in the design of any adhesive, and optimized to the particular chemistry, mechanical properties, and mechanical solicitation of the glued parts. In this work, we present different formulations of mechanically robust PEG-based hydrogels and explore their potential as alternative matrices for tissue gluing with considerably better mechanical strength than reported systems. We describe adhesive networks based on collagen and PEG–catechol hydrogel, and hydrogel composites containing hydroxyapatite particles to improve the mechanical stability of the hydrogel matrix. The adhesion performance to pork skin was evaluated under different conditions: dry, wet, and in the presence of blood, and compared with reported and commercial systems. A biocompatible hydrogel composition was identified, with cohesive properties far better than PEG–catechol, both on wet skin and in the presence of blood. This system attained adhesion strength of ca. 40 kPa on skin tissue covered by blood, 6-fold higher than cyanoacrylate and 12-fold higher than fibrin.

2. Materials and Methods

2.1. Reagents and Materials

Tissues from pork (fresh skin) were obtained from a local butcher. The cyanoacrylate-based tissue adhesive Leukosan® was obtained from BSN Medical (Hamburg, Germany). The fibrin-based tissue adhesive ARTISS® was purchased from Baxter (Unterschleißheim, Germany). Collagen (type I, from rat tail, here referred to as Coll) and hydroxyapatite (nanopowder, <200 nm, referred to as HAp) were purchased from Sigma-Aldrich (Taufkirchen, Germany). 4-arm PEG succinimidyl carboxymethyl ester (PEG-NHS, M_w of 5, 10, and 40 kDa) was purchased from Jenkem Technology (Plano, TX, USA). All other reactants were purchased from Sigma-Aldrich and used as received. Poly(ethylene glycol) O,O',O'',O'''-tetra(acetic acid dopamine) amide (PEG-Dop, Mw of 5, 10, and 40 kDa) were prepared following reported procedures from our group [40,41]. Details on the synthetic procedure and characterization (proton nuclear magnetic resonance (^{1}H NMR) spectra, Supplementary Figure S1) can be found in the Supplementary Materials.

2.2. Preparation of Adhesive Formulations

For the preparation of pure PEG-Dop hydrogel, PEG-Dop was dissolved in 0.01 M phosphate-buffered saline (PBS, pH 7.4) at 10%, 20%, 30%, and 40% (w/v). An equivolume of NaIO$_4$ solutions (concentration range 30–240 mM) in deionized (DI) water was added to induce gelation. The Dop:NaIO$_4$ molar ratio was adjusted according to the specific requirements of experiments. For the adhesive test, 20 µL of the mixture of polymer and oxidant were spread on the skin. This amount was sufficient to cover 113 mm^2 area of the tissue probe without significant overflow.

PEG-Dop/HAp: Nanocomposite hydrogels of PEG-Dop/HAp were prepared by adding the HAp nanoparticles suspension (prepared previously) to the PEG-Dop solution at 1%, 2%, 5%, and 10% (w/v). The mixture was vortexed and sonicated for 5 min to ensure homogeneity. The concentration of PEG-Dop was kept constant at 30% (w/v). Mixtures with NaIO$_4$ solution were prepared and used as described above.

PEG-Dop/Coll and PEG-Dop/HAp/Coll: for the preparation of these two precursors, 1 M NaOH solution, 10× PBS, and DI water should be precooled and the whole mixing process should be performed on an ice bath. A collagen solution in 0.1 M acetic acid was neutralized with 1 M NaOH and 10× PBS to collagen concentrations of 0.1%, 0.2%, and 0.4% (w/v). PEG-Dop was added to the collagen solution and vortexed for 10 min to obtain a precursor of PEG-Dop/Coll (30% (w/v) PEG-Dop). For the PEG-Dop/HAp/Coll mixture, the HAp particles were first suspended in DI water by vortexing and sonicating for 10 min. The HAp suspension and PEG-Dop solution were added simultaneously to a previously neutralized collagen solution, and the mixture was vortexed and sonicated for 5 min for homogeneity. The pH of the resulting mixture was 7.0–7.4 and the concentration of the precursor was 0.2% (w/v) for collagen, 5% (w/v) for HAp, 30% (w/v) for PEG-Dop, and 1× for PBS. All collagen-containing systems were subsequently incubated at 37 °C for 3 h to allow collagen fibrillogenesis. Finally, the PEG-Dop/Coll and PEG-Dop/HAp/Coll network were mixed with an equivolume solution of 120 mM NaIO$_4$ in DI water and applied to the tissue samples for adhesion tests as described above.

Collagen (0.2% (w/v)): Collagen (Type I, from rat tail) was diluted in 0.1 M acetic acid and neutralized with 1 M NaOH and 10× to a collagen concentration of 0.2% (w/v) in 1× PBS at the pH of 7.0–7.4 (Note: precooled NaOH solution, 10× PBS, and DI water should be used, and the mixing should be performed in an ice bath). The collagen solution was kept at 37 °C to allow fibrillogenesis. After 3 h, 20 µL of the collagen solution was sandwiched between two skin disks and cured for 30 min at 37 °C.

Commercial tissue adhesives: 20 µL of cyanoacrylate or fibrin adhesive was applied to uniformly coat the skin surface without significant overflow, and cured between two disks of skin.

2.3. Adhesion Measurements

Fresh porcine skin was obtained from a local butcher and stored at −20 °C for a few months. When needed for the adhesion measurements, a portion of the tissue was defrosted and kept at room temperature in aluminum paper to prevent from drying. The defrosted tissue was used within a day.

For the adhesion measurements, porcine skin was cut into disks of 13 or 12 mm diameter with a hollow punch and the fat was removed with a scalpel. For adhesion measurements on wet skin, the skin disks were immersed in PBS for 2 min before adding the tissue glue on the skin covered by a thin layer of PBS. For adhesion measurements in the presence of blood, the skin sample was immersed in blood before the test. Blood was collected under informed consent from a healthy human donor, following the regulations and protocol approved by the ethic commission of the Ärztekammer des Saarlandes (EK-0001) and used within 10 min, or stored with heparin anticoagulants and kept at 4 °C for no longer than two days. No significant differences were found when the tests were performed with fresh blood or with heparin anticoagulants-treated blood.

Adhesion measurements were performed with a Zwick Roell 1446 Universal Testing Machine (Zwick Roell, Ulm, Germany) equipped with a 200 N load cell (Supplementary Figure S2). The protocol of adhesion test was adapted from the American Society for Testing and Materials (ASTM) standard F2258-05, to better suit our measurements. The main differences between our procedure and the ASTM standard were the shape of the skin samples (circular instead of squared, as we found circles were easier to cut precisely and reproducibly with a hollow punch), the pulling speed (1 mm/min instead of 2 mm/min, as the lower speed proved to be more reproducible in our hands), and the method for keeping the samples moist during the curing (no wet gauze was used, but was replaced by the specific conditions detailed below, in order to mimic application in clinics).

For measurements performed in dry conditions, cut skin disks of diameter 12 or 13 mm (contact area 113.04 or 132.7 mm^2, respectively) were immobilized with superglue on a "T" holder (Supplementary Figure S2). The adhesive mixture was deposited between the skin disks, a load of 0.2 N was applied and the adhesive was left for 30 min in the machine to allow complete gelation. The glue was always covering the whole area of the skin sample. For adhesion measurements on wet samples cured underwater, the samples were prepared as described before, but (i) the skin samples were immersed in PBS for 2 h and the flowing PBS was removed from the skin disks by tissue paper before applying the glue; and (ii) the skin–glue–skin sandwich was immersed in PBS for 30 min under a weight of 20 g for gelation. Then, the cured sample was glued to the machine for measurement. For adhesion measurements in presence of blood, blood was used instead of PBS and the curing was performed at 37 °C under a weight of 20 g, immersed in blood (approx. 5 mL was employed). The skin disks were pulled off at a rate of 1 mm/min.

The adhesion strength and work of adhesion were calculated from the maximal of the stress vs. strain curve. Each measurement was performed eight to ten times. In the box plots of the adhesion strength values (kPa) shown below (Figures 1, 3 and 4), the average stress (square and obtained value), the median stress (line), and the type of failure (I, interfacial failure; C, cohesive failure; C + I, mixture of interfacial and cohesive failure; T + C, mixture of tissue and cohesive failure) are specified for each glue as well.

2.4. Statistical Analysis

Statistical analysis was performed by one-way analysis of variance (ANOVA) used to determine significant difference in the result for each group of measurement followed by a post-hoc Tukey's test (GraphPad software, La Jolla, CA, USA). A value of $p < 0.05$ was used for statistical analysis, and significance difference was set to $p < 0.05$, $p < 0.01$, and $p < 0.001$.

2.5. Tensile Measurements

Tensile measurements were performed with a Zwick Roell 1446 Universal Testing Machine (Zwick Roell, Ulm, Germany) equipped with a 200 N load cell. The precursor mixtures (PEG-Dop 30%, PEG-Dop 30%/Coll 0.2% or PEG-Dop 30%/Coll 0.2%/HAp 5% (w/v)) and an equivalent volume of oxidant solution (120 mM NaIO$_4$ in DI water) were injected in a rectangular mold (0.8 mm × 25 mm × 75 mm) and cured for 2 h. The mold was covered by a glass slide and kept in a closed humid box to prevent from drying. For the tensile measurements, rectangular hydrogel samples were prepared, and the exact dimensions and effective length of the samples were measured before each experiment with a digital vernier scale. Tensile measurements were performed at a strain rate of 20 mm/min until failure. The Young's modulus, the tensile strength, and strain were respectively calculated from the initial slope and the maximal stress upon failure of the stress vs. strain curve. Each measurement was performed in triplicate.

2.6. Scanning Electron Microscopy Imaging

The microstructures of hydrogel samples were characterized with a scanning electron microscopy (SEM) FEI Quanta 400 (FEI, Hillsboro, OR, USA). The samples PEG-Dop 15%, PEG-Dop 15%/HAp 2.5%, PEG-Dop 15%/Coll 0.1%, and PEG-Dop 15%/HAp 2.5%/Coll 0.1% (w/v) were cured between two disks of skin for adhesion measurements. After being cured for 30 min, the cured hydrogels were immersed for 10 min in DI water and then serially transitioned from water to ethanol for 20 min immersions in 20%, 40%, 60%, 80% solutions of ethanol in water, ending with overnight immersion in pure ethanol. Ethanol dehydrated hydrogels were dried in a K850 critical point dryer (LOT-QuantumDesign GmbH, Darmstadt, Germany). Prior to imaging, dried hydrogels were mounted on SEM sample stages using carbon tape and sputter-coated with a gold layer. The samples were imaged at 5 kV.

2.7. In Vitro Cytocompatibility Studies

Preparation of hydrogel samples: 13 mm diameter glass coverslips were cleaned with ethanol followed by 10 min oxidative plasma treatment. Cleaned substrates were immersed in a 10 mg/mL solution of dopamine buffered with Tris-HCl 10 mM to pH 8.5 for 30 min, rinsed with ultrapure water and dried under an N$_2$ stream. With this treatment, a thin poly(dopamine) layer is deposited onto the glass. A quantity of 20 μL of a PEG-Dop 30%, Coll 0.2%, Hap 5% (w/v), and 120 mM NaIO$_4$ mixture was deposited on a Sigmacote® (Sigma-Aldrich, Taufkirchen, Germany) treated glass slide (76 mm × 26 mm) [42] and covered with the polydopamine-coated glass slide. The Sigmacote® layer avoids reaction and attachment of the glue with the glass substrate. The hydrogel was cured for 1 h in a closed humid box. The formed hydrogel was then gently peeled-off from both glass surfaces, washed with PBS three times, and kept immersed in PBS until cell seeding. Live/dead and MST-1 assays were performed in triplicate.

Live/dead assay: The cytocompatibility of the substrates was performed with Live/Dead Cell Double Staining (Sigma-Aldrich) based on fluorescein diacetate (FDA) and propidium iodide (PI) by following the manufacturer's protocol. Briefly, L929 cells (50,000 cells/well) were seeded in 24-well plates (Greiner Bio-One, Kremsmünster, Austria) overnight. The hydrogel sample was sterilized by immersion in 75% EtOH for 3 min, then introduced to the well; the cells were then incubated in the presence of the hydrogel for 24 h. After 24 h culture, the cell medium was removed and the cells were washed once with PBS. Subsequently, a working solution (30 μg/mL FDA and 40 μg/mL PI) was then added directly to the cells for 5 min, the staining solution was removed, and the cells were washed with PBS once. Subsequent imaging was performed using epifluorescence microscopy. Live cells were stained green (excitation/emission ≈490/515 nm) while dead cells appeared in red color (excitation/emission ≈535/617 nm). As control experiments, untreated cells (cells incubated in the absence of the hydrogel) were used as PI negative control (viable cells), and 0.1% Triton X-100 (TX) treated cells for 5 min were used as PI positive control (dead cells).

Cell viability assay: Cell viability tests were performed using WST-1 staining (Sigma-Aldrich) following the manufacturer's protocol. Briefly, L929 cells (50,000 cells/well) were seeded in 24-well plates (Greiner Bio-One) overnight. The hydrogel sample was then introduced to the well and the cells were incubated in the presence of the hydrogel for 24 h. Untreated cells (cells incubated in absence of hydrogel) were used as control. After 24 h, the WST-1 reagent was diluted in cell culture medium in a 1:10 ratio and added to the cells. Metabolically active cells cleaved the stable tetrazolium salt WST-1 to a soluble formazan. After 4 h incubation, the 24-well plates were measured at wavelengths 450 and 690 nm with a plate reader (Infinite M200; Techan, Durham, NC, USA). The mean absorbance of untreated cells was defined as 100%, and the absorbance of the cells incubated with hydrogel was calculated relative to this value.

3. Results and Discussion

PEG-Dop with high substitution degree (>90%) and different molecular weights (5, 10, and 40 kDa) were synthesized in a 1 g scale and used for adhesion testing in different mixtures. Pull-off experiments on glued pork skin samples were established to characterize adhesion performance using a tensile testing machine [8,28]. The adhesion strength was calculated from the maximum force detected in the pull-off curves, at which the glue started detaching from the skin (interfacial failure) or breaking in two parts (cohesive failure, Supplementary Figure S3).

In preliminary experiments, the adhesion strength of PEG-Dop/NaIO$_4$ mixtures at conditions reported by other groups [43] was tested for reference. PEG-Dop (15% (w/v)) with 120 mM NaIO$_4$ showed an adhesive strength of 63.6 kPa. Interestingly, PEG-Dop glue predominantly failed cohesively, i.e., a crack propagated through the gluing material, while the glue–skin interface remained stable. These results suggest that the strength of the interface PEG-Dop–skin is stronger than the PEG-Dop hydrogel itself. Consequently, adhesion performance of the PEG-Dop–skin system might be improved by mechanically reinforcing the polymeric network.

The stability of PEG-Dop underwater was checked by immersing the glued skin in PBS for 30 min before performing the adhesive test. PEG-Dop showed average values of adhesion strength of 25.5 kPa, significantly lower than in dry conditions (63.6 kPa) but still remarkable when compared with commercial systems like cyanoacrylate or fibrin (Figure 1A). Mainly, the failure type was cohesive. Poly(ethylene glycol) is a highly hydrophilic polymer that can easily uptake water depending on its cross-linking degree. Water uptake softens the gel and leads to lower cohesive properties and adhesive strength.

Our results suggest that cohesively-failed PEG-Dop used in reported works [28,40,41] might offer improved adhesive performance if the mechanical properties of the cured gel were to be significantly improved without compromising the interfacial strength provided by the Dop modification. Encouraged by this finding, we explored this possibility by tuning the formulation of PEG-Dop gels following four different strategies: (i) changing the molecular weight of 4-arm PEG and polymer concentration in the gluing solution, (ii) adding HAp nanoparticles to PEG-Dop glues, (iii) mixing with Coll, and (iv) tuning the ratio between Dop and oxidant.

Adhesive tests on PEG-Dop gels with different molecular weights (Mw of 5, 10, and 40 kDa) were performed first. An increase of Mw from 5 to 10 kDa resulted in enhanced adhesion strength from 46.5 to 63.6 kPa (Figure 1B). The failure was cohesive in both cases, in spite of the net loss of the density of Dop groups available for bonding with tissue in the 10 kDa PEG. A higher Mw (40 kDa) led to a decrease in adhesion strength (47.7 kPa) and a mixture of cohesive and interfacial failure. This can be associated with the lower density of Dop groups in the 40 kDa PEG-Dop and consequent reduced number of the skin–glue covalent interactions and weakening of the interface. Therefore, for the following experiments we chose PEG-Dop (Mw of 10 kDa).

A further improvement of the adhesion strength of PEG-Dop mixtures was achieved by optimizing the final concentration of 10 kDa PEG-Dop. Tests with 5%, 10%, 15%, 20% (w/v) were performed (Figure 1C). The highest adhesion strength was found for 10% (w/v) PEG-Dop, with adhesion strength of 64.5 kPa and a cohesive failure; whereas 5% (w/v) PEG-Dop showed lower adhesion strength (43.0 kPa) and a mixture of interfacial and cohesive failure. PEG-Dop content higher than 15% did not lead to significant improvement of the adhesive properties. A PEG-Dop concentration of 15% (w/v) was selected for the following experiments, since consumption of catechol groups is expected to occur when mixing with HAp or Coll in later experiments and a minimum of 10% intact catechol groups was assumed for maximizing adhesive performance.

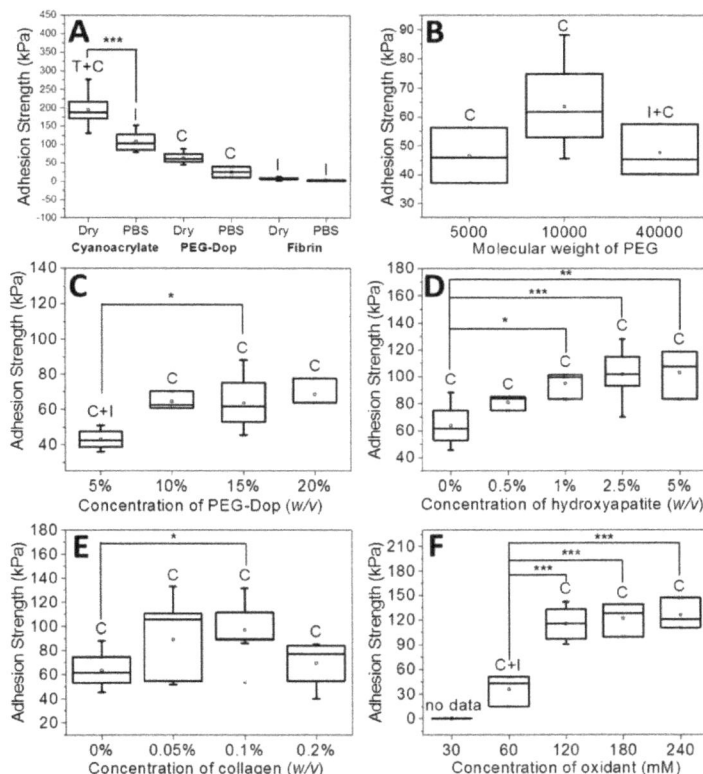

Figure 1. Adhesion strength of 10 kPa poly(ethylene glycol) O,O',O'',O'''-tetra(acetic acid dopamine) amide (PEG-Dop)/NaIO$_4$-based glues as function of the concentration of individual components in the gluing mixture. (**A**) Dry and wet test conditions (as compared with commercial tissue glues); (**B**) Molecular weight of poly(ethylene glycol) (PEG) (15% (w/v) PEG-Dop); (**C**) PEG-Dop concentration (120 mM NaIO$_4$); (**D**) Hydroxyapatite (Hap) nanoparticle concentration (15% (w/v) PEG-Dop, 120 mM NaIO$_4$); (**E**) Collagen (Coll) concentration (15% (w/v) PEG-Dop, 120 mM NaIO$_4$); (**F**) Oxidant concentration (15% (w/v) PEG-Dop). Failure type: I, interfacial; C, cohesive; C + I, mixture of interfacial and cohesive. * $p < 0.05$, ** $p < 0.01$, *** $p < 0.001$.

Addition of nanoparticles able to strongly interact with the polymer chains of the hydrogel is an effective way to reinforce mechanical properties and can lead to strong nanocomposite hydrogels [21,44–49]. When the interaction between the nanoparticle surface and the polymer chain is strong, the nanoparticles act as multivalent crosslinkers within the polymer network and enhance its mechanical strength. We applied this strategy to PEG-Dop hydrogels by mixing biocompatible HAp nanoparticles [50]. According to literature reports, the Dop end-groups in the PEG chain are expected to coordinate with the ionic surface of the HAp nanoparticles and reinforce the network [51–53]. In fact, tensile tests of PEG-Dop (15%)/HAp (2.5%) mixtures showed a >6-fold increase in the Young's modulus in comparison with the PEG-Dop hydrogel without added nanoparticles (Figure 2 and Table 1). Adhesion testing of PEG-Dop (15% (*w/v*)) formulations with increasing ratios of HAp particles (0, 0.5, 1, 2.5, and 5% (*w/v*)) gave increasing adhesion strengths (Figures 1D and 3). The type of adhesive failure remained cohesive for all samples. In conclusion, the addition of HAp nanoparticles led to a higher mechanical resistance of the hydrogel and to an increase in the final adhesive performance at concentrations up to 2.5%. It is important to note that the positive effects of the addition of HAp in terms of mechanical reinforcement are presumably counteracted by lower cross-linking degrees of the PEG matrix, since Dop units become coordinated to the HAp surface and are no longer available for self-reaction. This might be the reason why HAp concentrations higher than 2.5% did not further improve the adhesive performance.

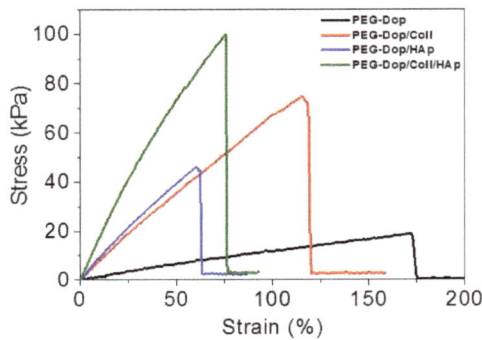

Figure 2. Representative tensile curves of PEG-Dop, PEG-Dop/Coll, composite hydrogel PEG-Dop/HAp, and composite mixture PEG-Dop/Coll/HAp. Concentration of the different components is as follows: 15% (*w/v*) PEG-Dop, 2.5% (*w/v*) HAp, 0.1% (*w/v*) Coll.

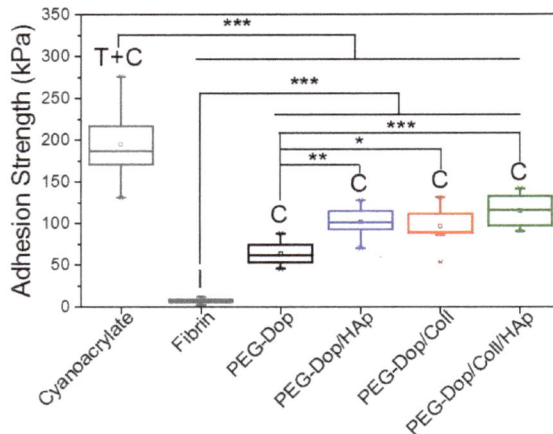

Figure 3. Adhesion strength values of PEG-Dop adhesives (15% (*w/v*) PEG-Dop, 2.5% (*w/v*) HAp, 0.1% (*w/v*) Coll), compared with commercial cyanoacrylate and fibrin glues on dry skin. Failure type: I, interfacial; C, cohesive; T + C, mixture of tissue and cohesive. * $p < 0.05$, ** $p < 0.01$, *** $p < 0.001$.

Table 1. Tensile tests of PEG-Dop, PEG-Dop/Coll, composite hydrogel PEG-Dop/HAp and composite mixture PEG-Dop/Coll/HAp.

	Young's Modulus (kPa)	Tensile Strength (kPa)	Strain (%)
PEG-Dop	15.2 ± 4.7	18 ± 1.3	182.3 ± 42.7
PEG-Dop/Coll	86.9 ±18.8	66.7 ± 6.5	100.9 ± 21.4
PEG-Dop/HAp	96.6 ± 4.9	45.2 ± 1.2	64.7 ± 3.0
PEG-Dop/Coll/HAp	181.5 ± 39.3	88.5 ± 9.5	61.3 ± 20.6

Concentration of the different components is as follows: 15% (w/v) PEG-Dop, 2.5% (w/v) HAp, 0.1% (w/v) Coll. Mean value ± standard deviations are given, n = 3.

We then tested the ability of PEG-Dop to form homogeneous mixtures with collagen type I at increasing concentrations (0.05–0.2% (w/v)). Collagen type I forms physical networks at pH 7–7.4 and 37 °C [27]. In the mixture PEG/Coll, the catechol units might also react with the nucleophiles (like amine groups) of the collagen and both networks are expected to be covalently interconnected. Homogeneous networks were obtained under conditions specified in the Materials and Methods section. The polymer mixture significantly enhanced the tensile strength of the PEG-Dop hydrogel, as reflected in the tensile test in Figure 2. The adhesive performance was also significantly improved. The adhesion strength of the PEG-Dop/Coll increased from 63.6 kPa (0% Coll) to 89.3 and 97.1 kPa when adding 0.05% and 0.1% Coll content (Figures 1E and 3). Pure collagen (0.2% (w/v)) gave a very low adhesion strength of 2.9 ± 1.4 kPa and failed interfacially (data not shown). The PEG-Dop/Coll mixtures including Coll content higher than 0.2% showed lower adhesion strengths (69.6 kPa), which we associate with the difficult solubilization of Coll at ≥0.4% concentrations, and consequent poor mixing and inhomogeneity of the PEG-Dop/Coll network. All adhesives showed cohesive failure, indicating that the interfacial interaction of Dop units with the skin was retained in the mixed hydrogel with collagen, in spite of catechol groups being eventually consumed by reaction with collagen and not by reaction with the tissue.

In an effort to push cohesiveness further, 2.5% (w/v) HAp nanoparticles were added to the PEG-Dop 15%/Coll 0.1% hydrogel. The Young's modulus obtained for PEG-Dop, PEG-Dop/Coll and PEG-Dop/Coll/HAp was 15.2, 86.9, and 181.5 kPa, respectively (Table 1), confirming the significant mechanical reinforcement by the mixing of HAp nanoparticles in the hydrogel. The PEG-Dop/Coll/HAp hydrogel showed a higher tensile strain (ultimate stress upon failure) than the PEG-Dop/HAp mixture due to the high elasticity of Coll (Table 1). In other words, collagen does not only enhance the strength of hydrogel, but also improves the strain loss induced by addition of HAp nanoparticles. The composite showed the highest adhesion strength (115.5 kPa). Note that this value is 15.4-fold higher than in fibrin, and represents ca. 60% of the adhesion strength of the cyanoacrylate glue (Figure 3). In summary, the incorporation of Coll and HAp nanoparticles in the PEG-Dop hydrogel significantly improved the cohesion and skin adhesion properties of the system. Importantly, the cohesive properties can be varied by tuning the Coll and HAp concentrations, to possibly match the properties of the different tissues for application. In all cases, cohesive failure was observed, indicating that there is still room for further improvement of the adhesive performance with other hydrogel compositions for load-bearing application, i.e., by using double networks incorporating dissipative elements [54–56]. This possibility will be studied in the future.

According to a previous report [43], the mechanical properties and curing speed of PEG-Dop are highly affected by both the alkalinity of the oxidant solution and the ratio of Dop vs. oxidant applied; this can impact the adhesive performance. In a final attempt to improve the performance of the hydrogels, we tested the adhesion strength of PEG-Dop glues at a Dop:oxidant ratio of 1:1; when the oxidant was added either in DI water or in 0.4 M NaOH solution. Adhesion strength in DI water was significantly higher (63.6 kPa, final pH of the mixture was 6) than in basic conditions (4.3 kPa, final pH of the mixture was 12). Failure type also changed from cohesive (in water) to interfacial (in NaOH) (Supplementary Figure S4). We attribute this difference to the fast curing speed observed when using the oxidant at a higher pH. The cross-linking of the system occurred immediately upon mixing, before the gel was placed on the skin and, therefore, a very small number of catechol groups were available for gluing to skin and reinforcing the interface.

The effect of the Dop:oxidant ratio was tested with the PEG-Dop/HAp/Coll system. Oxidant concentrations of 30, 60, 120, 180, 240 mM were used, which correspond to a Dop:oxidant ratio of 4:1, 2:1, 1:1, 1:1.5, and 1:2, respectively. At 30 mM oxidant concentration, no stable hydrogel was formed. At increasing oxidant concentrations of 60–240 mM, an increase in the adhesive strength up to 126.0 kPa and a change in the failure type from interfacial to cohesive were observed (Figure 1F). The oxidant concentration affects the ratio of catechol and quinone groups in the curing gel available for reaction with the different components, PEG, Coll, or HAp. The polymerization speed also increased with the oxidant concentration. Gelation time is a relevant parameter to medical applications, and has to be adjusted to the specific surgery conditions [40,43]. For the following experiments we used 120 mM oxidant.

In a real application, tissue adhesives need to perform in the presence of body fluids. Therefore, the adhesive strength of prospective tissue glues should be tested in the presence of PBS or blood. We characterized the adhesion performance of PEG-Dop/Coll/HAp on wet skin (after immersion in PBS for 2 min) and skin covered with blood (containing heparin anticoagulants) and compared with the performance on dry skin (Figure 4). PEG-Dop/Coll/HAp lost only 18% of its adhesion strength on wet skin (95 kPa). Importantly, this value was close to the adhesion strength of commercial cyanoacrylate glue to wet skin (108.1 kPa). It is worth mentioning that failure remained cohesive on our composite hydrogel on wet skin, while failure in cyanoacrylate changed to be interfacial. We attribute the loss of adhesive performance of cyanoacrylate in the presence of water to its hydrophobic nature, which prevents tissue contact and possibly penetration when water is in between. Notably, PEG-Dop/Coll/HAp significantly outperformed cyanoacrylate in adhesion tests in the presence of blood. Our composite hydrogel showed adhesion strength of 38.9 kPa, which is ≈5.6-fold higher than adhesion strength of cyanoacrylate (6.9 kPa) under comparable conditions. Failure remained cohesive in the composite gel, indicating that the strong interfacial interaction between the dopamine-modified system and the bloody tissue was retained, and possibly further reinforced by attractive interactions between the collagen and the blood proteins. This was also found in fibrin glue, which changed from interfacial to cohesive failure when measured on skin in the presence of PBS or blood. Fibrin–blood interactions during fibrinogenesis and clot formation seem to positively influence adhesion of fibrin glues to skin. However, the adhesion strength of fibrin was ≈13.4-fold lower than the adhesion strength of PEG-Dop/Coll/HAp under the same measuring conditions. In summary, PEG-Dop-based tissue glues can lead to customized systems with application-specific cohesive and adhesive properties, including mechanically demanding uses.

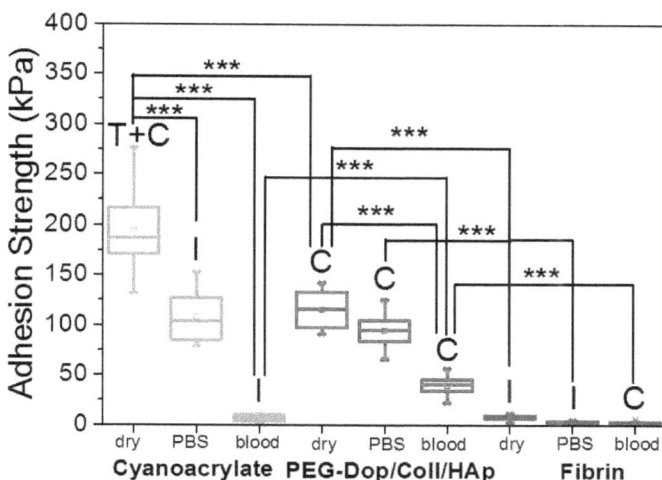

Figure 4. Adhesion strength values of PEG-Dop adhesives on "dry" skin, and "wet" skin (either with phosphate-buffered saline (PBS) or with blood) compared with commercial tissue glues. Failure type: I, interfacial; C, cohesive; T + C, mixture of tissue and cohesive. *** $p < 0.001$.

It should be noted that the reinforcement mechanism of collagen type I in our mixtures is not clear. An important feature of collagen type I is its possibility to form fibrils, and fibrillar protein structures can mechanically reinforce gels. In order to test if fibril formation occurred in our gels, the morphology of bulk PEG-Dop, PEG-Dop/HAp, PEG-Dop/Coll, and PEG-Dop/Coll/HAp hydrogels was characterized by SEM. Samples cured for 30 min were subjected to critical point drying and imaged by SEM. Porous structures were observed in all hydrogels (Figure 5). Changes in pore sizes were observed with added HAp, Coll, or both to the polymer mixture. Addition of HAp resulted in a more compact structure (smaller pores). In PEG-Dop/HAp hydrogel (Figure 5B), HAp appeared to be not homogeneously distributed, with the pores around HAp smaller than pores in other places. PEG-Dop/Coll hydrogels (Figure 5C) showed similar pore sizes to PEG-Dop, although size distribution was broader. Individual collagen fibers were not observed in the mixture, which typically have a striated morphology [57]. The final formulation showed an intermediate pore geometry (Figure 5D) which, according to the mechanical tests, was beneficial to the improvement of strength and maximum strain of the hydrogel. Future studies will change the gel preparation method and explore possible beneficial properties of fibril formation for mechanical reinforcement of the gels.

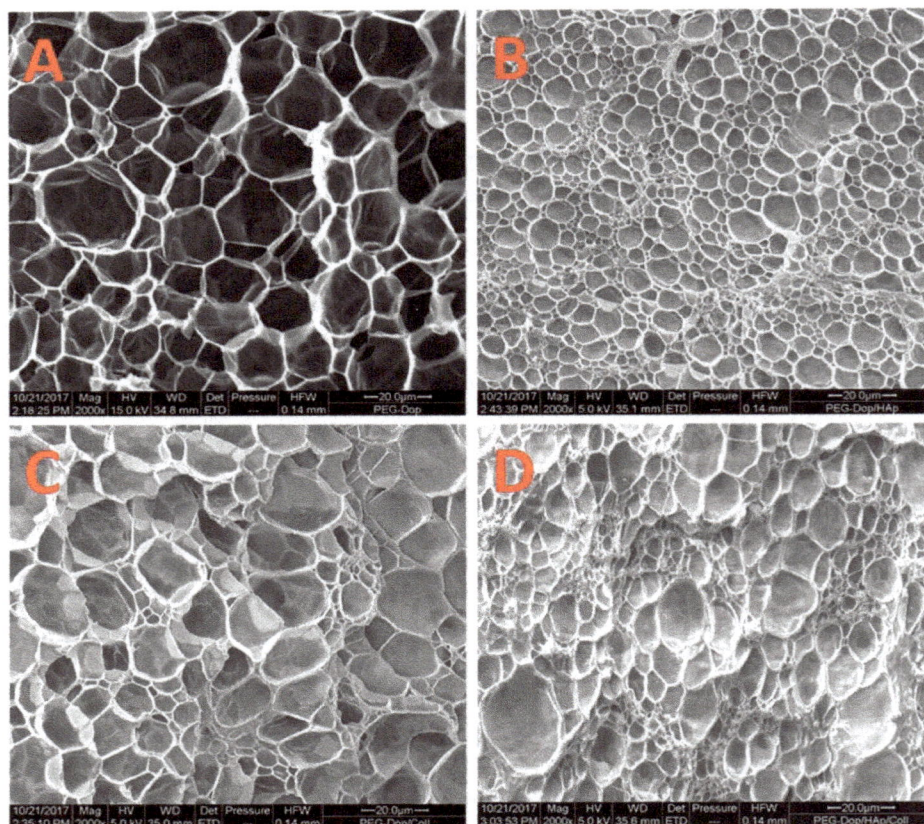

Figure 5. Scanning electron microscopy (SEM) images of the bulk structure of (**A**) PEG-Dop; (**B**) PEG-Dop/HAp; (**C**) PEG-Dop/Coll; (**D**) PEG-Dop/HAp/Coll. Concentration of the different components was: 15% (w/v) PEG-Dop, 2.5% (w/v) HAp, 0.1% (w/v) Coll. In all cases, scale bar: 20 µm.

Finally, we tested the cytocompatibility of the PEG-Dop/Coll/HAp hydrogel. Fibroblasts were cultured for 24 h either in the presence or absence of a thin hydrogel, and live/dead and WST-1 assays were performed to determine cell viability (Supplementary Figure S5). Cells incubated in presence of hydrogel remained viable (96%, in relation to a control of cells incubated in absence of hydrogel). These results are in agreement with those reported on the literature for PEG-Dop, which has proved biocompatible in vitro and in vivo [8,37].

4. Conclusions

Research in tissue glues has mainly been focused on improving the reactivity between artificial polymers and tissue with effective and biocompatible chemistries using in situ forming hydrogels. However, adhesive failure in hydrogels is often of a cohesive nature as a consequence of their swelling and low mechanical stability. Consequently, efforts in the design of more mechanically robust hydrogel backbones can significantly increase adhesive performance of hydrogel-based tissue glues. For this purpose, reinforced hydrogels offer interesting possibilities. In our work, mechanical reinforcement of PEG-Dop with collagen and by adding HAp nanoparticles lead to a significant increase in adhesive performance. It is important to note that these systems surpass by far the adhesion strength of commercially available products and reported adhesive systems in the presence of blood. The biocompatibility and the optional tunable degradability of PEG-Dop gels including peptide sequences in the backbone make this material's design promising for the real application of bioinspired tissue glues in medicine. Finally, the cohesive properties of the hydrogel might be further reinforced by introducing dissipative components in the backbone, as recently demonstrated for tough hydrogels, by integrating double network architectures, or by allowing fibrillar collagen to form an internal physical network. These possibilities will be the subject of further studies by our group.

Supplementary Materials: The following are available online at www.mdpi.com/2313-7673/2/4/23/s1. Synthesis of PEG-Dop, Figure S1: Proton nuclear magnetic resonance (^{1}H-NMR) spectra of PEG-Dop polymers, Figure S2: Pictures of the set-up used for the adhesion strength measurements, Figure S3: Picture of pork skin samples bonded with PEG-Dop adhesive hydrogel and fractured during the adhesion test, Figure S4: Adhesion strength values of 15% (w/v) PEG-Dop mixtures with 120 mM oxidant in 0.4 M NaOH solution vs. in DI water, Figure S5: Cytotoxicity experiments.

Acknowledgments: The authors thank Bilal Al-Nawas and Alexander Hoffmann from the University Medical Center in Mainz, Germany for useful discussions on tissue adhesives in surgery. The authors also acknowledge Karl-Peter Schmitt (INM) and Andreas Hannewald (Max-Planck-Institut für Polymerforschung) for their help with the adhesion and tensile measurements, Daniel Flormann (Saarland University) for providing access to blood for the experiments, Markus Koch (INM) for help with SEM imaging, Marcelo Salierno (Max-Planck-Institut für Polymerforschung) and Aleeza Farrukh (INM) for the cell experiments. J.F. and S.Z. acknowledge the financial support from China Scholarship Council. A.d.C. acknowledges financial support from BiomaTICS at University Medical Center Mainz and from the EU within BioSmartTrainee, the Marie Sklodowska-Curie Innovative Training School, No. 642861.

Author Contributions: J.F. and X.-A.T. equally contributed to this work. The manuscript was written through contributions of all authors. All authors have given approval to the final version of the manuscript.

Conflicts of Interest: The authors declare no conflict of interest.

References

1. Bochyńska, A.; Hannink, G.; Grijpma, D.W.; Buma, P. Tissue adhesives for meniscus tear repair: An overview of current advances and prospects for future clinical solutions. *J. Mater. Sci. Mater. Med.* **2016**, *27*, 85. [CrossRef] [PubMed]

2. O'Rorke, R.D.; Pokholenko, O.; Gao, F.; Cheng, T.; Shah, A.; Mogal, V.; Steele, T.W.J. Addressing unmet clinical needs with UV bioadhesives. *Biomacromolecules* **2017**. [CrossRef] [PubMed]

3. Sedó, J.; Saiz-Poseu, J.; Busqué, F.; Ruiz-Molina, D. Catechol-based biomimetic functional materials. *Adv. Mater.* **2013**, *25*, 653–701. [CrossRef] [PubMed]

4. Westwood, G.; Horton, T.N.; Wilker, J.J. Simplified polymer mimics of cross-linking adhesive proteins. *Macromolecules* **2007**, *40*, 3960–3964. [CrossRef]

5. Lee, B.P.; Messersmith, P.B.; Israelachvili, J.N.; Waite, J.H. Mussel-inspired adhesives and coatings. *Annu. Rev. Mater. Res.* **2011**, *41*, 99. [CrossRef] [PubMed]

6. Lee, B.P.; Dalsin, J.L.; Messersmith, P.B. Synthesis and gelation of DOPA-modified poly(ethylene glycol) hydrogels. *Biomacromolecules* **2002**, *3*, 1038–1047. [CrossRef] [PubMed]

7. Lee, B.P.; Huang, K.; Nunalee, F.N.; Shull, K.R.; Messersmith, P.B. Synthesis of 3,4-dihydroxyphenylalanine (DOPA) containing monomers and their co-polymerization with PEG-diacrylate to form hydrogels. *J. Biomater. Sci. Polym. Ed.* **2004**, *15*, 449–464. [CrossRef] [PubMed]

8. Brubaker, C.E.; Kissler, H.; Wang, L.-J.; Kaufman, D.B.; Messersmith, P.B. Biological performance of mussel-inspired adhesive in extrahepatic islet transplantation. *Biomaterials* **2010**, *31*, 420–427. [CrossRef] [PubMed]

9. Shao, H.; Stewart, R.J. Biomimetic underwater adhesives with environmentally triggered setting mechanisms. *Adv. Mater.* **2010**, *22*, 729–733. [CrossRef] [PubMed]

10. Ryu, J.H.; Lee, Y.; Kong, W.H.; Kim, T.G.; Park, T.G.; Lee, H. Catechol-functionalized chitosan/pluronic hydrogels for tissue adhesives and hemostatic materials. *Biomacromolecules* **2011**, *12*, 2653–2659. [CrossRef] [PubMed]

11. Fan, C.; Fu, J.; Zhu, W.; Wang, D.-A. A mussel-inspired double-crosslinked tissue adhesive intended for internal medical use. *Acta Biomater.* **2016**, *33*, 51–63. [CrossRef] [PubMed]

12. Bré, L.P.; Zheng, Y.; Pêgo, A.P.; Wang, W. Taking tissue adhesives to the future: From traditional synthetic to new biomimetic approaches. *Biomater. Sci.* **2013**, *1*, 239–253. [CrossRef]
13. Duarte, A.; Coelho, J.; Bordado, J.; Cidade, M.; Gil, M. Surgical adhesives: Systematic review of the main types and development forecast. *Prog. Polym. Sci.* **2012**, *37*, 1031–1050. [CrossRef]
14. Peak, C.W.; Wilker, J.J.; Schmidt, G. A review on tough and sticky hydrogels. *Colloid Polym. Sci.* **2013**, *291*, 2031–2047. [CrossRef]
15. Li, Q.; Barrett, D.G.; Messersmith, P.B.; Holten-Andersen, N. Controlling hydrogel mechanics via bio-inspired polymer–nanoparticle bond dynamics. *ACS Nano* **2016**, *10*, 1317–1324. [CrossRef] [PubMed]
16. Liu, L.; Li, L.; Qing, Y.; Yan, N.; Wu, Y.; Li, X.; Tian, C. Mechanically strong and thermosensitive hydrogels reinforced with cellulose nanofibrils. *Polym. Chem.* **2016**, *7*, 7142–7151. [CrossRef]
17. Yang, J.; Zhang, X.; Ma, M.; Xu, F. Modulation of assembly and dynamics in colloidal hydrogels via ionic bridge from cellulose nanofibrils and poly(ethylene glycol). *ACS Macro Lett.* **2015**, *4*, 829–833. [CrossRef]
18. Visser, J.; Melchels, F.P.W.; Jeon, J.E.; Van Bussel, E.M.; Kimpton, L.S.; Byrne, H.M.; Dhert, W.J.A.; Dalton, P.D.; Hutmacher, D.W.; Malda, J. Reinforcement of hydrogels using three-dimensionally printed microfibres. *Nat. Commun.* **2015**, *6*, 6933. [CrossRef] [PubMed]
19. Baniasadi, M.; Minary-Jolandan, M. Alginate-collagen fibril composite hydrogel. *Materials* **2015**, *8*, 799–814. [CrossRef] [PubMed]
20. Chen, S.H.; Tsao, C.T.; Chang, C.H.; Lai, Y.T.; Wu, M.F.; Liu, Z.W.; Chuang, C.N.; Chou, H.C.; Wang, C.K.; Hsieh, K.H. Synthesis and characterization of reinforced poly(ethylene glycol)/chitosan hydrogel as wound dressing materials. *Macromol. Mater. Eng.* **2013**, *298*, 429–438. [CrossRef]
21. Gaharwar, A.K.; Dammu, S.A.; Canter, J.M.; Wu, C.-J.; Schmidt, G. Highly extensible, tough, and elastomeric nanocomposite hydrogels from poly(ethylene glycol) and hydroxyapatite nanoparticles. *Biomacromolecules* **2011**, *12*, 1641–1650. [CrossRef] [PubMed]
22. Gaharwar, A.K.; Rivera, C.P.; Wu, C.-J.; Schmidt, G. Transparent, elastomeric and tough hydrogels from poly(ethylene glycol) and silicate nanoparticles. *Acta Biomater.* **2011**, *7*, 4139–4148. [CrossRef] [PubMed]
23. Liu, Y.; Meng, H.; Konst, S.; Sarmiento, R.; Rajachar, R.; Lee, B.P. Injectable dopamine-modified poly(ethylene glycol) nanocomposite hydrogel with enhanced adhesive property and bioactivity. *ACS Appl. Mater. Interfaces* **2014**, *6*, 16982–16992. [CrossRef] [PubMed]
24. Myung, D.; Waters, D.; Wiseman, M.; Duhamel, P.E.; Noolandi, J.; Ta, C.N.; Frank, C.W. Progress in the development of interpenetrating polymer network hydrogels. *Polym. Adv. Technol.* **2008**, *19*, 647–657. [CrossRef] [PubMed]
25. Da Cunha, C.B.; Klumpers, D.D.; Li, W.A.; Koshy, S.T.; Weaver, J.C.; Chaudhuri, O.; Granja, P.L.; Mooney, D.J. Influence of the stiffness of three-dimensional alginate/collagen-I interpenetrating networks on fibroblast biology. *Biomaterials* **2014**, *35*, 8927–8936. [CrossRef] [PubMed]
26. Chan, B.K.; Wippich, C.C.; Wu, C.J.; Sivasankar, P.M.; Schmidt, G. Robust and semi-interpenetrating hydrogels from poly(ethylene glycol) and collagen for elastomeric tissue scaffolds. *Macromol. Biosci.* **2012**, *12*, 1490–1501. [CrossRef] [PubMed]
27. Munoz-Pinto, D.J.; Jimenez-Vergara, A.C.; Gharat, T.P.; Hahn, M.S. Characterization of sequential collagen-poly(ethylene glycol) diacrylate interpenetrating networks and initial assessment of their potential for vascular tissue engineering. *Biomaterials* **2015**, *40*, 32–42. [CrossRef] [PubMed]
28. Kastrup, C.J.; Nahrendorf, M.; Figueiredo, J.L.; Lee, H.; Kambhampati, S.; Lee, T.; Cho, S.-W.; Gorbatov, R.; Iwamoto, Y.; Dang, T.T. Painting blood vessels and atherosclerotic plaques with an adhesive drug depot. *Proc. Natl. Acad. Sci. USA* **2012**, *109*, 21444–21449. [CrossRef] [PubMed]
29. Lee, C.; Shin, J.; Lee, J.S.; Byun, E.; Ryu, J.H.; Um, S.H.; Kim, D.-I.; Lee, H.; Cho, S.-W. Bioinspired, calcium-free alginate hydrogels with tunable physical and mechanical properties and improved biocompatibility. *Biomacromolecules* **2013**, *14*, 2004–2013. [CrossRef] [PubMed]
30. Hong, S.; Yang, K.; Kang, B.; Lee, C.; Song, I.T.; Byun, E.; Park, K.I.; Cho, S.W.; Lee, H. Hyaluronic acid catechol: A biopolymer exhibiting a pH-dependent adhesive or cohesive property for human neural stem cell engineering. *Adv. Funct. Mater.* **2013**, *23*, 1774–1780. [CrossRef]
31. Shin, J.; Lee, J.S.; Lee, C.; Park, H.J.; Yang, K.; Jin, Y.; Ryu, J.H.; Hong, K.S.; Moon, S.H.; Chung, H.M. Tissue adhesive catechol-modified hyaluronic acid hydrogel for effective, minimally invasive cell therapy. *Adv. Funct. Mater.* **2015**, *25*, 3814–3824. [CrossRef]

32. Park, H.-J.; Jin, Y.; Shin, J.; Yang, K.; Lee, C.; Yang, H.S.; Cho, S.-W. Catechol-functionalized hyaluronic acid hydrogels enhance angiogenesis and osteogenesis of human adipose-derived stem cells in critical tissue defects. *Biomacromolecules* **2016**, *17*, 1939–1948. [CrossRef] [PubMed]

33. Lee, M.; Kim, Y.; Ryu, J.H.; Kim, K.; Han, Y.-M.; Lee, H. Long-term, feeder-free maintenance of human embryonic stem cells by mussel-inspired adhesive heparin and collagen type I. *Acta Biomater.* **2016**, *32*, 138–148. [CrossRef] [PubMed]

34. Li, A.; Mu, Y.; Jiang, W.; Wan, X. A mussel-inspired adhesive with stronger bonding strength under underwater conditions than under dry conditions. *Chem. Commun.* **2015**, *51*, 9117–9120. [CrossRef] [PubMed]

35. Mehdizadeh, M.; Weng, H.; Gyawali, D.; Tang, L.; Yang, J. Injectable citrate-based mussel-inspired tissue bioadhesives with high wet strength for sutureless wound closure. *Biomaterials* **2012**, *33*, 7972–7983. [CrossRef] [PubMed]

36. Ai, Y.; Wei, Y.; Nie, J.; Yang, D. Study on the synthesis and properties of mussel mimetic poly(ethylene glycol) bioadhesive. *J. Photochem. Photobiol. B* **2013**, *120*, 183–190. [CrossRef] [PubMed]

37. Kivelio, A.; DeKoninck, P.; Perrini, M.; Brubaker, C.; Messersmith, P.; Mazza, E.; Deprest, J.; Zimmermann, R.; Ehrbar, M.; Ochsenbein-Koelble, N. Mussel mimetic tissue adhesive for fetal membrane repair: Initial in vivo investigation in rabbits. *Eur. J. Obstet. Gynecol. Reprod. Biol.* **2013**, *171*, 240–245. [CrossRef] [PubMed]

38. Brubaker, C.E.; Messersmith, P.B. Enzymatically degradable mussel-inspired adhesive hydrogel. *Biomacromolecules* **2011**, *12*, 4326–4334. [CrossRef] [PubMed]

39. Li, Y.; Meng, H.; Liu, Y.; Narkar, A.; Lee, B.P. Gelatin microgel incorporated poly(ethylene glycol)-based bioadhesive with enhanced adhesive property and bioactivity. *ACS Appl. Mater. Interfaces* **2016**, *8*, 11980–11989. [CrossRef] [PubMed]

40. Paez, J.I.; Ustahüseyin, O.; Serrano, C.; Ton, X.-A.; Shafiq, Z.; Auernhammer, G.K.; d'Ischia, M.; del Campo, A. Gauging and tuning cross-linking kinetics of catechol-PEG adhesives via catecholamine functionalization. *Biomacromolecules* **2015**, *16*, 3811–3818. [CrossRef] [PubMed]

41. García-Fernández, L.; Cui, J.; Serrano, C.; Shafiq, Z.; Gropeanu, R.A.; Miguel, V.S.; Ramos, J.I.; Wang, M.; Auernhammer, G.K.; Ritz, S.; et al. Antibacterial strategies from the sea: Polymer-bound Cl-catechols for prevention of biofilm formation. *Adv. Mater.* **2013**, *25*, 529–533. [CrossRef] [PubMed]

42. Farrukh, A.; Paez, J.I.; Salierno, M.J.; Fan, W.; Berninger, B.; del Campo, A. Bifunctional poly(acrylamide) hydrogels through orthogonal coupling chemistries. *Biomacromolecules* **2017**, *18*, 906–913. [CrossRef] [PubMed]

43. Cencer, M.; Liu, Y.; Winter, A.; Murley, M.; Meng, H.; Lee, B.P. Effect of pH on the rate of curing and bioadhesive properties of dopamine functionalized poly(ethylene glycol) hydrogels. *Biomacromolecules* **2014**, *15*, 2861–2869. [CrossRef] [PubMed]

44. Haraguchi, K.; Takehisa, T. Nanocomposite hydrogels: A unique organic–inorganic network structure with extraordinary mechanical, optical, and swelling/de-swelling properties. *Adv. Mater.* **2002**, *14*, 1120. [CrossRef]

45. Wang, Q.; Mynar, J.L.; Yoshida, M.; Lee, E.; Lee, M.; Okuro, K.; Kinbara, K.; Aida, T. High-water-content mouldable hydrogels by mixing clay and a dendritic molecular binder. *Nature* **2010**, *463*, 339–343. [CrossRef] [PubMed]

46. Olsson, R.T.; Samir, M.A.; Salazar-Alvarez, G.; Belova, L.; Ström, V.; Berglund, L.A.; Ikkala, O.; Nogues, J.; Gedde, U.W. Making flexible magnetic aerogels and stiff magnetic nanopaper using cellulose nanofibrils as templates. *Nat. Nanotechnol.* **2010**, *5*, 584–588. [CrossRef] [PubMed]

47. Weiner, S.; Wagner, H.D. The material bone: Structure-mechanical function relations. *Annu. Rev. Mater. Sci.* **1998**, *28*, 271–298. [CrossRef]

48. Balasundaram, G.; Webster, T.J. A perspective on nanophase materials for orthopedic implant applications. *J. Mater. Chem.* **2006**, *16*, 3737–3745. [CrossRef]

49. Murugan, R.; Ramakrishna, S. Development of nanocomposites for bone grafting. *Compos. Sci. Technol.* **2005**, *65*, 2385–2406. [CrossRef]

50. Kim, T.G.; Park, S.-H.; Chung, H.J.; Yang, D.-Y.; Park, T.G. Microstructured scaffold coated with hydroxyapatite/collagen nanocomposite multilayer for enhanced osteogenic induction of human mesenchymal stem cells. *J. Mater. Chem.* **2010**, *20*, 8927–8933. [CrossRef]

51. Bai, Y.; Chang, C.-C.; Choudhary, U.; Bolukbasi, I.; Crosby, A.J.; Emrick, T. Functional droplets that recognize, collect, and transport debris on surfaces. *Sci. Adv.* **2016**, *2*, e1601462. [CrossRef] [PubMed]

52. Chirdon, W.M.; O'Brien, W.J.; Robertson, R.E. Adsorption of catechol and comparative solutes on hydroxyapatite. *J. Biomed. Mater. Res. B Appl. Biomater.* **2003**, *66*, 532–538. [CrossRef] [PubMed]

53. Sebei, H.; Minh, D.P.; Lyczko, N.; Sharrock, P.; Nzihou, A. Hydroxyapatite-based sorbents: Elaboration, characterization and application for the removal of catechol from the aqueous phase. *Environ. Technol.* **2016**, 1–10. [CrossRef] [PubMed]

54. Yuk, H.; Zhang, T.; Lin, S.; Parada, G.A.; Zhao, X. Tough bonding of hydrogels to diverse non-porous surfaces. *Nat. Mater.* **2016**, *15*, 190–196. [CrossRef] [PubMed]

55. Li, J.; Illeperuma, W.R.; Suo, Z.; Vlassak, J.J. Hybrid hydrogels with extremely high stiffness and toughness. *ACS Macro Lett.* **2014**, *3*, 520–523. [CrossRef]

56. Sun, J.-Y.; Zhao, X.; Illeperuma, W.R.; Chaudhuri, O.; Oh, K.H.; Mooney, D.J.; Vlassak, J.J.; Suo, Z. Highly stretchable and tough hydrogels. *Nature* **2012**, *489*, 133–136. [CrossRef] [PubMed]

57. Raub, C.B.; Suresh, V.; Krasieva, T.; Lyubovitsky, J.; Mih, J.D.; Putnam, A.J.; Tromberg, B.J.; George, S.C. Noninvasive assessment of collagen gel microstructure and mechanics using multiphoton microscopy. *Biophys. J.* **2007**, *92*, 2212–2222. [CrossRef] [PubMed]

biomimetics

MDPI

Article

Cell-Adhesive Bioinspired and Catechol-Based Multilayer Freestanding Membranes for Bone Tissue Engineering

Maria P. Sousa and João F. Mano *

CICECO—Aveiro Institute of Materials, Department of Chemistry, University of Aveiro, 3810-193 Aveiro, Portugal; mariajsousa@ua.pt
* Correspondence: jmano@ua.pt; Tel.: +351-234-370-733

Academic Editors: Marco d'Ischia and Daniel Ruiz-Molina
Received: 7 July 2017; Accepted: 25 September 2017; Published: 5 October 2017

Abstract: Mussels are marine organisms that have been mimicked due to their exceptional adhesive properties to all kind of surfaces, including rocks, under wet conditions. The proteins present on the mussel's foot contain 3,4-dihydroxy-L-alanine (DOPA), an amino acid from the catechol family that has been reported by their adhesive character. Therefore, we synthesized a mussel-inspired conjugated polymer, modifying the backbone of hyaluronic acid with dopamine by carbodiimide chemistry. Ultraviolet–visible (UV–Vis) spectroscopy and nuclear magnetic resonance (NMR) techniques confirmed the success of this modification. Different techniques have been reported to produce two-dimensional (2D) or three-dimensional (3D) systems capable to support cells and tissue regeneration; among others, multilayer systems allow the construction of hierarchical structures from nano- to macroscales. In this study, the layer-by-layer (LbL) technique was used to produce freestanding multilayer membranes made uniquely of chitosan and dopamine-modified hyaluronic acid (HA-DN). The electrostatic interactions were found to be the main forces involved in the film construction. The surface morphology, chemistry, and mechanical properties of the freestanding membranes were characterized, confirming the enhancement of the adhesive properties in the presence of HA-DN. The MC3T3-E1 cell line was cultured on the surface of the membranes, demonstrating the potential of these freestanding multilayer systems to be used for bone tissue engineering.

Keywords: mussel-inspired; biomimetic; dopamine; multilayer freestanding membranes; adhesiveness; osteogenic differentiation; bone tissue engineering

1. Introduction

Catechol-based materials have been investigated quite a bit in the last decade, presenting an interesting structural and chemical versatility capable of being applied in different fields such as food and agrochemical engineering, green technology, analytical, materials science, biomedicine, and biotechnology [1–3]. In fact, catechols are aromatic derivatives with two contiguous (ortho-)hydroxyl groups that occur ubiquitously in nature, being a part of different biochemical processes and functions as simple molecular units or even as macromolecules [4,5]. These interesting molecules occur in different systems such as fruits, tea, and insects, but it is on marine mussels that catechols were identified as being responsible for their extremely adhesive properties under wet conditions [4,6]. This adhesiveness is mediated by a class of protein, the mussel adhesive proteins (MAPs), known for containing a high amount of the noncationic amino acid 3,4-dihydroxy-L-alanine (DOPA) [7,8]. Intense research has been conducted in this field over the last few decades and it has been proven that the catechol element of DOPA is mainly responsible for the strong adhesion to different type of wet substrates, from inorganic to organic ones [7,9]. To develop MAPs-based materials, different strategies have been employed [10]. Recombinant DNA technology permits the engineering of MAPs precursors purified from *Escherichia coli* and converting into DOPA-containing mimetic by tyrosinase treatment [11]. Another strategy is the chemical modification of polymer backbones with adhesive moieties; for instance, DOPA-modified poly(ethylene glycol) macromers were synthetized through standard peptide chemistry and suggested to produce catechol-based hydrogels [12]. Also DOPA-modified poly(vinyl alcohol) hydrogels were produced with self-healing and pH-responsiveness properties [13]. The chemical modification with DOPA or its derivatives (like dopamine (DN)) of different natural polymers has also been investigated in recent years, from chitosan [14] to alginate [15], or even dextran [16] or hyaluronic acid [17]. For instance, DOPA-modified alginates with different substitution degrees were synthesized and used to produce membranes with enhanced adhesive and biocompatible properties [18].

Natural-based polymers are generally recognized by their high biocompatibility when compared with synthetic ones, exhibiting some recognition domains for cell-binding or cell-mediated processes and making them very interesting materials for tissue engineering and biomedical applications [19]. However, some challenges remain; mechanical properties and controlled biodegradability have been gaining increasing importance. Therefore, some strategies have been envisaged to enhance the potential of tissue engineering and biomedical products and their bioactivity, modifying the physico-chemistry or even the topography of the materials [20,21]. To improve the adhesive properties of materials, mainly in a hydrated environment such as the human body, researchers have been investigating the marine mussel system and taking advantage of the chemical reactivity of catechol moieties [6,22,23].

In this study, a biomimetic approach was combined with the use of natural-based polymers to obtain a mussel-bioinspired system for tissue engineering purposes, focusing on bone tissue engineering. Nowadays, autologous or allogeneic bone transplantation are still the most employed strategies for bone defect treatment, but it entails some risks of causing secondary trauma or even immune system rejection [24]. To overcome these drawbacks, several efforts have been applied to find a bone substitute that is ideally composed of three elements: cells, support material, and bioactive agents. An ideal tissue substitute must mimic the extracellular matrix (ECM) and different material parameters should be taken into account, such as the chemical groups, the biochemical properties, and the topography at the material–cell interface [25,26]. To date, different processing methodologies have been suggested to produce bone substitutes where the support material mimics ECM; nano- and microscale control of different properties along with the deposition of hierarchical films have been suggested for this purpose [27,28].

Layer-by-layer (LbL) is a versatile and inexpensive technique that has been widely applied in this context, being based on the sequential of complementary multivalent molecules on a substrate via electrostatic and/or nonelectrostatic interactions [29–32]. Different authors have reported LbL strategies to mimic some aspect of ECM [33]; Mhanna et al. [34] coated polydimethylsiloxane substrates with specific ECM macromolecules using LbL technology; they used collagen type I, chondroitin sulfate, and heparin and, depending on the composition of the film, they studied specific ECM–cell interactions. Layer-by-Layer-based products have also been exploited for bone tissue engineering purposes, offering fine control over different parameters like film thickness, architecture, chemistry, and even mechanical and topographical properties [35,36]. Oliveira et al. [37] assembled 10 tetralayers of human platelet lysates and marine-origin polysaccharides by LbL technology and then shaped them into fibrils by freeze-drying; the resulting scaffolds could induce the differentiation of human adipose stem cells into mature osteoblasts. In turn, Crouzier et al. [38] showed that cross-linked poly(L-lysine)/hyaluronic acid (HA) can serve as a reservoir for recombinant human bone morphogenetic protein-2 (rhBMP-2) delivery to myoblasts and induce their differentiation into osteoblasts in a dose-dependent manner. Overall, the LbL technique allows us to produce bioinspired and tunable materials to local deliver immobilized growth factors or other bioactive agents and even to instruct stem cells towards osteogenic phenotypes. Moreover, these films can be deposited on an extensive range of substrates of different composition, size, and shape.

Herein, LbL methodology was used to produce bioinspired freestanding multilayer membranes containing catechol groups on the surface and the bulk to improve the adhesiveness properties of the material. In this sense, DN moieties were chemically grafted onto HA to develop multilayer membranes through electrostatic interactions with chitosan (CHT). The modification of HA was confirmed by nuclear magnetic resonance (NMR) and ultraviolet–visible (UV–Vis) spectroscopy. The ability to construct multilayer films was monitored using quartz crystal microbalance with dissipation (QCM-D). Adhesion mechanical tests and in vitro adhesion assays assessed the effect of having DN on the performance of the freestanding multilayer membranes.

Overall, the combination of the versatility of LbL methodology with the protein mussels' inspiration prompted us to exploit catechol-containing multilayer membranes to enhance the interfacial interaction between cells and materials, which takes advantage of the adhesive properties conferred by DN moieties. The potential to induce in vitro bone tissue regeneration was investigated, using a pre-differentiated MC3T3-E1 cell line.

2. Materials and Methods

2.1. Materials

Chitosan with a *N*-deacetylation degree of 80% and a molecular weight in the range of 190–310 kDa, HA as hyaluronic acid sodium salt from *Streptococcus equi* with a molecular weight in the range of 1500–1800 kDa, DN as dopamine hydrochloride with a molecular weight of 189.64 g mol^{-1} and *N*-(3-dimethylaminopropyl)-*N'*-ethylcarbodiimide hydrochloride (EDC) (purum $\geq$ 98.0% (AT)) were purchased from Sigma (St. Louis, MO, USA). These materials were used as received, except CHT, which was purified afterwards, following a standard procedure reported elsewhere [39].

2.2. Synthesis of Dopamine-Modified Hyaluronic Acid

The conjugate of HA modified with DN was synthesized using EDC as an activation agent of the carboxyl groups on HA chains, based on the procedure proposed by Lee et al. [40]. Basically, 1 g of HA was dissolved in 100 mL of phosphate-buffered saline (PBS, Sigma) solution and the pH was adjusted to 5.5 with a hydrochloric acid (HCl, 37%, reagent grade, Sigma) aqueous solution. Then, this solution was purged with nitrogen for 30 min and mixed with EDC and DN and maintained in reaction at 4 °C for 3 h. The pH was maintained at 5.5. Extensive dialysis was performed to remove unreacted chemicals and urea byproducts. After this step, the resulting conjugate was lyophilized for one week and then stored at −20 °C, protected from the light, to avoid oxidation.

2.2.1. Ultraviolet–Visible Spectrophotometry

Ultraviolet–visible spectrophotometer (Jasco V-560 PC) was used to confirm the substitution of dopamine in the conjugate. A solution of 1 mg mL^{-1} in sodium acetate buffer (Scharlab, Barcelona, Spain) with 0.15 M sodium chloride (NaCl, LabChem, Pittsburgh, PA, USA) at pH 5.5 was prepared for the UV–Vis analysis and placed in 1 cm quartz cells. The wavelength range used for this analysis was from 190 nm to 900 nm. Sodium acetate buffer with 0.15 M sodium chloride and at pH 5.5 was used as the reference solution.

2.2.2. Nuclear Magnetic Resonance

^{1}H-NMR analyses were made dissolving overnight the HA, DN and HA-DN in deuterated water (D$_2$O, Cambridge Isotope Laboratories, Inc., Andover, MA, USA) at a concentration of 1 mg mL^{-1} (^{1}H-NMR). The spectra were obtained using a spectrometer BioSpin 300 MHz (Bruker, Billerica, MA, USA). The spectra were recorded at 298 K and 300 MHz for ^{1}H.

2.3. Quartz Crystal Microbalance with Dissipation

The formation of the multilayers of CHT and HA-DN was followed in situ by QCM-D (Q-Sense, Biolin Scientific, Göteborg, Sweden). The mass change results from the variation of the normalized resonant frequency ($\Delta f / \upsilon$) of an oscillating quartz crystal when adsorption occurs on the surface and the dissipation factor (ΔD) provides a measure of the energy loss in the system. The measurements can be conducted at the fundamental frequency and at several overtones number ($\upsilon = 1, 3, 5, \ldots, 11$). Chitosan was used as the polycation while HA or HA-DN acted as the polyanion. Fresh polyelectrolyte solutions were prepared by dissolution of HA-DN, HA, and CHT in sodium acetate buffer containing 0.15 M of NaCl to yield a final concentration of 1 mg mL^{-1}, at pH 5.5. The sensor crystals used were AT-cut quartz (Q-Sense) with gold-plated polished electrodes. These crystals were excited at 5 MHz as well as at 15, 25, 35, 45, and 55 MHz corresponding to the 3rd, 5th, 7th, 9th, and 11th overtones. The crystals were previously cleaned with a pre-exposition to UV ozone (BioForce Nanosciences, Salt Lake City, UT, USA) irradiation during 10 min followed by an immersion on a 5:1:1-mixture of mQ-water, ammonia (25%, Sigma), and hydrogen peroxide (30%, Sigma) at 75 °C, during 5 min. Then the crystals were exposed to a sequential sonication for 3 min in acetone, ethanol, and isopropanol (all from Sigma) and then dried with flowing nitrogen gas avoiding contamination prior to use. To ensure that the crystals are perfectly clean and therefore show a null frequency, all the experiments started with a buffer/solvent baseline. Then, the polyelectrolyte solutions were injected into the cell during 10 min at a flow rate of 20 µL min^{-1}, starting with CHT. A rinsing step of 10 min with the solvent was included between the adsorptions of each polyelectrolyte. The multilayer systems were assembled at pH 5.5. The pH was adjusted with HCl or sodium hydroxide (NaOH, pellets, Fine Chemicals, Akzo Nobel Chemicals S.A., Mons, Belgium). Chitosan/HA films were prepared for comparison, to conclude about the DN effect onto the multilayer system. Films with eight bilayers were produced. All experiments were conducted at 25 °C. During the entire process $\Delta f / \upsilon$ and ΔD shifts were continuously recorded as a function of time. Thickness measurements were performed using the Voigt viscoelastic model implemented in the QTools software (Q-Sense, version 3.1.29.619). Changes in resonant frequency and dissipation of the fifth overtone were fitted. Based on the assumed growth models, the thickness of the multilayer films after 200 cycles was estimated for each system.

2.4. Freestanding Production and Characterization

Multilayer CHT/HA and CHT/HA-DN were built on polypropylene (PP) substrates (Auchan, Villeneuve-d'Ascq, France). Prior to the depositions, these surfaces were cleaned with ethanol and rinsed thoroughly with water before being dried with a stream of nitrogen. The polyelectrolyte solutions were freshly prepared at 1 mg mL^{-1} in a sodium acetate solution containing 0.15 M NaCl, being the pH adjusted to 5.5. The PP substrates were firstly dipped in CHT solution for 6 min and then rinsed twice in the washing solution (acetate buffer, pH 5.5) for 2 min each. Then, they were immersed in the polyanion solution (HA or HA-DN) for 6 min and again twice in the washing solution. This procedure was repeated 200 times with the help of a homemade dipping robot. After drying, the multilayer films were easily detached from the PP substrates without any damage resulting on the freestanding membranes [CHT/HA]$_{200}$ and [CHT/HA-DN]$_{200}$. After production, the membranes were characterized using different techniques and equipment.

2.4.1. Scanning Electron Microscopy and Energy-Dispersive X-ray Spectroscopy

The surface morphology of both sides of [CHT/HA]$_{200}$ and [CHT/HA-DN]$_{200}$ membranes was observed using a Hitachi SU-70 (Hitachi, Tokyo, Japan) scanning electron microscope. All samples were coated with a conductive layer of sputtered gold/palladium. The scanning electron microscopy (SEM) micrographs were taken at an accelerating voltage of 4 kV and at different magnifications. For the cross-section observation, the detached freestandings were immersed in liquid nitrogen until free fracture. After that, the free fracture was placed at 45° and observed by SEM. Energy-dispersive X-ray spectroscopy (EDS, Hitachi) was also used to determine the elemental components of the top surface and in the cross-section of the membranes. The samples were also sputtered with gold/palladium and the analysis was made at an accelerating voltage of 15 kV. The ratio between the oxygen (O) and the nitrogen (N) presented on the top surfaces was quantified in a representative area of the membrane (A = 0.136 mm^2).

2.4.2. Adhesive Mechanical Tests

The adhesion properties of the multilayer were firstly evaluated using a universal mechanical testing machine (Instron model 5966, High Wycombe, Buckinghamshire, UK), following the standard test method for shear strength of single-lap-joint adhesively bonded metal specimens by tension loading ASTM D1002 (ASTM International, West Conshohocken, PA, USA) with slightly modifications. All the adhesion experiments were conducted at 25 °C, at a cross-head speed of 5 mm min^{-1} and using a 1.0 kN static load cell. The lap shear adhesion specimens were squares (20 mm × 20 mm) of freestanding membranes that were incubated at 37 °C and equilibrated in a 50% humidity atmosphere prior to testing. Briefly, the samples were put between two glass slides and left in contact overnight. Then, the systems were stressed until enough force was applied to trigger their detachment and pull them apart, using the Instron apparatus. The lap shear bonding strength was then determined from the maximum of the force–deformation curve obtained. The average and standard deviations were determined using the results from five samples.

While lap shear strength gives a quantitative idea of the adhesive properties of the membranes, other nonconventional test was made to observe the bioadhesiveness potential of these systems. Briefly, the [CHT/HA]$_{200}$ and [CHT/HA-DN]$_{200}$ freestanding membranes were put in contact with a clean surface of porcine bone tissue in a 50% humidity controlled environment at 37 °C. Then, the freestanding membranes were pulled out of the bone with tweezers. This process was recorded by a video camera (Canon EOS 1200D, Tokyo, Japan).

2.5. In Vitro Cellular Tests

The sub-clone 4 of MC3T3-E1 cell line was obtained from the American Type Culture Collection (ATCC)-Laboratory of the Government Chemist (LGC) standards (ATCC® CRL-2593™) [41]. The cells were cultured with Minimum Essential Medium (MEM) Alpha Modification (1X), (α-MEM, Gibco, Thermo Fisher Scientific, Waltham, MA, USA) supplemented with sodium bicarbonate suitable for cell culture (Sigma), 10% fetal bovine serum (FBS, Life Technologies™, Thermo Fisher Scientific, Waltham, MA, USA), and 1% antibiotic–antimycotic (Gibco) at pH 7.4. The cells were cultured in 75 cm^2 tissue culture flasks and incubated at 37 °C in a humidified air atmosphere of 5% CO$_2$. The medium was changed every three–four days. At 80–90% of confluence, cells grown in tissue culture flasks were washed with Dulbecco's phosphate-buffered saline (DPBS, Corning, NY, USA) and then detached by a chemical procedure with trypLE™ express solution (Life Technologies™) for 5 min at 37 °C in a humidified air atmosphere of 5% CO$_2$. To inactivate the trypLE™ express effect, cell culture medium was added. The cells were then centrifuged at 300 × g and 25 °C for 5 min and the medium was decanted. Cells between passage 12 and 13 were used for this study. Prior to cell seeding, the samples were cut in small squares of 25 mm^2 or 1 cm^2, treated with UV ozone for 10 min and immersed in ethanol for 2 h. Then, 150 µL or 300 µL (depending on the size of the sample) of supplemented α-MEM containing a cell suspension with a density of 2 × 10^4 cells cm^{-2} was dropped above the surfaces of the [CHT/HA]$_{200}$ and the [CHT/HA-DN]$_{200}$ freestanding membranes, and the positive control tissue culture polystyrene surface (TCPS, Sarstedt AG & Co., Nümbrecht, Germany) (in triplicate). Then, the samples were incubated at 37 °C in a humidified air atmosphere of 5% CO$_2$. After 4 h, cells already started to adhere and fresh basal culture medium was added.

2.5.1. Metabolic Activity of MC3T3-E1 Cells

The samples were tested for cytotoxicity by analyzing their metabolic activity using the alamarBlue® reduction assay (Invitrogen™, Thermo Fisher Scientific, Waltham, MA, USA). Briefly, the samples (small squares of 25 mm^2) with adhered cells were placed in a nontreated surface 48-well cell culture plate (in triplicate) and incubated at 37 °C and 5% CO_2. At one, three and seven days of culture, the assay was performed, always protecting from light. Briefly, the culture medium was removed and 500 μL of supplemented α-MEM containing 10% (v/v) of alamarBlue solution was added to each well. The samples were then incubated in the dark, overnight, at 37 °C and 5% CO_2. After 12 h, 100 μL of each well (in triplicate) was transferred to a 96-well plate. The absorbance was monitored at 570 nm and 600 nm, using a microplate reader Synergy HTX (BioTek Instruments, Inc., Winooski, VT, USA).

2.5.2. DNA Quantification Assay

A DNA quantification assay (Quant-iT™ PicoGreen® dsDNA Assay Kit, Invitrogen™, Thermo Fisher Scientific) was also performed to evaluate cell proliferation when cultured on the samples' surface. All seeding procedure was repeated for this assay. For each culture time, the samples were washed with DPBS, and then, transferred with 1 mL of ultrapure sterile water to an Eppendorf flask. These Eppendorf flasks were placed at 37 °C for 1 h and then immediately stored at −80 °C until use. The quantification of total DNA was determined after cell lysis, according with the manufacturer's description. After transferring each solution to a 96-well white opaque plate (in triplicate), the plate was incubated at 25 °C, protected from the light, for 10 min. The standard curve for DNA analysis was generated with provided DNA from the assay kit. Fluorescence was read at excitation of 485/20 nm and emission of 528/20 nm using a microplate reader Synergy HTX (BioTek Instruments, Inc.).

2.5.3. Morphological Observation of MC3T3-E1 Cells

MC3T3-E1 cell morphology was observed using a fluorescence microscope (Axio Imager 2, Zeiss, Oberkochen, Germany). Briefly, the cells were seeded above the samples (squares 1 cm^2) at a density of 2×10^4 cells cm^{-2} and cultured for three and seven days, using basal culture conditions. After each time-point, the samples were gently washed with sterile DPBS and fixed with 10% (v/v) of formalin (Sigma) in DPBS solution for 30 min. To obtain morphological fluorescence images, a rhodamine phalloidin (Thermo Fisher Scientific) and 4′,6-diamidino-2-phenylindole (DAPI, Thermo Fisher Scientific) fluorescent assay was performed at each time culture period; DAPI stains preferentially nuclei and phalloidin the actin fibers of the cell cytoskeleton. Firstly, the fixed samples were permeabilized with 0.2% (v/v) of Triton X-100 (Sigma) in DPBS solution for 10 min and then blocked with 5% FBS (v/v) in DPBS solution for 30 min. Then, the samples were treated with rhodamine phalloidin for 45 min and consequently with DAPI for 15 min. Afterwards, the cell morphology was observed using the fluorescence microscope.

2.5.4. Osteogenic Potential of Dopamine-Modified Hyaluronic Acid Membranes and Differentiation of MC3T3-E1 Cells by Immunocytochemistry

To evaluate the osteogenic potential of these substrates, cells were cultured at 2×10^4 cells cm^{-2} in basal growth medium, at 37 °C and 5% of CO_2. After five days in basal conditions, the medium was changed for osteogenic medium (α-MEM containing 10% FBS, 10 mM β-glycerolphosphate disodium salt hydrate (Sigma), and 50 μg mL^{-1} L-ascorbic acid (Cayman Chemical, Ann Arbor, MI, USA)). The differentiation medium was changed every three days.

Intracellular osteopontin expression has been reported as a marker for osteogenic differentiation [42]. After 14 days in differentiation medium, the samples were fixed in 10% (v/v) formalin (Sigma) in DPBS. Following the fixation step, the fixed samples were permeabilized with 0.2% (v/v) of Triton X-100 in DPBS solution for 10 min and then blocked with 5% FBS (v/v) in DPBS solution for 60 min. Then, the samples were examined for protein expression visualization using a mouse antibody against osteopontin (BioLegend, San Diego, CA, USA), by incubation overnight, at 4 °C. Subsequently, the samples were treated with the corresponding secondary antibody anti-mouse Alexa Fluor 647 (Invitrogen™) for 1 h at 25 °C and consequently with rhodamine phalloidin and DAPI for 45 min and 15 min, respectively. Between each step the samples were extensively washed. Afterwards, the cell morphology was observed using fluorescence microscopy.

2.6. Statistical Analysis

All the experiments were performed with at least three replicates and were independently performed three times. Results are expressed as mean ± standard deviation. Differences between the experimental results were analyzed using the one-factor or two-factor analysis of variance (ANOVA), with the Bonferroni's multiple comparison test, defined with a statistical significance of $p < 0.05$.

3. Results

The HA was functionalized with DN and the resulting conjugate was combined with CHT to produce multilayer biomimetic membranes by LbL. The chemical structures of each compound, including the resulting chemical structure of HA-DN are represented in Figure 1. We hypothesize that the presence of DN moieties along the thickness of the films, especially on the last layer and top surface of the membrane, could enhance the interaction between cell and material and improve the osteogenic potential of MC3T3-E1 cells.

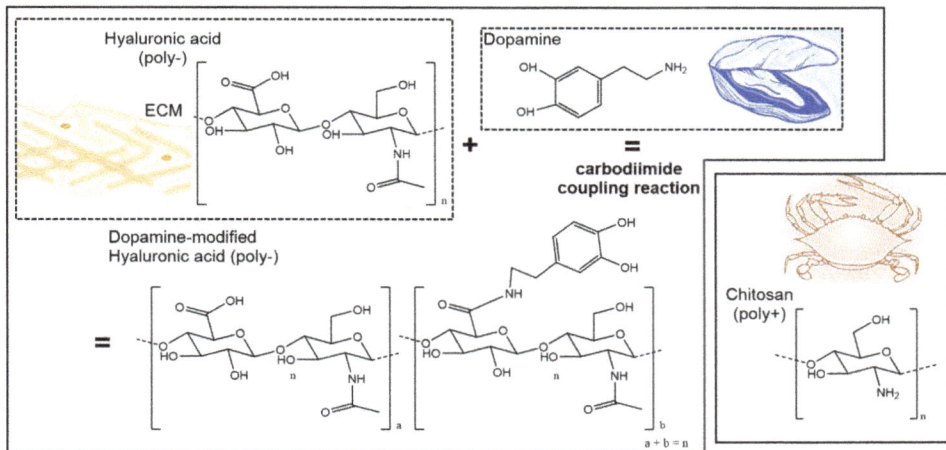

Figure 1. Chemical structure of hyaluronic acid (HA), dopamine (DN), and chitosan (CHT). Synthesis and chemical structure of dopamine-modified hyaluronic acid (HA-DN). ECM: extracellular matrix.

3.1. Synthesis and Characterization of Conjugated Dopamine-Modified Hyaluronic Acid

The conjugation of DN on HA backbone was achieved by the standard carbodiimide coupling method. Using EDC chemistry, the carboxyl group of HA was activated to react with the amine group of DN. After lyophilizing, the resulting conjugated was characterized by UV–Vis and NMR for ^{1}H.

Figure 2A shows the UV–Vis spectrum of each polymer solution, HA and HA-DN. The conjugation of DN onto the backbone of HA was confirmed by the presence of a typical peak around 280 nm, characteristic of dopamine. As expected, this peak did not appear for HA, confirming the presence of DN moieties on the final conjugated HA-DN [43].

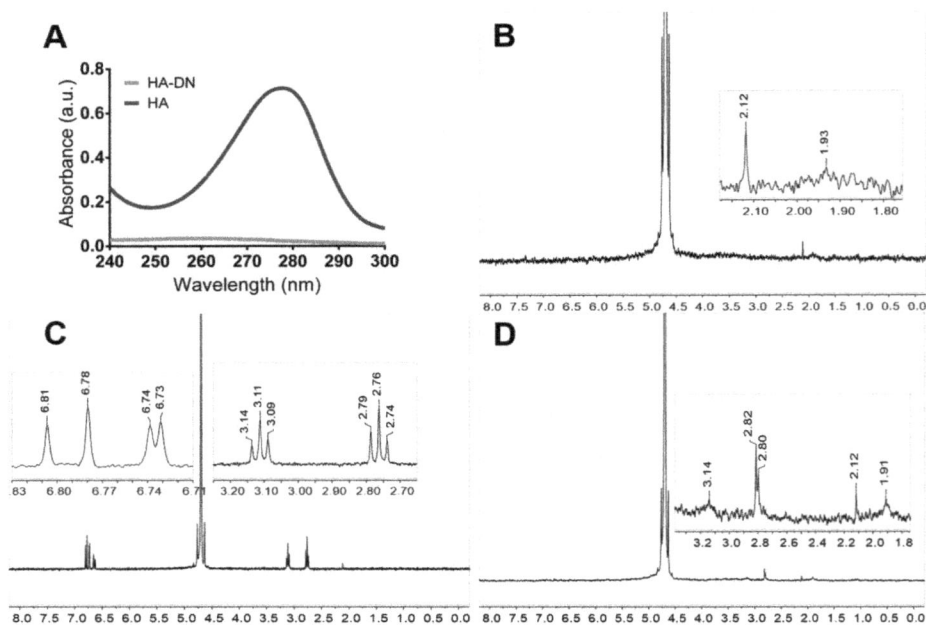

Figure 2. Characterization of conjugated dopamine-modified hyaluronic acid. (**A**) Ultraviolet–visible (UV–Vis) spectra of the control (HA) and the catechol-based conjugate (HA-DN). ^{1}H-nuclear magnetic resonance (NMR) spectra of (**B**) HA; (**C**) DN and (**D**) the synthesized conjugate HA-DN, all with an expanded view. a.u.: Arbitrary units.

Figure 2B–D shows the NMR spectra of HA, DN, and HA-DN for ^{1}H. Regarding the spectrum of HA, the peak at $\delta = 1.93$ ppm is associated with the protons of the methyl group [44]. The spectrum of DN was characterized by the triplets centered at $\delta = 2.76$ ppm and at $\delta = 3.11$ ppm that are associated with the protons of the aliphatic group [45]. In turn, the multiplets between $\delta = 6.73$ ppm and $\delta = 6.81$ ppm are related with the protons in ortho- and meta-coupling position of the ring [17]. The spectrum of HA-DN was consistent with the HA and DN ^{1}H-NMR spectra, as observed in Figure 2D. Both the results of UV–Vis spectroscopy and NMR confirmed that DN was successfully conjugated to HA.

3.2. Multilayer Construction and Thickness Estimation

The assembly of CHT and HA or HA-DN was first monitored on a gold quartz crystal by QCM-D. Figure 3A shows the LbL assembly of CHT and HA and Figure 3B shows the assembly of CHT and HA-DN. For both conditions, the first change in frequency happened when CHT was deposited on bare gold quartz. The second decrease in frequency corresponded to HA or HA-DN layers and this behavior was repeated during all the experiment (eight bilayers), indicating the successful of alternate adsorption of CHT and HA or HA-DN onto the quartz crystal surface. For both systems, the film seemed to have a stable growth for the first layers but it appeared more unstable with the addition of further layers. We hypothesize that it could be due to the formation of soluble macromolecular complexes between the previous layer and the new polymer solution [46]. Besides tmonitoring frequency variation (Δf), QCM-D technology also detects dissipation variation (ΔD), allowing to take assumptions of the multilayer hydration state. When the film is rigid, the Δf and ΔD for the fundamental frequency superpose with the signals recorded in the higher harmonic [47]; in the case of CHT/HA and CHT/HA-DN systems the soft nature of the layers adhering on the crystal leads to the dispersion of the different overtones. A more swollen multilayer film was achieved for polysaccharide-based films when compared with other LbL systems [48]. The assembly of CHT and HA or HA-DN induced similar frequency changes but different dissipation variations, with a bigger dissipation shift when CHT was deposited. This could be related to the fact that more water molecules were entrapped in the CHT layer [49].

Using the modeling tool Q-Tools [49], other parameters could be estimated like the thickness of the films. The Voigt model was chosen to estimate the film thickness of each system, requiring three parameters to be fixed: solvent density, solvent viscosity, and layer density. The solvent viscosity was fixed at 1 mPa s (the same as water) and the film density at 1100 kg m^{-3} (frequently assumed to return the lowest calculation error). The solvent density was changed by trial and error between 1000 and 1080 kg m^{-3} until error was minimized. Figure 3C,D represents the thickness estimation along with the number of bilayers for CHT/HA and CHT/HA-DN systems, respectively. For the CHT/HA system, the thickness of the films seemed to increase exponentially with the number of bilayers. On the other hand, CHT/HA-DN seemed to increase linearly with the number of bilayers. These results are in accordance with the related literature [17]; CHT/HA multilayer films are uniquely composed of polysaccharides and naturally more water-rich than the other system. In turn, the presence of DN on the CHT/HA-DN system seemed to change the growth regime to a linear model; this could indicate that DN conferred less water content on the films. After eight bilayers, the estimated thickness for the CHT/HA system is 157 nm and for the CHT/HA-DN it is 137 nm. Using the resulting linear model, after 200 bilayers we expected a thickness of the CHT/HA-DN system of about 4.0 μm. As HA-DN presents bigger chains, we expected that the thickness of the CHT/HA-DN could be higher than for the CHT/HA system. However, for the first eight bilayers, we observed the opposite trend. We hypothesize that it could be the result of the re-arrangement of the polymer chains when LbL happens; the HA-DN chains seemed to compact more than HA chains, for the first layers of the film.

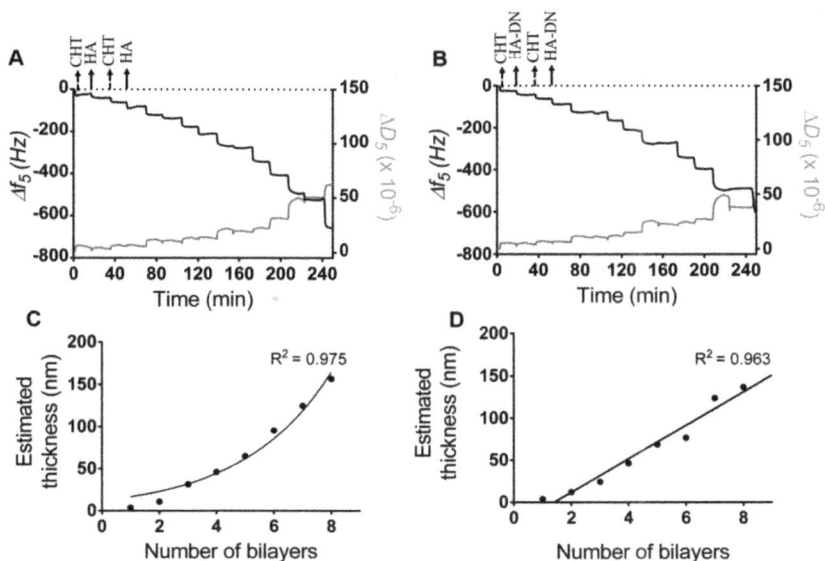

Figure 3. Build-up assemblies of (**A**) CHT and HA, and (**B**) CHT and HA-DN, monitored by quartz crystal microbalance with dissipation (QCM-D). Data shows the normalized frequency (Δf) and dissipation (ΔD) variations at the fifth overtone as a function of the time. Cumulative thickness evolution of the (**C**) CHT/HA and (**D**) CHT/HA-DN multilayer systems as a function of the number of deposition bilayers (Voigt model).

3.3. Production of the Freestanding Multilayer Membranes

From QCM-D monitoring results, CHT and HA or CHT and HA-DN could be combined to produce multilayer systems, with times of deposition of about 10 min; although from QCM-D data, using 6 min of deposition would be enough to construct such multilayer films. This observation was valuable to reduce the times of processing. Therefore, to produce freestanding polyelectrolyte multilayered membranes, we dipped an inert substrate successively on CHT and HA or on CHT and HA-DN solutions, for 6 min each immersion. The chosen underlying substrates were simple PP sheets that are widely available, inexpensive, and can be cut into a large range of shapes and sizes [50]. Between each dipping, we realized a washing step in the sodium acetate buffer (pH 5.5), to remove the excess of polymer. After 200 bilayers, the resulting multilayer films were dried at 25 °C and then easily detached from the underlying substrate, without requiring any post-treatment to dissolve the underlying substrate or any kind of mechanical force that could damage the membrane (see in Supplementary Figure S1A images representing the sequence of actions to detach the membrane from the PP substrate). The number of cycles has a direct influence on the thickness of the membranes and, indirectly, on their robustness and easiness to handle [50]. These membranes were designed as a support material for cells to adhere and differentiate, allowing bone tissue regeneration; this means that we needed the smallest thickness possible, without compromising the stability in physiological medium and the handling. The LbL parameters were optimized, choosing for instance a different number of cycles, but 200 bilayers seemed to be the best compromise between thickness and stability/handling.

The build-up of such LbL-based freestanding membranes has been reported in the literature [50,51], and even with polymers with a potentially adhesive character it was possible to detach the membranes; this may be due to the first layer being CHT. The photograph of both [CHT/HA]$_{200}$ and [CHT/HA-DN]$_{200}$ membranes is presented in Figure S1B, with some differences in color and transparency perceptible.

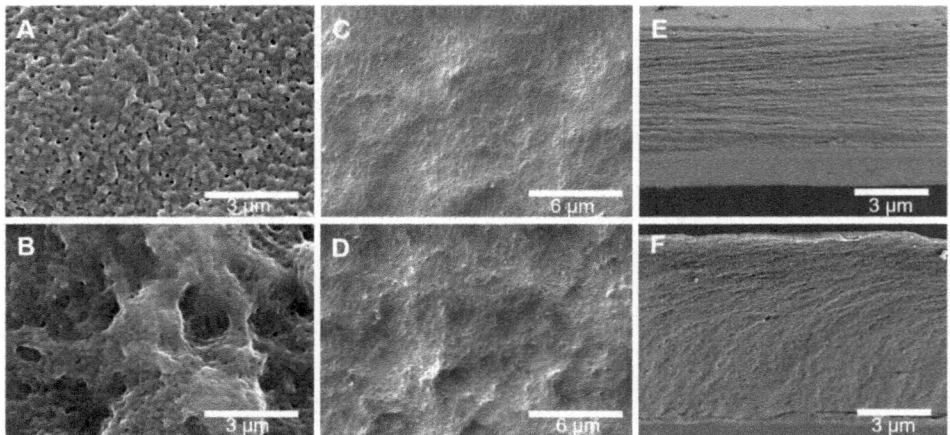

Figure 4. Representative scanning electron microscopy (SEM) images of the surfaces of (**A**) [CHT/HA]$_{200}$ (HA-ending side); (**B**) [CHT/HA-DN]$_{200}$ (HA-DN-ending side); (**C**) [CHT/HA]$_{200}$ (CHT-ending side); (**D**) [CHT/HA-DN]$_{200}$ (CHT-ending side). Representative SEM images of the cross-section of the (**E**) [CHT/HA]$_{200}$ and the (**F**) [CHT/HA-DN]$_{200}$ freestanding membranes.

3.3.1. Surface Morphology and Thickness of the Freestanding Membranes

The morphologies of the surface of the freestanding membranes were investigated by SEM. Figure 4A shows the top view of upper side of the [CHT/HA]$_{200}$ membrane (HA-ending layer); a closed network pore configuration and a smooth surface was observed. In contrast, the top-view of the upper side of the [CHT/HA-DN]$_{200}$ membrane (HA-DN-ending layer) shows an interesting pore network, with bigger pore diameters and a rougher surface, as shown in Figure 4B. Therefore, some differences were noted between the morphologies of the upper layer of the two systems; we hypothesize that the presence of DN in the last layer of the freestanding membrane conferred a higher pore network system and rougher structures than HA alone. Figure 4C,D shows a SEM image of the down side of the [CHT/HA]$_{200}$ and [CHT/HA-DN]$_{200}$ membranes, respectively. Similar morphologies were observed on the CHT side of the membranes, highlighting the rough nature conferred by CHT. Figure 4E,F represents the cross-section of the [CHT/HA]$_{200}$ and [CHT/HA-DN]$_{200}$ freestanding membranes after free-fracture, respectively; [CHT/HA]$_{200}$ cross-section shows a more homogeneous distribution of the polyelectrolytes layers along with the thickness of the membrane, being possible to observe a kind of LbL stratification. The thickness of the [CHT/HA]$_{200}$ membrane is around 5.5 ± 0.1 μm. A thickness of 7.7 ± 0.1 μm was achieved when DN is conjugated with HA and integrated in a multilayer system with CHT. The differences between the thickness of the membranes and the organization along the z-axis could be due to the arrangement of the polymer chains during the multilayer construction. Comparing the real thickness values with the ones estimated from QCM-D data, we obtained thicker [CHT/HA-DN]$_{200}$ films than expected. This fact could be due to some accumulation phenomenon and chain arrangement along with the thickness, which could happen after the construction of the initial layers of the multilayer. Curiously, the thickness of the [CHT/HA]$_{200}$ membranes was much lower than expected by the exponential growth. We hypothesize that for a higher number of bilayers, we cannot assume exponential growth but instead linear growth ($R^2 = 0.969$); therefore, the estimated thickness would be about 4.4 μm, which is approximated to be the real thickness of the freestanding [CHT/HA]$_{200}$ membrane.

3.3.2. Chemical Analysis of the Surface of the Freestanding Membranes

The surface chemical properties of the freestanding multilayer membranes were investigated by EDS analysis. Figure 5A,B shows EDS maps for [CHT/HA]$_{200}$ and [CHT/HA-DN]$_{200}$ membranes along the thickness (cross-section). Visually, both membranes revealed the presence of carbon, oxygen and nitrogen. All these elements are presented along with all the thickness of the membranes. In turn, Figure 5C,D shows EDS map images for the [CHT/HA]$_{200}$ and [CHT/HA-DN]$_{200}$ membranes' top surface and the respective quantification for the same area (Figure 5E and 5F, respectively). Both membranes revealed the presence of carbon, oxygen, sodium, and nitrogen. In fact, the ration between oxygen and nitrogen is significant higher for [CHT/HA-DN]$_{200}$ than for [CHT/HA]$_{200}$ membranes. We hypothesize that the nitrogen quantity could be higher for the [CHT/HA-DN]$_{200}$ due to the presence of DN in the last layer, which adds amine groups to the system.

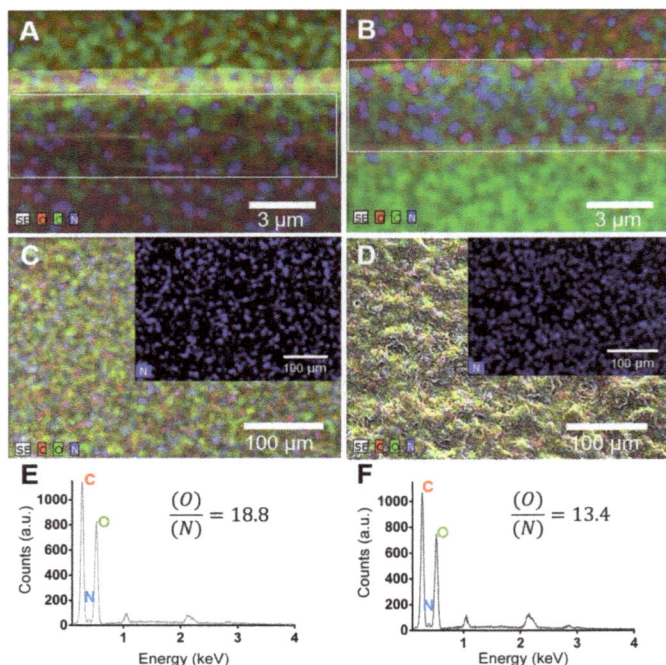

Figure 5. Mixed element map for carbon (C), oxygen (O) and nitrogen (N) of the cross-section of (**A**) [CHT/HA]$_{200}$ and (**B**) [CHT/HA-DN]$_{200}$ freestanding membranes. Mixed element map for carbon (C), oxygen (O), and nitrogen (N) of upper surface of (**C**) [CHT/HA]$_{200}$ and (**D**) [CHT/HA-DN]$_{200}$ freestanding membranes. Energy-dispersive X-ray spectra and ration quantification of O/N of (**E**) [CHT/HA]$_{200}$ and (**F**) [CHT/HA-DN]$_{200}$ freestanding membranes. a.u.: Arbitrary units.

3.3.3. Adhesive Properties of the Freestanding Membranes

In this work, freestanding membranes were used to glue two pieces of glass and this system was used as a proof of the concept for the adhesive strength of the [CHT/HA]$_{200}$ and [CHT/HA-DN]$_{200}$ membranes. The size of these membranes was around 20 mm × 20 mm, and the thickness was between 5 and 8 μm. Briefly, the freestanding membrane was sandwiched between two pieces of glass, in a controlled environment, trying to mimic the inner body conditions (37 °C in a moist environment), and incubated overnight. Most of the methods to determine the adhesive properties involve determination of the perpendicular force required to separate two surfaces. As biological tissues are more susceptible to shear stress than tensile stress, we decided to investigate the force that makes an adhesive slide on a surface in the direction parallel to the plane of contact [52]. The ability of [CHT/HA]$_{200}$ and [CHT/HA-DN]$_{200}$ to glue two glass substrates together was evaluated using a universal mechanical testing machine according to the standard procedure ASTM D1002, with subtle modifications. A heavy load (1.0 kN) could be held on a small joint area (20 mm × 20 mm) until it caused the detachment of the glass slides (see the mounting scheme in Figure 6A). Figure 6B presents the values of adhesive strength for each condition. From the lap shear adhesion tests, it can be seen that there were higher load values for the same displacement for [CHT/HA-DN]$_{200}$ membranes, indicating that more load is required to separate the glued glass slides. The calculation of lap shear adhesion strength was performed for each case; the [CHT/HA]$_{200}$ system presents an adhesion strength of 3.4 ± 0.6 MPa, while the [CHT/HA-DN]$_{200}$ system presents an adhesion strength of 8.6 ± 2.2 MPa. Such a difference highlights the adhesive strength of the membrane containing DN compared with the control. We hypothesize that the presence of the catechol moieties in the [CHT/HA-DN]$_{200}$ membranes increased the adhesion force between the two glass slides. The adhesive characteristics of catechol-based materials have been already investigated [17,53,54]. For instance, Ninan et al. [55] studied the adhesive properties of marine mussel adhesive extracts to bond porcine skin in controlled dry and humid conditions; the tissue joint strength was about 1 MPa for mussel extract joints under humid conditions. Also, Kim et al. [56] reported the development of a water-immiscible mussel protein-based adhesive, composed of a complex coacervate of HA and DOPA with strong underwater adhesion; an adhesive strength of about 0.14 ± 0.03 MPa was obtained, using rat bladder tissue as the contact substrate. Even using a different contact substrate, the guidelines for adhesive tests were the same as for the previous examples. Even so, we obtained significantly higher values of strength adhesion with [CHT/HA-DN]$_{200}$ membranes. Well-known bioadhesives are fibrin and cyanoacrylate-based materials, which present a shear adhesive strength around 0.013 MPa and 0.068 MPa, respectively [57]. Bioadhesive hydrogels have been reported mainly for topical wound dressings and sealants as their adhesive strength is still considered to be weak [57]. For tissue engineering applications, bioadhesive films are of more interest. Layer-by-layer technology has been used for this purpose; Neto et al. [17,58] already reported the ability to produce multilayer coatings composed of CHT and DOPA-modified HA with enhanced adhesive properties. The adhesive strength of the coating was about 2.3 ± 2.2 MPa, more than triple that of the control coating. Other LbL systems have been reported, but mostly in the form of coatings and not as freestanding membranes, which is more relevant for developing supports for tissue engineering purposes. To the best of our knowledge, there has already been one work on such adhesive freestanding multilayer membranes, but instead of dopamine they took advantage of the adhesive properties of levan derivatives [52]. Another strategy was envisaged to evaluate the bioadhesiveness of this membrane. After putting the [CHT/HA]$_{200}$ and [CHT/HA-DN]$_{200}$ membranes in contact with a clean surface of porcine bone at 37 °C, at 50% humidity overnight, it was possible to observe that more force is required to pull out the [CHT/HA-DN]$_{200}$ membrane (see representative images in Figure 6C (i) before and (ii) after applying a detachment force with tweezers, and the video recording in Supplementary Video S1). The [CHT/HA]$_{200}$ membrane was easier to detach from the surface of porcine bone than the [CHT/HA-DN]$_{200}$ membrane. This result is in accordance with the lap shear adhesion strength test, highlighting the adhesive potential of DN when incorporated in these multilayer freestanding systems.

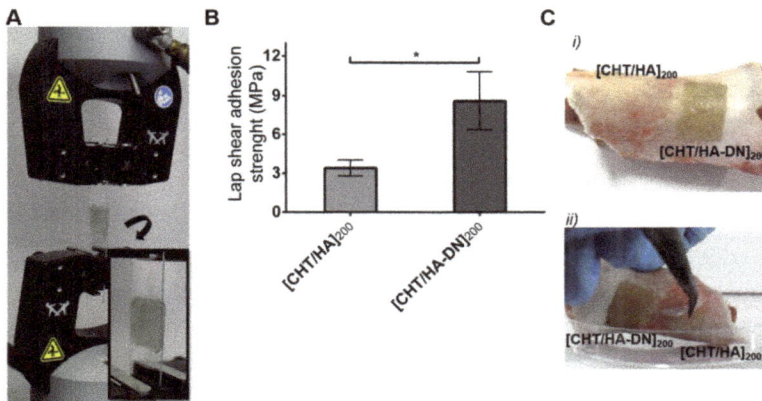

Figure 6. Adhesive properties of the freestanding membranes. (**A**) Mounting scheme for testing the lap shear adhesion strength on the Instron equipment; (**B**) lap shear adhesions strength values for each system. Significant differences were found for $p < 0.05$. (**C**) Representative images of the adhesiveness potential of [CHT/HA]$_{200}$ and [CHT/HA-DN]$_{200}$ freestanding membranes on a clean surface of porcine bone: (i) before and (ii) after applying a detachment force with tweezers.

3.4. In Vitro Cell Studies

MC3T3-E1 cells were seeded above [CHT/HA]$_{200}$ and [CHT/HA-DN]$_{200}$ freestanding membranes. The performance of such kind of adhesive substrates was evaluated for different cellular functions, namely the metabolic activity, cytotoxicity, proliferation, and morphology of MC3T3-E1. A preliminary immunofluorescent assay was performed to evaluate the osteogenic potential of these substrates. The MC3T3-E1 response for TCPS surfaces were used as reference and positive control.

3.4.1. Metabolic Activity, Cytotoxicity, Proliferation, and Morphology of MC3T3-E1 Cells

The metabolic activity of MC3T3-E1 cells and the cytotoxicity of [CHT/HA]$_{200}$ and [CHT/HA-DN]$_{200}$ membranes were evaluated by the Alamar blue colorimetric test. Figure 7A shows the results for the resulting Alamar blue absorbance. After one day of culture in basal growth medium, the cells seemed to adhere to the different substrates with no significant differences. Nevertheless, the values of absorbance of cells seeded above the freestanding membranes were comparable to the values for TCPS surfaces. This observation, combined with the increase in absorbance along with the days of culture for each condition, is an indication of the non-cytotoxicity of the materials. After three and seven days of culture, significant differences in metabolic activity were noted between [CHT/HA]$_{200}$ and [CHT/HA-DN]$_{200}$ freestanding membranes; higher absorbance values for cells seeded above [CHT/HA-DN]$_{200}$ freestanding membranes indicate enhanced metabolic activity and viability. Also, the proliferation of MC3T3-E1 was estimated using a standard DNA quantification assay. Figure 7B shows the content of DNA for each condition up to seven days of culture. After one day of culture the DNA content of MC3T3-E1 cells seeded above [CHT/HA]$_{200}$, [CHT/HA-DN]$_{200}$, and TCPS was quite similar, while for three and seven days of culture there were significant differences between these conditions. Following the same trend observed for Alamar blue results, the highest rates of proliferation were found for [CHT/HA-DN]$_{200}$ freestanding membranes, indicating a better cell response to the catechol-containing membranes. Additionally, the morphology of the cells was observed using a fluorescence assay (see Figure 7C). After three and seven days of culture in basal growth medium, the cells seeded above the different surfaces were fixed and stained with specific markers: phalloidin (in red) to label the actin cytoskeleton and DAPI (in blue) to label the nuclei of the cells. The representative images, presented in Figure 7C, corroborate the results of metabolic activity and DNA quantification: after three and seven days of culture, the density of cells adhered on the [CHT/HA-DN]$_{200}$ membranes was significantly higher than for the [CHT/HA]$_{200}$ membranes and close to the cell density on the TCPS surfaces. Moreover, the morphology of the MC3T3-E1 cells also differed; cells adhered above the [CHT/HA-DN]$_{200}$ started to establish cell–cell contact with each other after just one day of culture, while for cells seeded on [CHT/HA]$_{200}$ membranes this cell–cell contact was only perceptible after three days of culture; these observations reinforced the proliferation results as cell–cell contact promotes cell proliferation. We hypothesize that DN presence along the thickness of the multilayer membranes and concretely on the surface improved their biological performance at different stages: adhesion, viability, communication, and proliferation. These results could have potential applicability in the tissue engineering field, as adhesive and biocompatible substrates to support cells' functions. There are other works [15,59–61] reporting the positive effect of catechol-based materials in promoting cellular adhesion and good function. Polycaprolactone scaffolds were modified using a mussel-inspired approach; polydopamine coating and hyaluronic acid immobilization seemed to be an effective way to improve cellular performance [62]. Zhang et al. [63] suggested LbL methodology to simply coat titanium implants with HA-DN and CHT, aiming to enhance the osteoblast proliferation. The greatest advantage of our system over the reported examples and the existing literature is related to the combination of a freestanding substrate composed of natural-based materials with enhanced adhesive strength and improved cell response. Moreover, this kind of flexible substrate could be produced, handled, and applied in quite a simple and adaptable way.

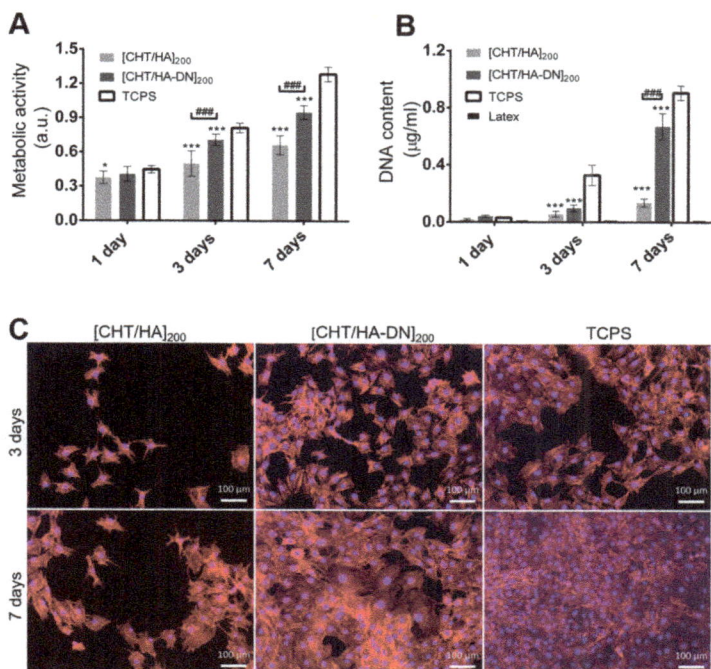

Figure 7. In vitro cell studies. (**A**) **Metabolic activity** of MC3T3- E1 cells seeded on the membranes (Alamar Blue assay). Significant differences were found between membranes and TCPS conditions (for * $p < 0.05$; *** $p < 0.001$) and between the two kind of systems (### $p < 0.001$); (**B**) DNA content of MC3T3- E1 seeded above the membranes (PicoGreen Kit). Significant differences were found between membranes and tissue culture polystyrene surface (TCPS) conditions (*** $p < 0.001$) and between the two kind of systems (### $p< 0.001$). (**C**) Fluorescence images of MC3T3- E1 cells stained with phalloidin (red) and 4′,6-diamidino-2-phenylindole (DAPI) (blue), at three and seven days of culture on the [CHT/HA]$_{200}$ and [CHT/HA-DN]$_{200}$ membranes and the TCPS (positive control). a.u.: Arbitrary units.

3.4.2. Differentiation of MC3T3-E1 Cells

The differentiation of mouse MC3T3-E1 pre-osteoblasts toward the formation of a mineralized extracellular matrix was also evaluated. This cell line was chosen as the model for our system due to its usually compressed level of differentiation and the ability to form a mineralized bone-like extracellular matrix [64]. Figure 8 shows the immunofluorescence images of cells seeded on the different materials, using osteopontin as the osteogenic marker. Typical osteogenic differentiation of MC3T3-E1 cells occurs in three phases: proliferation, extracellular matrix deposition and maturation, and finally mineralization. Each phase corresponds to higher expressions of certain genes; osteopontin is expressed near the later stages of osteogenic differentiation. Therefore, after 14 days in a differentiation medium containing ascorbic acid and β-glycerolphosphate, different behaviors could be observed for the freestanding systems. The cells cultured on [CHT/HA-DN]$_{200}$ showed stronger immunofluorescence for osteopontin protein staining than the cells cultured on [CHT/HA]$_{200}$ membranes, and very similar fluorescence to the TCPS positive control (see Figure 8A–C). We hypothesize that catechol-based moieties provided the multilayer films with important properties to improve their osteogenic potential. The characteristic chemical groups of [CHT/HA]$_{200}$ and [CHT/HA-DN]$_{200}$ membranes could be considered osteogenic differentiation promoters; besides CH$_2$ and CH$_3$ groups, these surfaces also present NH_2 and OH groups and for all of them a positive effect on endorsing osteogenic differentiation has been reported [65]. Few works have investigated the potential of mussel-inspired adhesive proteins for in vitro bone formation. For instance, Yu et al. [66] coated titanium substrates with polydopamine to facilitate the homogeneous covalent immobilization of collagen on their surface and promote the osteogenic differentiation of MC3T3-E1 cells. Another group [67] synthesized a conjugate of alginate and dopamine and produced alginate–dopamine gels, which seemed to promote the osteogenic differentiation of mesenchymal stem cells. The adhesive character of DN allowed the author to coat the gel with silver, providing antibacterial properties. Moreover, there are some studies reporting that the polydopamine coating could enhance hydroxyapatite nucleation and then promote mineralization; Lee et al. [68] coated 3D-printed polycaprolactone scaffolds with polydopamine to easily graft rhBMP-2. In addition, a higher amount of BMP-2 resulted in better bone tissue formation; even with small doses of this protein, they could induce osteogenic differentiation. In contrast, even in the absence of any other grafted protein or growth factor, our multilayer system based on mussel-inspired catechol groups could enhance the potential to differentiate MC3T3-E1 cells into osteoblasts. Figure 8D is a merged image of the MC3T3-E1 cells cultured on the $[CHT/HA-DN]_{200}$ membrane, overlaying osteopontin (green), phalloidin (red), and DAPI (blue) markers. As phalloidin stains the actin cytoskeleton of the cells and DAPI stains the nucleus, the appearance of osteopontin as an intracytoplasmic marker is clear from Figure 8D. We hypothesize that this positive intracellular activity could be related to the cellular calcification induced by the substrate and the culture conditions during the active stage of the differentiation of MC3T3-E1 cells in osteoblasts [69].

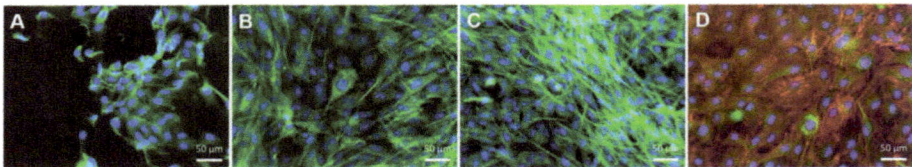

Figure 8. Osteopontin immunofluorescence images of MC3T3-E1 cells stained in green and with DAPI (in blue), after 14 days in osteogenic medium and cultured on the (**A**) $[CHT/HA]_{200}$ and (**B**) $[CHT/HA-DN]_{200}$ membranes and (**C**) TCPS (positive control); (**D**) Merged image of MC3T3-E1 cells cultured on the $[CHT/HA-DN]_{200}$ membrane is shown in the overlay with osteopontin (green), phalloidin (red) and DAPI (blue) markers.

Note that even after 21 days immersed in a physiological medium, the membranes seemed to be stable, but presented some signs of degradation.

Overall, the $[CHT/HA-DN]_{200}$ freestanding membranes showed enhanced adhesive strength properties, as well as improved cell adhesion, proliferation, and differentiation, making them a good candidate to regenerate bone tissue.

4. Conclusions

The notable ability of DOPA and its analogues to form strong interactions with both organic and inorganic surfaces was inspired by the process of wet adhesion in mussels and has been used to produce materials with unique adhesion properties. Bioadhesive materials have started to gain importance in bone tissue engineering strategies, which have been appearing to overcome some issues, often related to implant failure, by creating a system that directly promotes bone ingrowth into the material's structure and helps with tissue regeneration. Other features have been reported as significant for the success of the bone tissue engineering system. Using the LbL technique offers a unique vehicle to create a biomimetic environment due to the ability to incorporate materials that are presented in the ECM or have a bioactive role and assemble them into a functional tissue-like unit.

In conclusion, we covalently bonded DN on the backbone of HA with success, conferring important properties to this glycosaminoglycan. Taking advantage of the conjugate HA-DN and their negative nature at pH 5.5, we produced thin multilayer freestanding membranes composed only of CHT and HA-DN, using a simple LbL technology based on electrostatic interactions. Interestingly, when comparing membranes without DN with membranes where DN was conjugated with HA, we clearly enhanced the adhesive strength. MC3T3-E1 cell adhesion, viability, proliferation, and density were enhanced when cultured on [CHT/HA-DN]$_{200}$. We hypothesize that the mechanical and morphological differences between the different multilayer systems had a positive impact on cellular behavior. Additionally, our preliminary results for MC3T3-E1 differentiation studies suggest that the presence of HA-DN on the multilayer membranes provided better differentiation signals. We assume that enhanced differentiation of MC3T3-E1 cells is related to the morphology, chemistry, and mechanical properties conferred by this catechol-based material. Therefore, our investigation suggests a cheap, scalable, and versatile technology to produce biocompatible and osteophilic [CHT/HA-DN]$_{200}$ multilayer membranes with interesting adhesive properties that could potentially be applied in bone regeneration.

Supplementary Materials: The following are available online at http://www.mdpi.com/2313-7673/2/4/19/s1: Figure S1: Production of the freestanding multilayer membranes; Video S1: Adhesiveness of the freestanding membranes to a clean surface of porcine bone.

Acknowledgments: M.P.S. acknowledges the Portuguese Foundation for Science and Technology (FCT) for financial support through Grant No. SFRH/BD/97606/2013. This work was supported by the European Research Council grant agreement ERC-2014-ADG-669858 for the ATLAS project. The authors acknowledge Carmen Freire (CICECO, University of Aveiro, Aveiro, Portugal) for providing the Instron equipment to carry out the lap shear adhesion tests.

Author Contributions: M.P.S. and J.F.M. conceived and designed the experiments; M.P.S. performed the experiments and analyzed the data; M.P.S. wrote the paper.

Conflicts of Interest: The authors declare no conflict of interest.

References

1. Lee, H.; Dellatore, S.M.; Miller, W.M.; Messersmith, P.B. Mussel-inspired surface chemistry for multifunctional coatings. *Science* **2007**, *318*, 426–430. [CrossRef] [PubMed]
2. Sedó, J.; Saiz-Poseu, J.; Busqué, F.; Ruiz-Molina, D. Catechol-based biomimetic functional materials. *Adv. Mater.* **2013**, *25*, 653–701. [CrossRef] [PubMed]
3. Liu, Y.; Ai, K.; Lu, L. Polydopamine and its derivative materials: Synthesis and promising applications in energy, environmental, and biomedical fields. *Chem. Rev.* **2014**, *114*, 5057–5115. [CrossRef] [PubMed]
4. Faure, E.; Falentin-Daudré, C.; Jérôme, C.; Lyskawa, J.; Fournier, D.; Woisel, P.; Detrembleur, C. Catechols as versatile platforms in polymer chemistry. *Prog. Polym. Sci.* **2013**, *38*, 236–270. [CrossRef]
5. Busqué, F.; Sedó, J.; Ruiz-Molina, D.; Saiz-Poseu, J. Catechol-based biomimetic functional materials and their applications. In *Bio- and Bioinspired Nanomaterials*; Ruiz-Molina, D., Novio, F., Roscini, C., Eds.; Wiley-VCH Verlag GmbH & Co. KGaA: Weinheim, Germany, 2014; pp. 277–308.
6. Lee, B.P.; Messersmith, P.B.; Israelachvili, J.N.; Waite, J.H. Mussel-inspired adhesives and coatings. *Annu. Rev. Mater. Res.* **2011**, *41*, 99–132. [CrossRef] [PubMed]
7. Lee, H.; Scherer, N.F.; Messersmith, P.B. Single-molecule mechanics of mussel adhesion. *Proc. Natl. Acad. Sci. USA* **2006**, *103*, 12999–13003. [CrossRef] [PubMed]
8. Silverman, H.G.; Roberto, F.F. Understanding marine mussel adhesion. *Mar. Biotechnol.* **2007**, *9*, 661–681. [CrossRef] [PubMed]
9. Lee, H.; Lee, B.P.; Messersmith, P.B. A reversible wet/dry adhesive inspired by mussels and geckos. *Nature* **2007**, *448*, 338–341. [CrossRef] [PubMed]
10. Cha, H.J.; Hwang, D.S.; Lim, S. Development of bioadhesives from marine mussels. *Biotechnol. J.* **2008**, *3*, 631–638. [CrossRef] [PubMed]

11. Choi, Y.S.; Yang, Y.J.; Yang, B.; Cha, H.J. In vivo modification of tyrosine residues in recombinant mussel adhesive protein by tyrosinase co-expression in *Escherichia coli*. *Microb. Cell Fact.* **2012**, *11*, 139. [CrossRef] [PubMed]
12. Lee, B.P.; Dalsin, J.L.; Messersmith, P.B. Synthesis and gelation of DOPA-modified poly(ethylene glycol) hydrogels. *Biomacromolecules* **2002**, *3*, 1038–1047. [CrossRef] [PubMed]
13. Shi, D.; Liu, R.; Dong, W.; Li, X.; Zhang, H.; Chen, M.; Akashi, M. pH-dependent and self-healing properties of mussel modified poly(vinyl alcohol) hydrogels in a metal-free environment. *RSC Adv.* **2015**, *5*, 82252–82258. [CrossRef]
14. Kim, K.; Ryu, J.H.; Lee, D.Y.; Lee, H. Bio-inspired catechol conjugation converts water-insoluble chitosan into a highly water-soluble, adhesive chitosan derivative for hydrogels and LbL assembly. *Biomater. Sci.* **2013**, *1*, 783–790. [CrossRef]
15. Lee, C.; Shin, J.; Lee, J.S.; Byun, E.; Ryu, J.H.; Um, S.H.; Kim, D.-I.; Lee, H.; Cho, S.-W. Bioinspired, calcium-free alginate hydrogels with tunable physical and mechanical properties and improved biocompatibility. *Biomacromolecules* **2013**, *14*, 2004–2013. [CrossRef] [PubMed]
16. Park, J.Y.; Yeom, J.; Kim, J.S.; Lee, M.; Lee, H.; Nam, Y.S. Cell-repellant dextran coatings of porous titania using mussel adhesion chemistry. *Macromol. Biosci.* **2013**, *13*, 1511–1519. [CrossRef] [PubMed]
17. Neto, A.I.; Cibrão, A.C.; Correia, C.R.; Carvalho, R.R.; Luz, G.M.; Ferrer, G.G.; Botelho, G.; Picart, C.; Alves, N.M.; Mano, J.F. Nanostructured polymeric coatings based on chitosan and dopamine-modified hyaluronic acid for biomedical applications. *Small* **2014**, *10*, 2459–2469. [CrossRef] [PubMed]
18. Scognamiglio, F.; Travan, A.; Borgogna, M.; Donati, I.; Marsich, E.; Bosmans, J.W.A.M.; Perge, L.; Foulc, M.P.; Bouvy, N.D.; Paoletti, S. Enhanced bioadhesivity of dopamine-functionalized polysaccharidic membranes for general surgery applications. *Acta Biomater.* **2016**, *44*, 232–242. [CrossRef] [PubMed]
19. Mano, J.F.; Silva, G.A.; Azevedo, H.S.; Malafaya, P.B.; Sousa, R.A.; Silva, S.S.; Boesel, L.F.; Oliveira, J.M.; Santos, T.C.; Marques, A.P.; et al. Natural origin biodegradable systems in tissue engineering and regenerative medicine: Present status and some moving trends. *J. R. Soc. Interface* **2007**, *4*, 999–1030. [CrossRef] [PubMed]
20. Cai, K.; Rechtenbach, A.; Hao, J.; Bossert, J.; Jandt, K.D. Polysaccharide-protein surface modification of titanium via a layer-by-layer technique: Characterization and cell behaviour aspects. *Biomaterials* **2005**, *26*, 5960–5971. [CrossRef] [PubMed]
21. Alves, N.M.; Pashkuleva, I.; Reis, R.L.; Mano, J.F. Controlling cell behavior through the design of polymer surfaces. *Small* **2010**, *6*, 2208–2220. [CrossRef] [PubMed]
22. Moulay, S. Dopa/catechol-tethered polymers: Bioadhesives and biomimetic adhesive materials. *Polym. Rev.* **2014**, *54*, 436–513. [CrossRef]
23. Kaushik, N.K.; Kaushik, N.; Pardeshi, S.; Sharma, J.G.; Lee, S.H.; Choi, E.H. Biomedical and clinical importance of mussel-inspired polymers and materials. *Mar. Drugs* **2015**, *13*, 6792–6817. [CrossRef] [PubMed]
24. Amini, A.R.; Laurencin, C.T.; Nukavarapu, S.P. Bone tissue engineering: Recent advances and challenges. *Crit. Rev. Biomed. Eng.* **2012**, *40*, 363–408. [CrossRef] [PubMed]
25. Geckil, H.; Xu, F.; Zhang, X.; Moon, S.; Demirci, U. Engineering hydrogels as extracellular matrix mimics. *Nanomedicine* **2010**, *5*, 469–484. [CrossRef] [PubMed]
26. Gong, T.; Xie, J.; Liao, J.; Zhang, T.; Lin, S.; Lin, Y. Nanomaterials and bone regeneration. *Bone Res.* **2015**, *3*, 15029. [CrossRef] [PubMed]
27. Nassif, N.; Gobeaux, F.; Seto, J.; Belamie, E.; Davidson, P.; Panine, P.; Mosser, G.; Fratzl, P.; Giraud Guille, M.-M. Self-assembled collagen-apatite matrix with bone-like hierarchy. *Chem. Mater.* **2010**, *22*, 3307–3309. [CrossRef]
28. Yang, W.; Xi, X.; Si, Y.; Huang, S.; Wang, J.; Cai, K. Surface engineering of titanium alloy substrates with multilayered biomimetic hierarchical films to regulate the growth behaviors of osteoblasts. *Acta Biomater.* **2014**, *10*, 4525–4536. [CrossRef] [PubMed]
29. Decher, G. Fuzzy nanoassemblies: Toward layered polymeric multicomposites. *Science* **1997**, *277*, 1232–1237. [CrossRef]
30. Tang, Z.; Wang, Y.; Podsiadlo, P.; Kotov, N.A. Biomedical applications of layer-by-layer assembly: From biomimetics to tissue engineering. *Adv. Mater.* **2006**, *18*, 3203–3224. [CrossRef]

31. Ariga, K.; Yamauchi, Y.; Rydzek, G.; Ji, Q.; Yonamine, Y.; Wu, K.C.-W.; Hill, J.P. Layer-by-layer nanoarchitectonics: Invention, innovation, and evolution. *Chem. Lett.* **2014**, *43*, 36–68. [CrossRef]
32. Gentile, P.; Carmagnola, I.; Nardo, T.; Chiono, V. Layer-by-layer assembly for biomedical applications in the last decade. *Nanotechnology* **2015**, *26*, 422001. [CrossRef] [PubMed]
33. Silva, J.M.; Reis, R.L.; Mano, J.F. Biomimetic extracellular environment based on natural origin polyelectrolyte multilayers. *Small* **2016**, *12*, 4308–4342. [CrossRef] [PubMed]
34. Mhanna, R.F.; Vörös, J.; Zenobi-Wong, M. Layer-by-layer films made from extracellular matrix macromolecules on silicone substrates. *Biomacromolecules* **2011**, *12*, 609–616. [CrossRef] [PubMed]
35. Gribova, V.; Auzely-Velty, R.; Picart, C. Polyelectrolyte multilayer assemblies on materials surfaces: From cell adhesion to tissue engineering. *Chem. Mater.* **2012**, *24*, 854–869. [CrossRef] [PubMed]
36. Shah, N.J.; Hyder, M.N.; Quadir, M.A.; Dorval Courchesne, N.-M.; Seeherman, H.J.; Nevins, M.; Spector, M.; Hammond, P.T. Adaptive growth factor delivery from a polyelectrolyte coating promotes synergistic bone tissue repair and reconstruction. *Proc. Natl. Acad. Sci. USA* **2014**, *111*, 12847–12852. [CrossRef] [PubMed]
37. Oliveira, S.M.; Reis, R.L.; Mano, J.F. Assembling human platelet lysate into multiscale 3D scaffolds for bone tissue engineering. *ACS Biomater. Sci. Eng.* **2015**, *1*, 2–6. [CrossRef]
38. Crouzier, T.; Ren, K.; Nicolas, C.; Roy, C.; Picart, C. Layer-by-layer films as a biomimetic reservoir for rhBMP-2 delivery: Controlled differentiation of myoblasts to osteoblasts. *Small* **2009**, *5*, 598–608. [CrossRef] [PubMed]
39. Correia, C.; Caridade, S.; Mano, J. Chitosan membranes exhibiting shape memory capability by the action of controlled hydration. *Polymers* **2014**, *6*, 1178–1186. [CrossRef]
40. Lee, H.; Lee, Y.; Statz, A.R.; Rho, J.; Park, T.G.; Messersmith, P.B. Substrate-independent layer-by-layer assembly by using mussel-adhesive-inspired polymers. *Adv. Mater.* **2008**, *20*, 1619–1623. [CrossRef] [PubMed]
41. MC3T3-E1 Subclone 4 (ATCC® CRL-2593™). Available online: https://www.lgcstandards-atcc.org/Products/All/CRL-2593.aspx (accessed on 4 October 2017).
42. Baler, K.; Ball, J.P.; Cankova, Z.; Hoshi, R.A.; Ameer, G.A.; Allen, J.B. Advanced nanocomposites for bone regeneration. *Biomater. Sci.* **2014**, *2*, 1355–1366. [CrossRef]
43. Chen, S.-M.; Peng, K.-T. The electrochemical properties of dopamine, epinephrine, norepinephrine, and their electrocatalytic reactions on cobalt(II) hexacyanoferrate films. *J. Electroanal. Chem.* **2003**, *547*, 179–189. [CrossRef]
44. Pomin, V.H. NMR chemical shifts in structural biology of glycosaminoglycans. *Anal. Chem.* **2014**, *86*, 65–94. [CrossRef] [PubMed]
45. Mueller, D.D.; Morgan, T.D.; Wassenberg, J.D.; Hopkins, T.L.; Kramer, K.J. Proton and carbon-13 NMR of 3-O and 4-O conjugates of dopamine and other catecholamines. *Bioconjugate Chem.* **1993**, *4*, 47–53. [CrossRef]
46. Croll, T.I.; O'Connor, A.J.; Stevens, G.W.; Cooper-White, J.J. A blank slate? Layer-by-layer deposition of hyaluronic acid and chitosan onto various surfaces. *Biomacromolecules* **2006**, *7*, 1610–1622. [CrossRef] [PubMed]
47. Kujawa, P.; Schmauch, G.; Viitala, T.; Badia, A.; Winnik, F.M. Construction of viscoelastic biocompatible films via the layer-by-layer assembly of hyaluronan and phosphorylcholine-modified chitosan. *Biomacromolecules* **2007**, *8*, 3169–3176. [CrossRef] [PubMed]
48. Picart, C. Polyelectrolyte multilayer films: From physico-chemical properties to the control of cellular processes. *Curr. Med. Chem.* **2008**, *15*, 685–697. [CrossRef] [PubMed]
49. Alves, N.M.; Picart, C.; Mano, J.F. Self assembling and crosslinking of polyelectrolyte multilayer films of chitosan and alginate studied by QCM and IR spectroscopy. *Macromol. Biosci.* **2009**, *9*, 776–785. [CrossRef] [PubMed]
50. Caridade, S.G.; Monge, C.; Gilde, F.; Boudou, T.; Mano, J.F.; Picart, C. Free-standing polyelectrolyte membranes made of chitosan and alginate. *Biomacromolecules* **2013**, *14*, 1653–1660. [CrossRef] [PubMed]
51. Sousa, M.P.; Cleymand, F.; Mano, J.F. Elastic chitosan/chondroitin sulfate multilayer membranes. *Biomed. Mater.* **2016**, *11*, 035008. [CrossRef] [PubMed]
52. Costa, R.R.; Neto, A.I.; Calgeris, I.; Correia, C.R.; Pinho, A.C.M.; Fonseca, J.; Oner, E.T.; Mano, J.F. Adhesive nanostructured multilayer films using a bacterial exopolysaccharide for biomedical applications. *J. Mater. Chem. B* **2013**, *1*, 2367–2374. [CrossRef]

53. Yamada, K.; Chen, T.; Kumar, G.; Vesnovsky, O.; Topoleski, L.D.; Payne, G.F. Chitosan based water-resistant adhesive. Analogy to mussel glue. *Biomacromolecules* **2000**, *1*, 252–258. [CrossRef] [PubMed]
54. Zhou, J.; Defante, A.P.; Lin, F.; Xu, Y.; Yu, J.; Gao, Y.; Childers, E.; Dhinojwala, A.; Becker, M.L. Adhesion properties of catechol-based biodegradable amino acid-based poly(ester urea) copolymers inspired from mussel proteins. *Biomacromolecules* **2015**, *16*, 266–274. [CrossRef] [PubMed]
55. Ninan, L.; Monahan, J.; Stroshine, R.L.; Wilker, J.J.; Shi, R. Adhesive strength of marine mussel extracts on porcine skin. *Biomaterials* **2003**, *24*, 4091–4099. [CrossRef]
56. Kim, H.J.; Hwang, B.H.; Lim, S.; Choi, B.-H.; Kang, S.H.; Cha, H.J. Mussel adhesion-employed water-immiscible fluid bioadhesive for urinary fistula sealing. *Biomaterials* **2015**, *72*, 104–111. [CrossRef] [PubMed]
57. Lauto, A.; Mawad, D.; Foster, L.J.R. Adhesive biomaterials for tissue reconstruction. *J. Chem. Technol. Biotechnol.* **2008**, *83*, 464–472. [CrossRef]
58. Neto, A.I.; Vasconcelos, N.L.; Oliveira, S.M.; Ruiz-Molina, D.; Mano, J.F. High-throughput topographic, mechanical, and biological screening of multilayer films containing mussel-inspired biopolymers. *Adv. Funct. Mater.* **2016**, *26*, 2745–2755. [CrossRef]
59. Lynge, M.E.; van der Westen, R.; Postma, A.; Stadler, B. Polydopamine—A nature-inspired polymer coating for biomedical science. *Nanoscale* **2011**, *3*, 4916–4928. [CrossRef] [PubMed]
60. Ryu, J.H.; Lee, Y.; Kong, W.H.; Kim, T.G.; Park, T.G.; Lee, H. Catechol-functionalized chitosan/pluronic hydrogels for tissue adhesives and hemostatic materials. *Biomacromolecules* **2011**, *12*, 2653–2659. [CrossRef] [PubMed]
61. Madhurakkat Perikamana, S.K.; Lee, J.; Lee, Y.B.; Shin, Y.M.; Lee, E.J.; Mikos, A.G.; Shin, H. Materials from mussel-inspired chemistry for cell and tissue engineering applications. *Biomacromolecules* **2015**, *16*, 2541–2555. [CrossRef] [PubMed]
62. Jo, S.; Kang, S.M.; Park, S.A.; Kim, W.D.; Kwak, J.; Lee, H. Enhanced adhesion of preosteoblasts inside 3D PCL scaffolds by polydopamine coating and mineralization. *Macromol. Biosci.* **2013**, *13*, 1389–1395. [CrossRef] [PubMed]
63. Zhang, X.; Li, Z.; Yuan, X.; Cui, Z.; Yang, X. Fabrication of dopamine-modified hyaluronic acid/chitosan multilayers on titanium alloy by layer-by-layer self-assembly for promoting osteoblast growth. *Appl. Surf. Sci.* **2013**, *284*, 732–737. [CrossRef]
64. Boskey, A.L.; Roy, R. Cell culture systems for studies of bone and tooth mineralization. *Chem. Rev.* **2008**, *108*, 4716–4733. [CrossRef] [PubMed]
65. Barradas, A.M.; Lachmann, K.; Hlawacek, G.; Frielink, C.; Truckenmoller, R.; Boerman, O.C.; van Gastel, R.; Garritsen, H.; Thomas, M.; Moroni, L.; et al. Surface modifications by gas plasma control osteogenic differentiation of MC3T3-E1 cells. *Acta Biomater.* **2012**, *8*, 2969–2977. [CrossRef] [PubMed]
66. Yu, X.; Walsh, J.; Wei, M. Covalent immobilization of collagen on titanium through polydopamine coating to improve cellular performances of MC3T3-E1 cells. *RSC Adv.* **2013**, *4*, 7185–7192. [CrossRef] [PubMed]
67. Zhang, S.; Xu, K.; Darabi, M.A.; Yuan, Q.; Xing, M. Mussel-inspired alginate gel promoting the osteogenic differentiation of mesenchymal stem cells and anti-infection. *Mater. Sci. Eng. C* **2016**, *69*, 496–504. [CrossRef] [PubMed]
68. Lee, S.J.; Lee, D.; Yoon, T.R.; Kim, H.K.; Jo, H.H.; Park, J.S.; Lee, J.H.; Kim, W.D.; Kwon, I.K.; Park, S.A. Surface modification of 3D-printed porous scaffolds via mussel-inspired polydopamine and effective immobilization of rhBMP-2 to promote osteogenic differentiation for bone tissue engineering. *Acta Biomater.* **2016**, *40*, 182–191. [CrossRef] [PubMed]
69. Tsutsumi, K.; Saito, N.; Kawazoe, Y.; Ooi, H.-K.; Shiba, T. Morphogenetic study on the maturation of osteoblastic cell as induced by inorganic polyphosphate. *PLoS ONE* **2014**, *9*, e86834. [CrossRef] [PubMed]

biomimetics

MDPI

Article

Size Control and Fluorescence Labeling of Polydopamine Melanin-Mimetic Nanoparticles for Intracellular Imaging

Devang R. Amin [1,2], Caroline Sugnaux [1], King Hang Aaron Lau [3] and Phillip B. Messersmith [1,*]

[1] Departments of Bioengineering and Materials Science and Engineering, University of California, Berkeley, 210 Hearst Mining Building, Berkeley, CA 94720, USA; damin@berkeley.edu (D.R.A.); csugnaux@berkeley.edu (C.S.)

[2] Department of Biomedical Engineering, Northwestern University, 2145 Sheridan Rd., Evanston, IL 60208, USA

[3] WestCHEM/Department of Pure and Applied Chemistry, University of Strathclyde, 295 Cathedral St., Glasgow G1 1XL, UK; aaron.lau@strath.ac.uk

*   Correspondence: philm@berkeley.edu; Tel.: +1-510-643-9631

Academic Editors: Marco d'Ischia and Daniel Ruiz-Molina
Received: 19 June 2017; Accepted: 30 August 2017; Published: 6 September 2017

Abstract: As synthetic analogs of the natural pigment melanin, polydopamine nanoparticles (NPs) are under active investigation as non-toxic anticancer photothermal agents and as free radical scavenging therapeutics. By analogy to the widely adopted polydopamine coatings, polydopamine NPs offer the potential for facile aqueous synthesis and incorporation of (bio)functional groups under mild temperature and pH conditions. However, clear procedures for the convenient and reproducible control of critical NP properties such as particle diameter, surface charge, and loading with functional molecules have yet to be established. In this work, we have synthesized polydopamine-based melanin-mimetic nanoparticles (MMNPs) with finely controlled diameters spanning ≈25 to 120 nm and report on the pH-dependence of zeta potential, methodologies for PEGylation, and the incorporation of fluorescent organic molecules. A comprehensive suite of complementary techniques, including dynamic light scattering (DLS), cryogenic transmission electron microscopy (cryo-TEM), X-ray photoelectron spectroscopy (XPS), zeta-potential, ultraviolet–visible (UV–Vis) absorption and fluorescence spectroscopy, and confocal microscopy, was used to characterize the MMNPs and their properties. Our PEGylated MMNPs are highly stable in both phosphate-buffered saline (PBS) and in cell culture media and exhibit no cytotoxicity up to at least 100 µg mL^{-1} concentrations. We also show that a post-functionalization methodology for fluorophore loading is especially suitable for producing MMNPs with stable fluorescence and significantly narrower emission profiles than previous reports, suggesting they will be useful for multimodal cell imaging. Our results pave the way towards biomedical imaging and possibly drug delivery applications, as well as fundamental studies of MMNP size and surface chemistry dependent cellular interactions.

Keywords: catechol; melanin; nanoparticle; dopamine

1. Introduction

Nanotechnology has garnered tremendous attention from the biomedical community over the past decade due to its potential to revolutionize cancer treatment by delivering targeted packages of chemotherapeutic drugs, thereby minimizing their adverse side-effects and boosting bioavailability [1–8]. Despite intense research, however, few nanotechnology-based solutions are clinically-approved as cancer therapeutics [8,9]. An improved understanding of nanoparticle–cell and nanoparticle–body interactions is essential for the optimization of nanoparticle (NP) design to improve therapeutic outcomes. Fluorescent NPs enable in-depth study of these phenomena, as illustrated in the recent use of quantum dots (QDs) by Chan et al. to study the fundamental mechanisms of hard NP clearance by the liver [10]. Improving the design of organic NP-based therapeutics requires study of soft organic NPs rather than hard inorganic NPs like QDs, spurring interest in fluorescent organic NPs (FONs) [11,12]. A variety of approaches including emulsion polymerization, block copolymer self-assembly, and nanoprecipitation in the presence of a fluorophore have been employed in the synthesis of FONs [11,12]. However, many FON synthesis techniques require the use of toxic organic solvents or surfactants, which must be removed following synthesis.

The recent development of polydopamine melanin-like NPs has created an opportunity to generate organic NPs in a non-toxic, straightforward strategy. Inspired by the presence of sepia melanin NPs in cuttlefish ink, Ju et al. first reported a synthesis of non-toxic melanin-like NPs composed of polydopamine [13]. Nanoparticles were prepared by dissolving dopamine·HCl in water at basic pH to generate polydopamine NPs, which had melanin-like free radical scavenging activity. Although the exact mechanism and species involved in the formation of polydopamine are still actively under investigation, there is mounting experimental and computational evidence that polydopamine forms via oxidation of dopamine, producing a complex series of subsequent reactions that are not fully understood [14,15]. This material has properties similar to the biological pigment eumelanin, which has functions including protection against harmful ultraviolet (UV) light, free radical scavenging, heavy metal sequestration, and structural roles, as in the *Glycera dibranchiata* bloodworm jaw [16–19].

One unique property of polydopamine is its chemical versatility. Studies on polydopamine surface coatings have shown that it can subsequently be modified via covalent bonding with amines and thiols, hydrogen bonding, π–π stacking, metal coordination, and electrostatic interactions [20,21]. This characteristic of polydopamine may be leveraged to form multifunctional polydopamine-based melanin-mimetic NPs (MMNPs) for biomedical applications without the use of coupling reagents or biomolecular modification, unlike many existing fluorescent NP systems [11,12,22].

Since the initial synthesis of MMNPs by Ju et al. [13], alternate methods of creating MMNPs have also been developed [23,24]. Polydopamine NPs have shown promising results when studied as melanin-like UV-protective materials in cells [25], anticancer photothermal agents in vivo [23], and as magnetic resonance imaging (MRI) contrast agents [26]. Despite these reports, neither the reproducibility of NP size control over a broad range of sub-200 nm diameters nor the NP surface charge have yet been studied in depth.

Related work has focused on synthesis of fluorescent microcapsules or plate-like nanostructures of polydopamine via oxidation by H_2O_2 [26–28] or by combination of polyethyleneimine and polydopamine [29,30]. These novel nanomaterials have promise as potential fluorescent organic NPs or microparticles, but they possess several shortcomings. First, the sizes and morphologies of these fluorescent materials limit conclusions that could be drawn with regard to the uptake and trafficking of spherical NPs with diameters below 100–200 nm, which is the most relevant size ranges for injectable nanotherapeutics. Second, each of these published methods of fluorescent polydopamine preparation results in broad fluorescence excitation and emission spectral peaks that may interfere with dyes to be used as co-stains for in vitro studies. Full-width half maximum (FWHM) of the peaks in these studies are on the order of 100 nm, limiting simultaneous use of other fluorophores. In contrast, quantum dots have spectral linewidths of just 12 nm [31]. Third, fluorescence excitation and emission peaks cannot be tuned using this approach. Modification of NPs with different fluorophores would permit this fine-tuning. We sought to address these three issues through our research.

In this work, we demonstrate a novel, straightforward method by which MMNPs with reproducibly tunable diameters under 100–200 nm can be synthesized and modified by two fluorescent rhodamine dyes, rhodamine 123 (RA123) and rhodamine B (RAB). We have developed procedures by which MMNPs may be rhodamine-labeled either in situ during MMNP formation or by post-functionalization of PEGylated MMNPs. Neither of these methods require any toxic or expensive chemical coupling reagents, organic solvents, surfactants, or oxidizing agents. We demonstrate that these materials have narrower fluorescence excitation and emission peaks (FWHM $\approx$40 nm) relative to other methods of fluorescent polydopamine NP preparation. As a proof of concept, we show that PEGylated fluorescent MMNPs are taken up by cells and accumulate in the perinuclear region, where they can be visualized by confocal microscopy.

2. Materials and Methods

2.1. Materials

Dopamine·HCl (DA, >98% purity), rhodamine 123 (RA123), and rhodamine B base (RAB) were purchased from Sigma-Aldrich (St. Louis, MO, USA). Methyl ether poly(ethylene glycol)-thiol (5 kDa, mPEG-SH) was purchased from Laysan Bio, Inc. (Arab, AL, USA). Neutral red dye was purchased from Amresco (Solon, OH, USA). Ultrapure (UP) water was obtained by purification of deionized water with a Barnstead Ultrapure Water Purification System (Thermo Fisher Scientific, Waltham, MA, USA) to a resistivity of at least 18.0 MΩ cm. Amicon®Ultra centrifugal filters with 10 kDa and 100 kDa molecular weight cut-off (MWCO) were obtained from EMD Millipore (Billerica, MA, USA). Dialysis cassettes (10 kDa MWCO) and cell culture reagents were obtained from Thermo Fisher Scientific. NIH/3T3 fibroblasts were purchased from the American Type Culture Collection (ATCC, Manassas, VA, USA).

2.2. Nanoparticle Synthesis and Modification

2.2.1. MMNP Synthesis

Melanin-mimetic nanoparticle synthesis was adapted from Ju et al. [13]. For a typical synthesis (1:1 NaOH:DA, 1 mg mL^{-1} DA), 22.68 mL of UP water and 1.32 mL of 0.1 M NaOH were added to a 50 mL round bottom flask and heated to 50 °C under vigorous stirring. Then, 1 mL of a 25 mg mL^{-1} DA solution was added. The flask was tightly capped, and the solution was vigorously stirred for 5 h at 50 °C. After 5 h, the reaction mixture was purified by centrifugal filtration (10 kDa MWCO), washing with UP water. Then, aggregates were removed by centrifugation at between 2000 and 6000 g followed by 0.45 μm filtration. The hydrodynamic diameters (D_h) of MMNPs were adjusted by controlling DA concentration (1 to 4 mg mL^{-1}) and NaOH:DA molar ratio (0.5:1 to 1:1).

2.2.2. In Situ Modification of MMNPs with Rhodamine B or Rhodamine 123

For in situ fluorophore modification, MMNPs synthesis was performed as noted above in a growth solution of 1 mg mL^{-1} DA and 1:1 NaOH:DA supplemented with 50 μg mL^{-1} RAB or RA123.

2.2.3. PEGylation of MMNPs

Melanin-mimetic nanoparticles were treated with 10 mM 5 kDa mPEG-SH overnight in 10 mM NaOH. Unbound mPEG-SH was removed by centrifugal filtration (100 kDa MWCO) at 2000× g with washing.

2.2.4. Post-Functionalization of MMNP@PEG with Rhodamine B or Rhodamine 123

Purified MMNP@PEG were post-functionalized for 24 h with RAB or RA123 in either UP water or aqueous pH 8.5 bicine buffer containing 40 μg mL^{-1} MMNP@PEG and 50 μg mL^{-1} RA123 or RAB. The post-functionalized NPs (MMNP@PEG@RAB and MMNP@PEG@RA123) were initially purified by at least six rounds of centrifugal filtration (100 kDa MWCO) with washing to remove unbound fluorophore. Nanoparticles were dialyzed (10 kDa MWCO) for four days in UP water before use in cell culture, replacing the dialysis bath at least five times.

2.2.5. Fluorophore Release Testing

In order to evaluate the release of fluorophore, 50–100 μg fluorophore-labeled MMNPs were dialyzed (10 kDa MWCO) in 200 mL UP water or 1× phosphate-buffered saline (PBS). MMNP@RA123@PEG and MMNP@RAB@PEG were dialyzed for seven days in PBS, and MMNP@PEG@RA123 and MMNP@PEG@RAB were dialyzed sequentially in UP water for three days and in 1× PBS for three days. The dialysis baths were replaced every 4–6 h for the first 12 h to preserve sink conditions and at least every 24 h for the next several days. Fluorophore release was quantified by measuring fluorescence of aliquots of the dialysis baths (Tecan Infinite M200, Männendorf, Switzerland) and quantifying fluorophore content using standard curves prepared from RA123 and RAB stock solutions.

2.3. Nanoparticle Characterization

2.3.1. Extinction Coefficient Calculation

An ultraviolet–visible (UV–Vis) plate reader (Synergy H1, BioTek, Winooski, VT, USA) was used to determine absorbances of solutions with known concentrations of MMNPs at wavelengths from 300 to 1000 nm. At least three batches of MMNPs from each set of synthesis conditions was used to calculate the extinction coefficients. Initial concentration of MMNPs was calculated by lyophilizing 1 mL suspensions of MMNPs. Exponential decay curves were fit to extinction coefficient data using OriginPro 2017 software (Student version, OriginLab, Northampton, MA, USA).

2.3.2. Dynamic Light Scattering and Zeta Potential Analysis

Dynamic light scattering (DLS) and zeta potential analysis of NPs was conducted using a Malvern Zetasizer Nano ZS instrument (Malvern, Worcestershire, UK). The z-average NP diameters of NP batches were calculated using cumulants analysis and was reported as the D_h. The polydispersity index (PDI) was also measured by DLS. During zeta potential measurements, pH was controlled during measurements by using 10 mM citrate buffer (pH 2.5–6.5), 10 mM N-2-hydroxyethylpiperazine-N-2-ethane sulfonic acid (HEPES) buffer (pH 7.0–7.5), or 10 mM bicine buffer (pH 8.0–9.0). Unless otherwise noted, zeta potential was measured at pH 7.4. At least three independently prepared batches of NPs were used for every reported D_h or zeta potential value.

2.3.3. Electron Microscopy Imaging

Scanning electron microscopy (SEM) images were obtained using an FEI Quanta 3D FEG SEM (Hillsboro, OR, USA). Conventional transmission electron microscopy (TEM) imaging was performed on a JEOL 1400 TEM (Tokyo, Japan) or FEI Tecnai 12 TEM (Hillsboro, OR, USA). Samples were prepared either with or without uranyl acetate staining. Cryogenic TEM (cryo-TEM) imaging was performed using a JEOL 1230 TEM (Tokyo, Japan). Nanoparticle size analysis was performed using ImageJ software [32] to measure diameters of at least 35 NPs from representative cryo-TEM images of each size range. All NPs in representative images were included in size analysis.

2.3.4. Spectroscopic Characterization

Fluorescence spectra of fluorescent NPs, fluorophores, and unmodified MMNPs were taken using a FluoroMax-4 spectrophotometer (Horiba Scientific, Irvine, CA, USA) with 5 nm slit widths. Samples containing RA123 and RA123 were excited at $\lambda_{ex} = 500$ nm, and those containing RAB and RAB were excited at $\lambda_{ex} = 555$ nm. The UV–Vis absorbance spectra were taken using a PerkinElmer Lambda UV/Vis/NIR (Waltham, MA, USA) or a UV2600 spectrophotometer (Shimadzu Scientific Instruments, Kyoto, Japan). Fluorescent spectra were not normalized, but absorbance spectra were normalized by multiplying each curve by a constant factor.

2.3.5. X-ray Photoelectron Spectroscopy Characterization

Gold-coated silicon substrates were first cleaned by sonication in UP water, acetone, and isopropanol for 10 min each. Then, after drying them with a flow of nitrogen, the substrates were exposed to a plasma discharge at 60 W for 10 min (Harrick Plasma Cleaner, Ithaca, NY, USA). A 50 µL drop of each NP suspension was then placed onto the surface of the substrates and left to dry overnight. Substrates were completely dried under vacuum prior to analysis using a PHI 5600 spectrometer (PerkinElmer) equipped with an Al monochromated 2 mm filament and a built-in charge neutralizer. The X-ray source operated at 350 W, 14.8 V, and 40° take-off angle. The atomic concentrations of sulfur, nitrogen, oxygen, and carbon of drop-casted MMNP and MMNP@PEG samples by performing survey scans between 0 and 1100 eV electron binding energies. Charge correction was performed setting the C 1s peak at 285.0 eV. Data analysis was conducted using MultiPak software version 9.6.015 (Physical Electronics, Chanhassen, MN, USA).

2.4. In Vitro Uptake and Cytocompatibility Evaluation

2.4.1. MMNP@PEG Cytocompatibility Study

The procedure used for MMNP@PEG cytocompatibility quantification by neutral red uptake was adapted from Repetto et al. [33]. NIH/3T3 fibroblasts were seeded onto a 96-well plate (10,000 cells/well) and incubated overnight in Dulbecco's Modified Eagle's medium (DMEM, Life Technologies Corporation, Carlsbad, CA, USA) supplemented with 5% newborn calf serum (NBCS, Fisher Scientific, Chicago, IL, USA) and 1% penicillin/streptomycin (Life Technologies Corporation). The cell media were then removed, and 0.2 µm filtered 42, 83, and 146 nm diameter MMNP@PEG samples were introduced to the wells in DMEM supplemented with 5% NBCS with penicillin/streptomycin. Dead cell control wells were treated with 0.2 mg mL^{-1} sodium lauryl sulfate-containing media, and live cell control wells were treated with media without MMNP@PEG. Each treatment was performed in three wells. After incubating the fibroblasts with MMNP@PEG for 24 h, cell media were removed from all wells, and the cells were rinsed with PBS. A 40 µg mL^{-1} neutral red solution was added to the wells in DMEM, and the cells were incubated for 3 h. The DMEM was then aspirated off the cells, and the cells were rinsed with PBS. Subsequently, a solution of 50% ethanol/49% UP water/1% glacial acetic acid was added to the wells. The absorbance of each well was read at λ_{abs} = 540 nm in a plate reader (Synergy H1, BioTek). The data were normalized as follows:

$$\text{Relative Cell Viability} = \frac{OD540_{treated} - OD540_{dead}}{OD540_{untreated}} \tag{1}$$

$OD540_{treated}$ represents the optical density of treated cells at 540 nm, $OD540_{dead}$ represents the optical density of dead control cells killed with 0.2 mg mL^{-1} sodium lauryl sulfate at 540 nm, and $OD540_{untreated}$ represents the optical density of live control cells treated with MMNP@PEG-free media at 540 nm. The OD540 of live cells treated with MMNP@PEG without neutral red was negligible following the PBS rinse step, obviating the need to correct OD540 of neutral red-treated cells further.

2.4.2. MMNP@PEG@RA123 Uptake Study

NIH/3T3 fibroblasts were seeded onto 35 mm tissue culture dishes (FluoroDish, World Precision Instruments, Sarasota, FL, USA) and incubated for 24 h with 0.2 µm filtered 20 µg mL^{-1} MMNP@PEG@RA123 in DMEM supplemented with 10% fetal bovine serum (FBS, Life Technologies) and 1% penicillin/streptomycin. MMNP@PEG@RA123 were prepared by post-functionalization of MMNP@PEG with RA123 in water. The cells were then rinsed with PBS and treated with Hoechst nuclear stain (Life Technologies). Standard and z-stack images of the live cells were taken using a Zeiss LSM 510 inverted confocal microscope (Carl Zeiss AG, Oberkochen, Germany). Hoechst staining was observed at λ_{em} = 410 nm and two-photon excitation (λ_{ex} = 760 nm), and MMNP@PEG@RA123 were visualized using λ_{em} = 525 nm and λ_{ex} = 488 nm. z-Stack images were taken with slices spaced evenly over 8–15 μm z-stack heights. Images were processed using ZEN 2.3 Lite software (Blue edition, Carl Zeiss Microscopy GmbH, Munich, Germany).

2.5. Statistical Analysis

Statistical analysis was performed in Minitab 17 (Minitab Inc., State College, PA, USA) by conducting analysis of variance (ANOVA) followed by post-hoc Tukey tests. Error bars in figures represent standard deviations (SD) or standard errors as specified.

3. Results and Discussion

3.1. MMNP Formation via Dopamine Autoxidation, Comparisons with Melanin, and Characterization of Surface Charge

After adding DA to aqueous NaOH, the solution gradually changed from colorless to yellow to dark brown-black within one hour. The purified MMNP suspension was black and demonstrated a smooth, monotonically decaying broadband UV–Vis absorbance like melanin, with highest absorbance in the UV region (Figure 1a). In contrast, measurements of the supernatant solution removed during growth show absorbance peaks at 280 and 398 nm superimposed upon this monotonically decaying curve. These peaks have been attributed to the (precursor) DA monomer and its oxidation product dopamine *o*-quinone, respectively. Since no oxidant was added to the growth solution except for ambient dissolved oxygen, the presence of both DA and dopamine *o*-quinone in the solution phase confirm that MMNP formation follows an autoxidation route [34,35]. The monotonically decaying UV–Vis absorbance in both the filtrate and pure MMNP spectra are consistent with the formation of polydopamine, some of which may be present as pre-formed oligomers below 10 kDa in the raw product (ibid.).

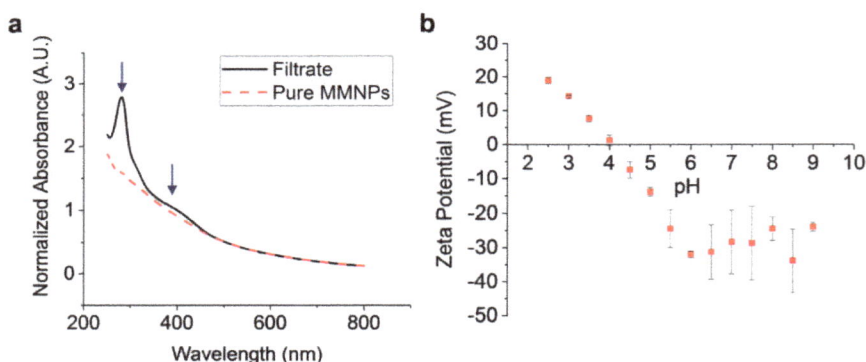

Figure 1. Ultraviolet–visible (UV–Vis) absorbance and surface charge of polydopamine-based melanin-mimetic nanoparticles (MMNPs). (**a**) UV-Vis spectra of purified MMNPs and filtrate removed from crude product via 10 kDa centrifugal filtration. Arrows in (**a**) indicate the two peaks observed at 280 and 398 nm in the filtrate absorbance spectrum that are absent in the purified MMNP absorbance spectrum. A.U.: Arbitrary units. (**b**) Zeta potential of MMNPs at pH 2.5–9.0. The isoelectric point is approximately pH 4.0–4.1.

For each synthetic condition, MMNP extinction coefficients were calculated at wavelengths from 300 to 1000 nm (Supplementary Figure S1). The fit of these coefficients to a single exponential decay function of wavelength was excellent (r^2 > 0.998; Supplementary Figure S1 and Table S1). Notably, these extinction coefficients match closely with values reported by Sarna et al. for melanin, especially for MMNPs with D_h < 50 nm (Supplementary Figure S1) [36,37]. These calculated extinction coefficients enabled rapid quantification of MMNP concentrations in our study and could be used in future work. Finally, we investigated the surface charge of MMNPs as a function of pH between pH 2.5 and 9.0 (Figure 1b) to determine the potential role of electrostatic interactions in MMNP surface loading (see Sections 3.3–3.5). An isoelectric point of approximately pH 4.0–4.1 was observed, which is in agreement with previous findings on polydopamine films [38]. X-ray photoelectron spectroscopy data also shows a carbon to oxygen ratio of the MMNPs that is essentially identical with polydopamine (see Section 3.3).

The chemical similarity of the MMNPs and polydopamine coatings indicates that common methods of polydopamine functionalization may be applied to the MMNPs. Moreover, the clear trend in surface ionization of MMNPs with pH suggests that electrostatic attraction may be utilized for MMNP modification in a pH-dependent manner (see Section 3.5).

3.2. MMNP Size Control

By varying the DA concentration (1 to 4 mg mL^{-1}) and NaOH:DA molar ratio (0.5:1 to 1:1) in the synthesis of MMNPs, nanoparticle D_h could be adjusted from 28 to 117 nm. Figure 2 shows the diameters of MMNPs as measured by DLS. Highly reproducible results were obtained by our synthetic methodology. All reported results represent the average of at least three (and up to 19) independent sample preparations.

Note that the commercially available DA is an HCl salt, and the NaOH serves to neutralize this salt as well as to increase the pH to facilitate polydopamine formation. Second, increasing DA concentration at constant NaOH:DA molar ratio resulted in larger NPs (Figure 2a and Supplementary Figure S2). Holding the NaOH:DA ratio constant at 1:1, increasing DA from 1 to 2 mg mL^{-1} resulted in an increase of D_h from 28.1 ± 8.8 nm to 49.5 ± 12.3 nm (mean ± SD). We attribute this to the increased quantity of dopamine available to bind to each nucleated NP. These trends demonstrate significantly finer control of NP diameter over the sub-100 nm scale compared to previous attempts to control MMNP sizes [13].

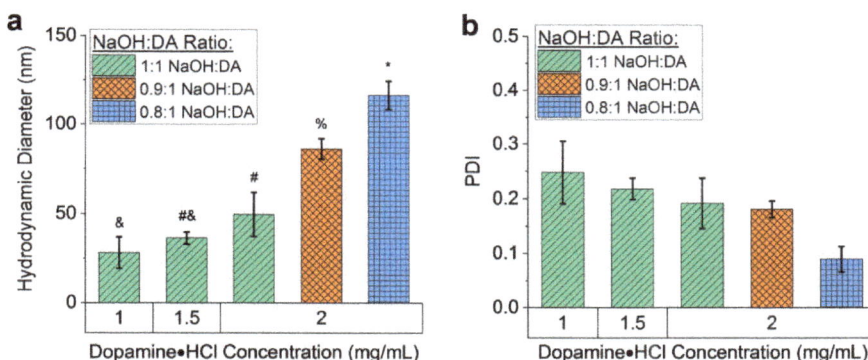

Figure 2. Dynamic light scattering (DLS) analysis of MMNPs. (**a**) Mean hydrodynamic diameters and (**b**) polydispersity indices (PDI) of multiple batches of MMNPs prepared at various dopamine·HCl (DA) concentrations and NaOH:DA ratios. $n = 3-19$ independently prepared batches of MMNPs were analyzed for each synthetic condition. Error bars represent standard deviations. Bars not sharing symbols in (**a**) differ significantly with $p < 0.001$.

Although the average D_h values were highly consistent from batch to batch, this consistency must be distinguished from the variance in MMNP diameter within each batch, which was assessed initially by PDI. The average PDI of NP batches prepared at each condition ranged from 0.09 to 0.25 (Figure 2b). This result indicates that individual MMNP batches are relatively monodisperse for organic NPs but that size analysis beyond cumulants analysis of DLS data is required [39]. As such, the polydispersity and morphology of MMNPs were further assessed by SEM, TEM, and cryo-TEM (Figure 3; see also Supplementary Figures S3 and S4, and Table S2). Spherical NPs were always observed. The imaging data also corroborate the DLS data demonstrating that NP size increased as DA concentration increased and as NaOH:DA ratio decreased (Figure 3).

Unimodal size distributions were observed for synthesis conditions that produced MMNPs up to a diameter of $\approx$50 to 60 nm (Figure 3). Minimal MMNP aggregation was observed in images obtained by cryo-TEM, which does not suffer from the drying artifacts of conventional TEM and SEM, confirming that the products mainly consisted of dispersed NPs. In particular, low polydispersities were obtained for MMNPs produced both at 1 mg mL^{-1} and 2 mg mL^{-1} DA with 1:1 NaOH:DA (SD of 9.1 nm and 12.4 nm were observed, respectively; Figure 3h,i and Supplementary Table S2). However, the NP size distribution was bimodal at 2 mg mL^{-1} DA with only 0.8:1 NaOH (Figure 3g), with distinct NP populations centering around D_h = 65 nm and 100 nm. Because the DLS signal intensity is related to the 6th power of the particle diameter (i.e., weighted more heavily toward the larger nanoparticles), this cryo-TEM result is consistent with the DLS data shown in Figure 2 indicating an average D_h = 120 nm at this condition.

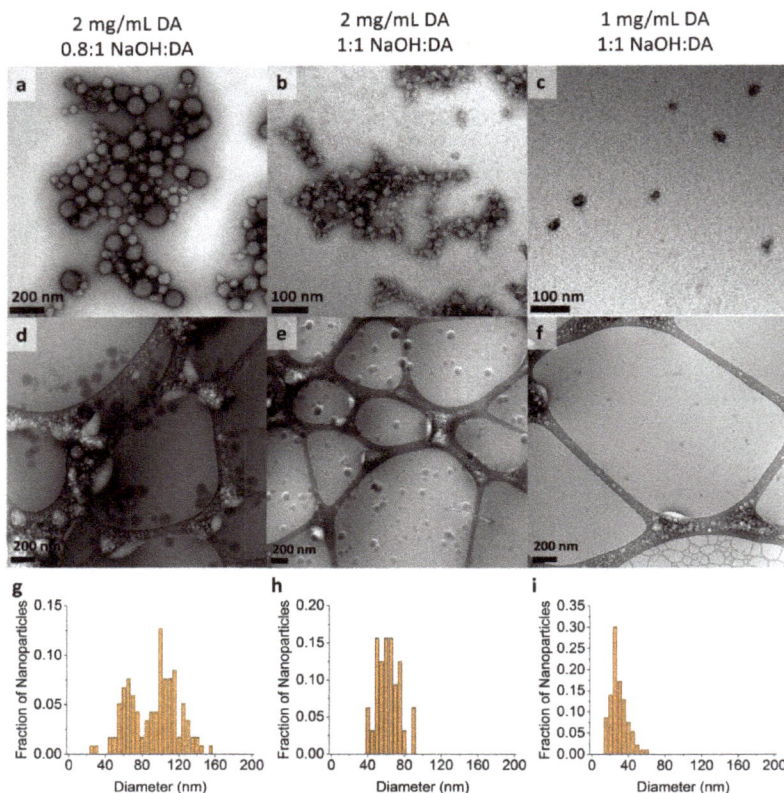

Figure 3. Transmission electron microscopy (TEM) images of MMNPs and quantitative analysis of nanoparticle diameter grown at the conditions specified at the top of each column. (**a–c**) TEM images with uranyl acetate negative stain. (**d–f**) Cryo-TEM images were taken without staining. Nanoparticles are spherical but have rougher appearances as diameter decreases. (**g–i**) Distribution of MMNP diameters in cryo-TEM images.

3.3. PEGylation to Produce MMNP@PEG

We focused on the 49.5 nm MMNPs for PEGylation studies because these NPs would remain within a biologically useful size regime following modification (i.e., D_h <100 nm). PEGylation was achieved by overnight treatment with 10 mM 5 kDa mPEG-SH in 10 mM NaOH, and the NPs were evaluated using zeta potential, DLS, and XPS data. Control batches of MMNPs were treated with 10 mM NaOH base without mPEG-SH.

The zeta potential of the resulting MMNP@PEG was -20.9 ± 0.5 mV at pH 7.4, which was significantly higher than those of both untreated MMNPs (-32.5 ± 0.1 mV) and base treated MMNPs (-33.8 $\pm$ 1.0 mV) at pH 7.4 (Figure 4a), indicating shielding of the negatively charged polydopamine MMNP surface. Consistent with the zeta potential results, DLS shows PEGylation increased the D_h of the MMNPs by 24 nm to 71.5 ± 1.1 nm (Figure 4b). Melanin-mimetic nanoparticles treated with base alone did not have a significantly greater D_h than untreated MMNPs, confirming that the NP diameter increase was not caused by MMNP growth or aggregation in basic conditions. Additionally, TEM imaging shows that MMNP@PEG have spherical morphology similar to that of MMNPs (Figure 4c), validating the use of the standard spherical NP analysis of the DLS data.

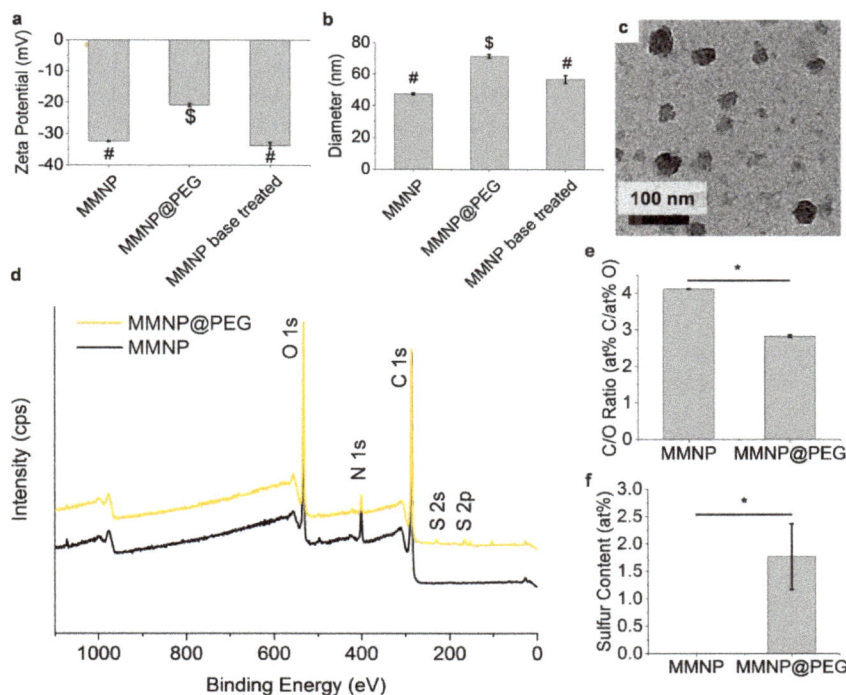

Figure 4. MMNP@PEG vs. MMNP zeta potential, hydrodynamic diameter, morphology, and atomic composition. (**a**) Zeta potentials and (**b**) hydrodynamic diameters of MMNPs, MMNP@PEG, and control MMNPs treated with 10 mM NaOH base. Samples not sharing symbols are significantly different ($p < 0.05$). (**c**) TEM image of MMNP@PEG. (**d**) XPS survey scans of MMNP and MMNP@PEG with assignments for O 1s, N 1s, C 1s, and S 2s, and S 2p peaks. (**e**) C/O atomic ratios in MMNP vs. MMNP@PEG calculated from C 1s and O 1s signal ratios (* $p < 0.01$). at%: Atomic percent relative to total C, N, O, and S content. (**f**) Sulfur content in MMNP vs. MMNP@PEG calculated from S 2p signal intensity expressed as at% S (* $p < 0.01$). Error bars represent standard errors.

The thickness of the PEG layer is 12 nm (half of the change in D_h), twice the thickness that would be expected from the mushroom regime [40,41], providing evidence that PEG is packed in the brush regime rather than the mushroom regime at the surface of MMNPs. A PEG brush causing a similar diameter increase has also been reported previously to provide sufficient resistance against protein adsorption and phagocytosis on other organic and inorganic NP cores, including poly(lactic acid) (PLA), poly(lactic-co-glycolic acid) (PLGA), polycaprolactone (PCL), and gold [42–47].

PEGylation was further corroborated by XPS analysis of drop-casted NPs suspensions (Figure 4d). In addition to O 1s, N 1s, and C 1s signals, the survey spectrum of MMNP@PEG reveals the presence S 2s and S 2p signals, indicating the presence of sulfur from mPEG-SH. Sulfur peaks were absent in unmodified MMNP controls. Moreover, the C/O atomic ratios decreased from 4.12 ± 0.01 for unfunctionalized MMNP controls to 2.82 ± 0.03 for MMNP@PEG. Since MMNP@PEG is compositionally a mixture of polydopamine and PEG, this latter C/O ratio is consistent with successful PEGylation because it is intermediate between the theoretical ratio of 4 for dopamine and its oxidation products and a ratio of 2 for mPEG-SH. Furthermore, while high resolution C 1s and O 1s spectra for MMNPs show π–π^*, C=O, C–O/C–N, C–H_x/C–C chemical shifts corresponding to previously reported polydopamine coatings (Supplementary Figure S5) [48,49], the spectra of MMNP@PEG show large increases of C–O components, demonstrating the presence of PEG on the MMNP@PEG.

3.4. Stability of MMNP@PEG

The stabilities of MMNP and MMNP@PEG samples were compared by immersion in $1\times$ PBS and in cell culture media (DMEM + 10% serum) for 24 h. Before PEGylation, MMNPs were stable in UP water but aggregated in $1\times$ PBS (Supplementary Figure S6a). Thus, the electrostatic repulsion between unfunctionalized MMNPs was insufficient in maintaining colloidal stability with screening at physiologic ionic strength. Incidentally, in cell culture media with serum, MMNPs do not visibly aggregate and appeared to be stable (Supplementary Figure S6b). Potentially, this effect is due to the sterically stabilizing effect of serum proteins bound to the polydopamine surface of MMNPs, as it is known that amine groups in proteins can covalently bind to the polydopamine surfaces at physiologic pH [50]. In contrast, MMNP@PEG remained stable for 24 h in both $1\times$ PBS and cell culture media (Supplementary Figure S6c,d). Even after fluorophore modification, it was noted that MMNP@PEG remained stable in $1\times$ PBS for up to seven days (see Sections 3.5.1 and 3.5.2).

3.5. Fluorescence Functionalization

We compared two methods of MMNP functionalization with small organic fluorophores (Figure 5): an in situ method in which the fluorescent molecules were mixed and incorporated with dopamine during MMNP formation, and a post-functionalization method in which the fluorophores were added onto purified MMNP@PEG. Both RA123 and RAB were used as model fluorophores. Both may interact with polydopamine via π–π stacking, hydrogen bonding, or electrostatic interactions. In addition, RA123 has a primary amine that may behave as a weak nucleophile to covalently bind to oxidized quinones in polydopamine (Figure 5a). The predominant structure of RAB is the fluorescent zwitterion, but a significant fraction of RAB also exists as a non-fluorescent lactone, with K_{eq} = [zwitterion]/[lactone] = 4.4 at 25 °C in water (Figure 5b) [51]. The acidic cation of RAB has $pK_a \approx 3.2$ but has been reported to increase up to 5.70 in the presence of microheterogeneities in solution, such as the interfaces formed by surfactants, thus stabilizing it at higher pH than in homogeneous solutions [52]. These molecules and their modes of binding may also be viewed as models for the incorporation of other functionalities, such as chemotherapeutics.

3.5.1. In situ Incorporation

Melanin-mimetic nanoparticles were labeled in situ by growing MMNPs in 1 mg mL^{-1} DA and 1:1 NaOH:DA in the presence of 50 μg mL^{-1} RA123 or RAB to prepare MMNP@RA123 or MMNP@RAB, respectively (Figure 5c). These NPs were then modified with 5 kDa mPEG-SH to form MMNP@RA123@PEG and MMNP@RAB@PEG. In order to remove loosely bound dye, samples were centrifugally filtered with extensive washing and then dialyzed for seven days in 1× PBS. Approximately 90% of the physisorbed dye remaining after centrifugal filtration was released within the first 24 h of dialysis (Supplementary Figure S7a). No aggregation was observed during immersion in PBS for one week, indicating good steric stability imparted by the PEG coating.

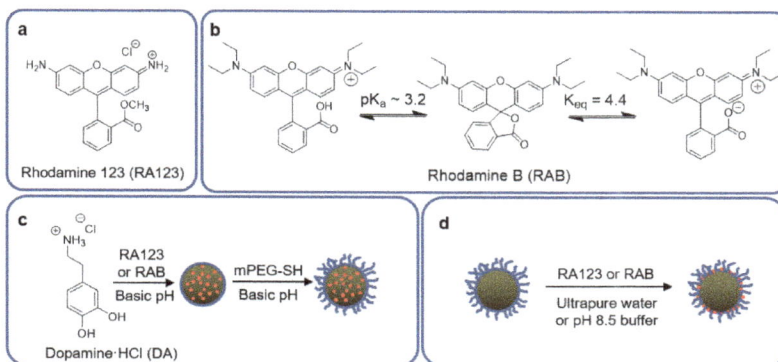

Figure 5. Approaches to synthesis of fluorescent MMNPs. (**a**) Structure of rhodamine 123 (RA123). (**b**) Structures of rhodamine B (RAB), including the fluorescent cationic acid, non-fluorescent neutral lactone, and fluorescent zwitterionic structures. (**c**) In situ approach and subsequent PEGylation: MMNP@RA123 and MMNP@RAB are first synthesized by DA polymerization in the presence of RA123 or RAB. These fluorescent NPs are subsequently PEGylated, forming MMNP@RA123@PEG and MMNP@RAB@PEG. (**d**) Post-functionalization approach: MMNP@PEG@RA123 and MMNP@PEG@RAB are formed by treatment of MMNP@PEG with RA123 or RAB in unbuffered ultrapure water or pH 8.5 buffer.

In the first step of MMNP growth in solution mixtures of DA and rhodamine, both UV–Vis absorbance spectroscopy and fluorimetry provided evidence that RA123 and RAB were successfully incorporated into the in situ labeled NPs and retained after centrifugal filtration (Supplementary Figures S8 and S9a,b). The fluorescence emission peaks of MMNP@RA123@PEG and MMNP@RAB@PEG were centered at 520 nm and 573 nm, respectively, similar to the free dyes, and remained at those locations following extensive dialysis (Figure 6a,c). The 10 nm red-shift in the MMNP@RA123@PEG absorbance peak (λ_{abs} = 510 nm) relative to free RA123 (λ_{abs} = 500 nm) may indicate some dye aggregation in the NPs. After seven day dialysis in 1× PBS, the fluorescent signal of a 25 µg mL^{-1} solution of MMNP@RA123@PEG approximately corresponds to that of 5.4 ng mL^{-1} free RA123, and the fluorescence of a 25 µg mL^{-1} solution of MMNP@RAB@PEG approximately corresponds to that of 3.6 ng mL^{-1} free RAB. These in situ-modified NPs also have full-width half-maximum (FWHM) of approximately 45 nm, which is two to three times narrower than previously reported fluorescent polydopamine NP systems [26–28,30]. The broadband UV–Vis absorbance pattern also verified that polydopamine growth could proceed in the presence of rhodamine (Supplementary Figure S8). Furthermore, the largely negative zeta potentials of both MMNP@RA123 and MMNP@RAB were not significantly different from those of MMNPs without rhodamine, indicating that the dyes were chiefly incorporated into the interior of the NPs (Supplementary Figure S10a). Otherwise, the positively charged RA123 or various forms of RAB would have increased the zeta potential significantly versus vs. MMNPs by masking the negative MMNP surface charge or by reversing it, especially if the fluorophores segregated to the NP surface.

After PEGylation, zeta potential measurement provided evidence that grafting the polydopamine NP surfaces with mPEG-SH was successful, as the zeta potentials became significantly less negative

(-9.9 ± 1.4 mV for MMNP@RA123@PEG and -7.6 ± 1.9 mV for MMNP@RAB@PEG, vs. -34.4 ± 0.8 mV for MMNP@RA123 and -39.3 ± 1.3 mV for MMNP@RAB; Supplementary Figure S10a). However, a high polydispersity interfered with quantitative use of DLS data (Supplementary Figure S10b), and an increase in D_h following PEGylation could not be confirmed. In fact, TEM shows that these NPs were more polydisperse and less well-defined than MMNPs grown without dye (Figure 6b,d). It is possible that a lower level of rhodamine incorporation could restore normal MMNP growth and this could be worth pursuing in future work given the encouraging fluorescence profile, colloidal stability, and straightforward synthesis of the in situ modified NPs.

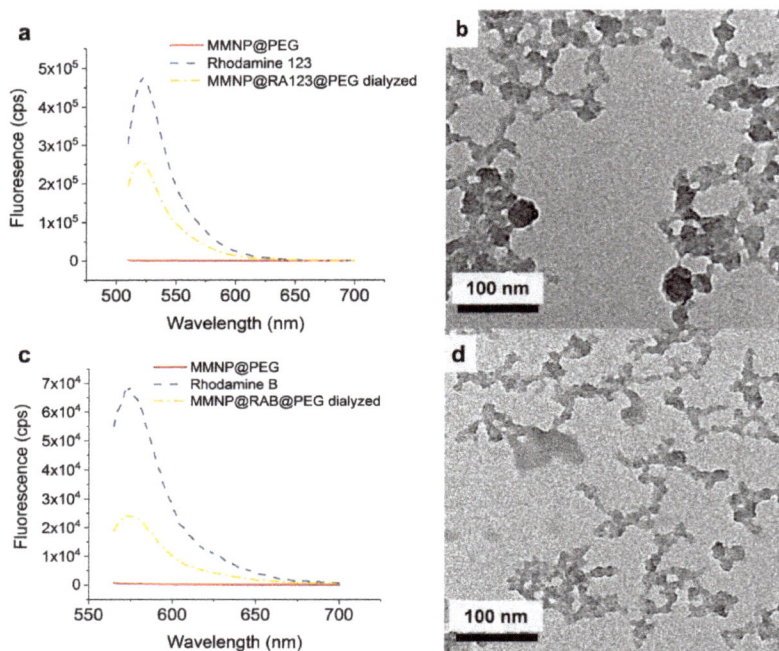

Figure 6. Fluorescence emission spectra and TEM images of in situ labeled MMNP@RA123 and MMNP@RAB. (**a**) Fluorescent emission spectra (λ_{ex} = 500 nm) of MMNP@RA123@PEG after seven day dialysis in 1× phosphate-buffered saline (PBS), rhodamine 123, and MMNP@PEG. (**b**) TEM image of MMNP@RAB. (**c**) Fluorescent emission spectra (λ_{ex} = 555 nm) of MMNP@RAB@PEG after seven day dialysis in 1× PBS, rhodamine B, and MMNP@PEG. (**d**) TEM image of MMNP@RA123.

3.5.2. Post-Functionalization

MMNP@PEG (40 µg mL^{-1}; D_h = 71.5 $\pm$ 0.6 nm) were post-functionalized by incubating in 50 µg mL^{-1} RA123 or RAB dye to form MMNP@PEG@RA123 or MMNP@PEG@RAB, respectively (Figure 5d). Two solution conditions were tested: functionalization in UP water and in buffer at pH 8.5. Both RA123 and RAB may modify the free polydopamine surface remaining in between the PEG chains via non-covalent interactions such as π–π stacking or hydrogen bonding. The positive charge of RA123 could also promote more electrostatic attraction to the negatively charged polydopamine surface than RAB. The primary amine on the RA123 could undergo Michael addition for covalent binding to polydopamine especially at the pH 8.5 basic condition as well [50], although this coupling may not be prominent, since the aromatic primary amine is a weak nucleophile [53].

Rhodamine functionalization was first confirmed by the appearance of prominent absorption peaks in UV–Vis spectra and fluorescence emission spectra taken directly following extensive centrifugal filtration to remove the dissolved free dye in the solution used for functionalization (Supplementary Figures S9c,d and S11). The red-shifted absorbance peaks on MMNP@PEG@RA123 (λ_{abs} = 520 nm) vs. free RA123 (λ_{abs} = 500 nm) indicate that the RA123 has aggregated on the NP surface, potentially due to high loading. No obvious differences were noted between samples modified in UP water or at pH 8.5. Moreover, both MMNP@PEG@RAB and MMNP@PEG@RA123 have significantly higher zeta potentials than MMNP@PEG (Figure 7a), further indicating coverage of and binding to the polydopamine NP surface underlying the PEG brush. The finding that the zeta potentials for all of the post-functionalized NPs were similar may indicate that the cationic form of RAB is stabilized at the negatively charged MMNP surface, as observed in microheterogeneous solutions containing surfactant micelles [52]. Dynamic light scattering measurements show that the NP diameter generally did not increase after dye functionalization, except for a <10% increase for MMNP@PEG@RAB modified at pH 8.5 (Figure 7b). It is thus unlikely that polydopamine growth or NP aggregation occurred during fluorophore loading.

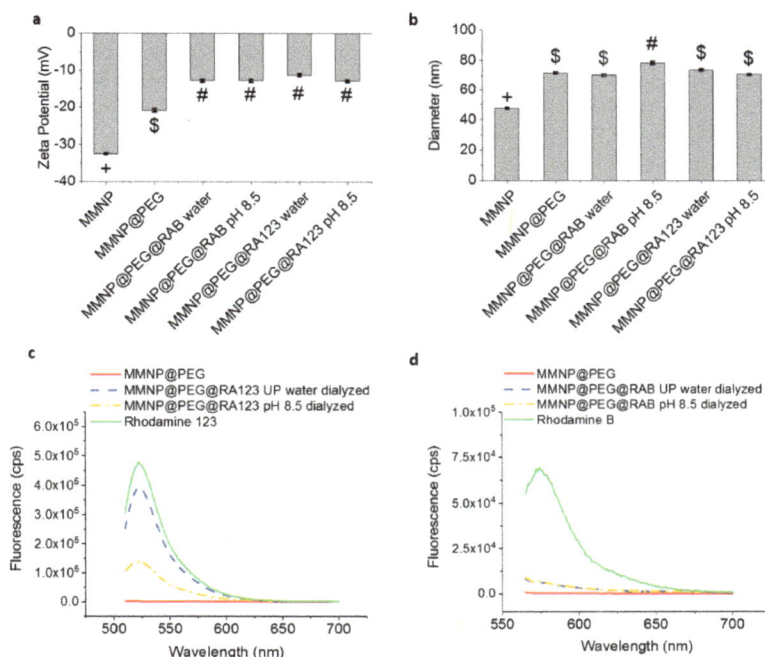

Figure 7. Fluorescence, hydrodynamic diameter, and zeta potential of rhodamine post-functionalized MMNP@PEG. (**a**) Zeta potentials and (**b**) hydrodynamic diameters of rhodamine post-functionalized MMNP@PEG samples prepared in water or at pH 8.5 vs. unmodified MMNPs and MMNP@PEG. Groups not sharing symbols have significantly different values ($p < 0.05$). (**c**) Fluorescence emission spectra (λ_{ex} = 500 nm) of 25 μg mL^{-1} samples of MMNP@PEG before and after modification with RA123 in water or at pH 8.5 followed by serial dialysis in ultrapure (UP) water for 72 h and 1× PBS for 72 h. Emission spectrum of RA123 was taken at 10 ng mL^{-1}. (**d**) Fluorescence emission spectra (λ_{ex} = 555 nm) of 25 μg mL^{-1} samples of MMNP@PEG before and after modification with RAB in water or at pH 8.5 followed by serial dialysis in UP water for 72 h and 1× PBS for 72 h. Emission spectrum of RAB was taken at 10 ng mL^{-1}.

To ensure that the fluorescent emission of MMNP@PEG@RA123 and MMNP@PEG@RAB was due to the fluorophores bound to the MMNPs and that this emission would be stable, additional dialysis was performed after centrifugal filtration—72 h in UP water followed by a further 72 h in $1\times$ PBS—to remove dye molecules that could be desorbed from the MMNPs. The dialysis process was successful in removing this loosely bound fraction (over 80% of removable fraction of dyes was released within the first 24 h of the first UP water dialysis) (Supplementary Figure S7b,c). Although measurements of the emission levels during dialysis do show that a large portion of the initially measured fluorescence was due to loosely bound dyes that desorb from the NP surface (Supplementary Figure S7b,c), the fluorescence of MMNP@PEG@RA123 was still detectable (Figure 7c). The emission peaks of MMNP@PEG@RA123 centered at λ_{em} = 520–524 nm, which are essentially unchanged from the free dye. The fluorescence remaining in 25 μg mL^{-1} samples of MMNP@PEG@RA123 corresponded to 2.9 ng mL^{-1} RA123 for the pH 8.5 modification condition and 8.3 ng mL^{-1} RA123 for the UP water modification condition. Thus, the RA123 remained strongly bound to the MMNP@PEG@RA123 surface, and the pH 8.5 condition did not enhance interactions between aromatic amines on RA123 and polydopamine vs. UP water.

On the other hand, no emission peaks are observed in dialyzed MMNP@PEG@RAB samples (Figure 7d). Taken together with zeta potential results, which suggest the presence of the acidic cation of RAB at the MMNP surface, the almost total removal of RAB after dialysis also suggests an electrostatic binding mechanism: During dialysis, the bound fluorescent RAB cation may equilibrate with the non-fluorescent lactone and fluorescent zwitterionic RAB forms, which may subsequently desorb from the NP surface due to less electrostatic attraction to polydopamine.

We also observed that during the dialysis process, more RA123 was released in the first UP water dialysis step for samples prepared at pH 8.5 than in UP water. This result indicates that covalent bonding is not preferred at pH 8.5 and that electrostatic attraction between RA123 and polydopamine may be the preferred mechanism of RA123 loading onto the MMNP surface. Regardless, the fluorescence of the MMNP@PEG@RA123 after extensive dialysis also shows that this physical binding is sufficient to obtain stable fluorescent NPs. More RA123 was released from samples prepared in UP water in the second dialysis step in $1\times$ PBS. The origin of this effect is unclear. Nevertheless, the level of RA123 fluorescence retained on NPs functionalized in UP water was significantly higher, and this approach was used to generate MMNP@PEG@RA123 for cell work.

3.6. In Vitro Cytocompatibility of MMNP@PEG and Imaging of MMNP@PEG@RA123

The viability of NIH/3T3 fibroblasts incubated in media loaded with MMNP@PEG was evaluated. A range of NPs with D_h = 42 nm to 146 nm were tested (the diameters refer to the values measured for the specific batch of NPs used for each viability assay rather than the averages shown in Figure 2). No toxicity was observed over a duration of 24 h at all tested concentrations (1–100 μg mL^{-1}; Figure 8a). In fact, some increase in relative cell viability was observed for cells treated with MMNP@PEGs, most notably for the smallest 42 nm diameter tested (up to 40% higher). This effect was previously observed in HeLa cells at 6–75 μg mL^{-1} treatments, but not in 4T1 cells [13,23]. It is possible that this dose-dependent effect stems from the known antioxidant capacity of MMNPs [13], which may alter cellular proliferation by limiting oxidative stress in some cells. Finally, confocal microscopy was used to characterize the cell uptake of MMNP@PEG@RA123. Figure 8b shows a representative three-dimensional (3D) z-stack composite reconstruction of the NIH/3T3 fibroblasts treated with both Hoechst dye and MMNP@PEG@RA123 (separately imaged with 760 nm two-photon and regular 488 nm excitation, respectively). After 24 h incubation, the fluorescence associated with the MMNP@PEG@RA123 could be clearly observed, even at the relatively low incubation concentration of 20 μg mL^{-1}. Additional z-stack confocal images and 3D reconstructions of treated cells vs. untreated control cells confirmed that the observed fluorescence was located within the cells, indicating MMNP internalization (Figure 8b and Supplementary Figures S12–S16). Co-staining the cells with the Hoechst dye used for nuclear staining revealed that the MMNP@PEG@RA123 was concentrated in the perinuclear region—they were excluded from both the cell nuclei and the filapodia regions. It was also observed that MMNP@PEG@RA123 have a punctate distribution within cells. From these confocal microscopy images, it is evident that MMNP@PEG@RA123 is sufficiently stable to be utilized in high-resolution, multimodal cell imaging.

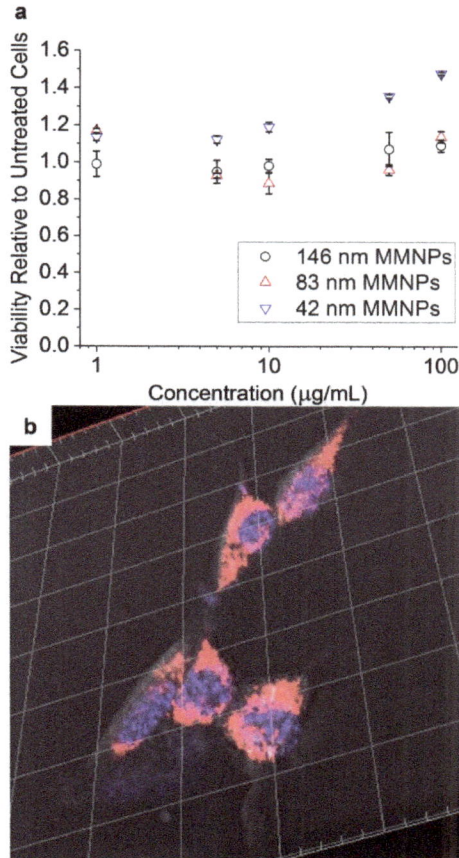

Figure 8. In vitro investigation of MMNP-cell interactions. (**a**) MMNP cytocompatibility with NIH/3T3 fibroblasts as measured by neutral red uptake viability assay. Error bars represent standard errors of triplicate experiments. (**b**) Representative confocal microscopy three-dimensional (3D) z-stack reconstruction image of Hoechst-stained NIH/3T3 fibroblasts treated with 20 μg mL^{-1} MMNP@PEG@RA123. Hoechst stain (blue) and rhodamine fluorescence (red/pink) are shown here; Scale bar: 20 μm between gridlines.

4. Conclusions

We have demonstrated spherical MMNPs labeled with fluorescent dyes with controlled diameters. Careful measurements based on multiple independent batches of NP preparation (up to 19) showed how adjustment of DA concentration and NaOH:DA ratio in MMNP synthesis could be used to achieve fine control of batch mean MMNP diameter in the sub-100 nm range. Similarly, our conditions for MMNP PEGylation produced particles with a high degree of stability in both 1× PBS and in cell culture media. Melanin-mimetic nanoparticle formation was shown to follow an autoxidation route,

Biomimetics **2017**, *2*, 17

and the similarities of MMNPs with polydopamine and melanin in terms of chemical identity and surface charge were shown by UV–Vis absorption, zeta potential, and XPS analysis. No cytotoxicity was observed over the entire range of diameters from ≈40 to 150 nm tested.

We also compared two approaches of loading MMNPs with aromatic fluorescent dyes—in situ dye loading during MMNP formation and post-functionalization after MMNP formation and PEGylation. The fluorescence spectra of MMNPs obtained using both protocols produced emission peak widths ≈40 nm FWHM, similar to the free dye and less than half that of previous reports of fluorescent polydopamine NPs. The in situ approach, however, modified the shape of the MMNPs, but post-functionalization could be used to produce spherical MMNPs with stable fluorescence suitable for high-resolution multimodal confocal live cell imaging.

The reproducible diameter control and facile methodologies for functionalizing and loading the MMNPs are highly applicable to fundamental studies and further refinement of organic NP–cell interactions, such as size-dependent cellular uptake and intracellular trafficking using targeting moieties. The stable and spectrally narrow fluorescence measured indicate that our protocol could be beneficial for incorporating dyes into MMNPs (or other polydopamine matrices) for multimodal imaging studies, or for delivery of therapeutic molecules with weak interactions to MMNPs.

Supplementary Materials: The following are available online at http://www.mdpi.com/2313-7673/2/3/17/s1: Figure S1: Extinction coefficients of melanin and MMNPs prepared at various synthetic conditions, Figure S2: Melanin-mimetic nanoparticle diameter dependence on [NaOH]:[DA] and DA concentration, Figure S3: Scanning electron microscopy image of MMNPs prepared in 2 mg mL^{-1} DA with 1:1 NaOH:DA, Figure S4: Transmission electron microscopy characterization of MMNPs, Figure S5: Melanin-mimetic nanoparticle and MMNP@PEG high-resolution C 1s and O 1s XPS peaks with peak deconvolutions, Figure S6: Evaluation of MMNP and MMNP@PEG stability in 1× PBS and DMEM + 10% serum, Figure S7: Dialysis of rhodamine-labeled MMNPs in water and PBS following synthesis and centrifugal filtration, Figure S8: Ultraviolet–visible absorbance spectra of in situ labeled MMNPs, Figure S9: Fluorescence emission spectra of rhodamine 123 and rhodamine B-labeled MMNPs following centrifugal filtration and before dialysis, Figure S10: Zeta potential and dynamic light scattering characterization of in situ labeled MMNPs, Figure S11: Ultraviolet–visible absorbance spectra of rhodamine-treated MMNP@PEG, Figure S12: Confocal z-stack images of control Hoechst-stained NIH/3T3 fibroblasts untreated with MMNP@PEG@RA123, Figure S13: Confocal z-stack images of Hoechst-stained NIH/3T3 fibroblasts treated for 24 h with 20 μg mL^{-1} MMNP@PEG@RA123, Figure S14: Three-dimensional reconstruction of confocal z-stack images of control Hoechst-stained NIH/3T3 fibroblasts untreated with MMNP@PEG@RA123, Figure S15: Three-dimensional reconstructions of confocal z-stack images of Hoechst-stained NIH/3T3 fibroblasts treated for 24 h with 20 μg mL^{-1} MMNP@PEG@RA123, Figure S16: A second area covered by three-dimensional reconstruction of confocal z-stack images of Hoechst-stained NIH/3T3 fibroblasts treated for 24 h with 20 μg mL^{-1} MMNP@PEG@RA123, Table S1: Parameters for exponential decay fitting of extinction coefficient vs. wavelength in Supplementary Figure S1, Table S2: Comparison of MMNP size distributions calculated by DLS and cryo-TEM for the samples analyzed in Figure 3d–i.

Acknowledgments: This work was supported by the National Institutes of Health (NIH) grant R37 DE014193, NIH Grant R01 DE021215, and by the International Institute for Nanotechnology (IIN) at Northwestern University. D.R.A. acknowledges support from NIH grants T32GM008152 and T32GM008449. C.S. acknowledges support from the Swiss National Foundation of Science Fellowship 165149. Confocal microscopy was performed at the Northwestern University Quantitative Bio-element Imaging Center generously supported by the National Science Foundation CHE-9810378/005. This work made use of the EPIC and Keck-II facilities of the Northwestern University Atomic and Nanoscale Characterization Experimental (NUANCE) Center, which has received support from the Soft and Hybrid Nanotechnology Experimental (SHyNE) Resource (NSF ECCS-1542205); the MRSEC program (NSF DMR-1121262) at the Materials Research Center; the IIN; the Keck Foundation; and the State of Illinois, through the IIN. Electron microscopy was also performed at the Bio-Imaging Facility (BIF) and Structural Biology CryoEM Facility at Northwestern University and at the Electron Microscopy Lab and California Institute for Quantitative Biosciences (QB3) at the University of California, Berkeley.

Author Contributions: D.R.A., K.H.A.L., and P.B.M. conceived and designed the experiments. D.R.A. performed the experiments, analyzed the data, and wrote the paper. C.S. performed XPS experiments, analyzed XPS data, and wrote about the XPS results. K.H.A.L. and P.B.M. contributed to the revision of the paper.

Conflicts of Interest: The authors declare no conflict of interest. The founding sponsors had no role in the design of the study; in the collection, analyses, or interpretation of data; in the writing of the manuscript, and in the decision to publish the results.

References and Note

1. Petros, R.A.; DeSimone, J.M. Strategies in the design of nanoparticles for therapeutic applications. *Nat. Rev. Drug Discov.* **2010**, *9*, 615–627. [CrossRef] [PubMed]
2. Peer, D.; Karp, J.M.; Hong, S.; FaroKhzad, O.C.; Margalit, R.; Langer, R. Nanocarriers as an emerging platform for cancer therapy. *Nat. Nanotechnol.* **2007**, *2*, 751–760. [CrossRef] [PubMed]
3. Heath, J.R.; Davis, M.E. Nanotechnology and cancer. *Ann. Rev. Med.* **2008**, *59*, 251–265. [CrossRef] [PubMed]
4. Gu, F.X.; Karnik, R.; Wang, A.Z.; Alexis, F.; Levy-Nissenbaum, E.; Hong, S.; Langer, R.S.; Farokhzad, O.C. Targeted nanoparticles for cancer therapy. *Nano Today* **2007**, *2*, 14–21. [CrossRef]
5. De, M.; Ghosh, P.S.; Rotello, V.M. Applications of nanoparticles in biology. *Adv. Mater.* **2008**, *20*, 4225–4241. [CrossRef]
6. Davis, M.E.; Chen, Z.; Shin, D.M. Nanoparticle therapeutics: An emerging treatment modality for cancer. *Nat. Rev. Drug Discov.* **2008**, *7*, 771–782. [CrossRef] [PubMed]
7. Byrne, J.D.; Betancourt, T.; Brannon-Peppas, L. Active targeting schemes for nanoparticle systems in cancer therapeutics. *Adv. Drug Deliv. Rev.* **2008**, *60*, 1615–1626. [CrossRef] [PubMed]
8. Suhair, S.; Rania, H.; Heba, A.-H.; Ola, T. Synergistic Interplay of medicinal chemistry and formulation strategies in nanotechnology—From drug discovery to nanocarrier design and development. *Curr. Top. Med. Chem.* **2017**, *17*, 1451–1468. [CrossRef]
9. Wilhelm, S.; Tavares, A.J.; Dai, Q.; Ohta, S.; Audet, J.; Dvorak, H.F.; Chan, W.C.W. Analysis of nanoparticle delivery to tumours. *Nat. Rev. Mater.* **2016**, *1*, 16014. [CrossRef]
10. Tsoi, K.M.; MacParland, S.A.; Ma, X.Z.; Spetzler, V.N.; Echeverri, J.; Ouyang, B.; Fadel, S.M.; Sykes, E.A.; Goldaracena, N.; Kaths, J.M.; et al. Mechanism of hard-nanomaterial clearance by the liver. *Nat. Mater.* **2016**, *15*, 1212–1221. [CrossRef] [PubMed]
11. Reisch, A.; Klymchenko, A.S. Fluorescent polymer nanoparticles based on dyes: Seeking brighter tools for bioimaging. *Small* **2016**, *12*, 1968–1992. [CrossRef] [PubMed]
12. Peng, H.S.; Chiu, D.T. Soft fluorescent nanomaterials for biological and biomedical imaging. *Chem. Soc. Rev.* **2015**, *44*, 4699–4722. [CrossRef] [PubMed]
13. Ju, K.-Y.; Lee, Y.; Lee, S.; Park, S.B.; Lee, J.-K. Bioinspired polymerization of dopamine to generate melanin-like nanoparticles having an excellent free-radical-scavenging property. *Biomacromolecules* **2011**, *12*, 625–632. [CrossRef] [PubMed]
14. Chen, C.-T.; Martin-Martinez, F.J.; Jung, G.S.; Buehler, M.J. Polydopamine and eumelanin molecular structures investigated with ab initio calculations. *Chem. Sci.* **2017**. [CrossRef] [PubMed]
15. Hong, S.; Na, Y.S.; Choi, S.; Song, I.T.; Kim, W.Y.; Lee, H. Non-covalent self-assembly and covalent polymerization co-contribute to polydopamine formation. *Adv. Funct. Mater.* **2012**, *22*, 4711–4717. [CrossRef]
16. D'Ischia, M.; Wakamatsu, K.; Cicoira, F.; Di Mauro, E.; Garcia-Borron, J.C.; Commo, S.; Galván, I.; Ghanem, G.; Kenzo, K.; Meredith, P.; et al. Melanins and melanogenesis: From pigment cells to human health and technological applications. *Pigment Cell Melanoma Res.* **2015**, *28*, 520–544. [CrossRef]
17. Meredith, P.; Sarna, T. The physical and chemical properties of eumelanin. *Pigment Cell Res.* **2006**, *19*, 572–594. [CrossRef] [PubMed]
18. Bustamante, J.; Bredeston, L.; Malanga, G.; Mordoh, J. Role of melanin as a scavenger of active oxygen species. *Pigment Cell Res.* **1993**, *6*, 348–353. [CrossRef] [PubMed]
19. Moses, D.N.; Mattoni, M.A.; Slack, N.L.; Waite, J.H.; Zok, F.W. Role of melanin in mechanical properties of Glycera jaws. *Acta Biomater.* **2006**, *2*, 521–530. [CrossRef] [PubMed]
20. Lee, H.; Dellatore, S.M.; Miller, W.M.; Messersmith, P.B. Mussel-inspired surface chemistry for multifunctional coatings. *Science* **2007**, *318*, 426–430. [CrossRef] [PubMed]
21. Lynge, M.E.; van der Westen, R.; Postma, A.; Stadler, B. Polydopamine—A nature-inspired polymer coating for biomedical science. *Nanoscale* **2011**, *3*, 4916–4928. [CrossRef] [PubMed]
22. Howes, P.D.; Chandrawati, R.; Stevens, M.M. Colloidal nanoparticles as advanced biological sensors. *Science* **2014**, *346*. [CrossRef] [PubMed]
23. Liu, Y.; Ai, K.; Liu, J.; Deng, M.; He, Y.; Lu, L. Dopamine-melanin colloidal nanospheres: An efficient near-infrared photothermal therapeutic agent for in vivo cancer therapy. *Adv. Mater.* **2013**, *25*, 1353–1359. [CrossRef] [PubMed]

24. Yan, J.; Yang, L.; Lin, M.-F.; Ma, J.; Lu, X.; Lee, P.S. Polydopamine spheres as active templates for convenient synthesis of various nanostructures. *Small* **2013**, *9*, 596–603. [CrossRef] [PubMed]

25. Huang, Y.; Li, Y.; Hu, Z.; Yue, X.; Proetto, M.T.; Jones, Y.; Gianneschi, N.C. Mimicking melanosomes: Polydopamine nanoparticles as artificial microparasols. *ACS Cent. Sci.* **2017**. [CrossRef] [PubMed]

26. Lin, J.-H.; Yu, C.-J.; Yang, Y.-C.; Tseng, W.-L. Formation of fluorescent polydopamine dots from hydroxyl radical-induced degradation of polydopamine nanoparticles. *Phys. Chem. Chem. Phys.* **2015**, *17*, 15124–15130. [CrossRef] [PubMed]

27. Zhang, X.; Wang, S.; Xu, L.; Feng, L.; Ji, Y.; Tao, L.; Li, S.; Wei, Y. Biocompatible polydopamine fluorescent organic nanoparticles: Facile preparation and cell imaging. *Nanoscale* **2012**, *4*, 5581–5584. [CrossRef] [PubMed]

28. Chen, X.; Yan, Y.; Müllner, M.; van Koeverden, M.P.; Noi, K.F.; Zhu, W.; Caruso, F. Engineering fluorescent poly(dopamine) capsules. *Langmuir* **2014**, *30*, 2921–2925. [CrossRef] [PubMed]

29. Zhao, C.; Zuo, F.; Liao, Z.; Qin, Z.; Du, S.; Zhao, Z. Mussel-inspired one-pot synthesis of a fluorescent and water-soluble polydopamine–Polyethyleneimine copolymer. *Macromol. Rapid Commun.* **2015**, *36*, 909–915. [CrossRef] [PubMed]

30. Liu, M.; Ji, J.; Zhang, X.; Zhang, X.; Yang, B.; Deng, F.; Li, Z.; Wang, K.; Yang, Y.; Wei, Y. Self-polymerization of dopamine and polyethyleneimine: Novel fluorescent organic nanoprobes for biological imaging applications. *J. Mater. Chem. B* **2015**, *3*, 3476–3482. [CrossRef]

31. Chan, W.C.W.; Nie, S. Quantum dot bioconjugates for ultrasensitive nonisotopic detection. *Science* **1998**, *281*, 2016–2018. [CrossRef] [PubMed]

32. Schneider, C.A.; Rasband, W.S.; Eliceiri, K.W. NIH image to ImageJ: 25 years of image analysis. *Nat Meth* **2012**, *9*, 671–675. [CrossRef]

33. Repetto, G.; del Peso, A.; Zurita, J.L. Neutral red uptake assay for the estimation of cell viability/cytotoxicity. *Nat. Protoc.* **2008**, *3*, 1125–1131. [CrossRef] [PubMed]

34. Bisaglia, M.; Mammi, S.; Bubacco, L. Kinetic and structural analysis of the early oxidation products of dopamine: Analysis of the interactions with α-synuclein. *J. Biol. Chem.* **2007**, *282*, 15597–15605. [CrossRef] [PubMed]

35. Herlinger, E.; Jameson, R.F.; Linert, W. Spontaneous autoxidation of dopamine. *J. Chem. Soc. Perkin Trans. 2* **1995**, 259–263. [CrossRef]

36. Sarna, T.; Swartz, H.A. The physical properties of melanins. In *The Pigmentary System*; Blackwell Publishing Ltd.: Hoboken, NJ, USA, 2007; pp. 311–341.

37. Sarna, T.; Sealy, R.C. Photoinduced oxygen consumption in melanin systems. Action spectra and quantum yields for eumelanin and synthetic melanin. *Photochem. Photobiol.* **1984**, *39*, 69–74. [CrossRef] [PubMed]

38. Ball, V. Impedance spectroscopy and zeta potential titration of dopa-melanin films produced by oxidation of dopamine. *Colloids Surf. A* **2010**, *363*, 92–97. [CrossRef]

39. Stetefeld, J.; McKenna, S.A.; Patel, T.R. Dynamic light scattering: A practical guide and applications in biomedical sciences. *Biophys. Rev.* **2016**, *8*, 409–427. [CrossRef] [PubMed]

40. In the mushroom regime of PEG packing at the MMNP surface, the diameter increase (Δd) would be expected to be double the Flory radius ($2 \times R_f$). The Flory radius is $R_f = \alpha n^{3/5}$, where n is the number monomers per polymer chain, and α is the segmental length of one monomer [41]. For PEG, α = 0.35 nm, and 5 kDa PEG has n $\approx$ 113 monomers per chain, resulting in R_f = 6.0 nm and Δd = 12.0 nm.

41. Allen, C.; Dos Santos, N.; Gallagher, R.; Chiu, G.N.C.; Shu, Y.; Li, W.M.; Johnstone, S.A.; Janoff, A.S.; Mayer, L.D.; Webb, M.S.; et al. Controlling the physical behavior and biological performance of liposome formulations through use of surface grafted poly(ethylene glycol). *Biosci. Rep.* **2002**, *22*, 225–250. [CrossRef] [PubMed]

42. Perry, J.L.; Reuter, K.G.; Kai, M.P.; Herlihy, K.P.; Jones, S.W.; Luft, J.C.; Napier, M.; Bear, J.E.; DeSimone, J.M. PEGylated PRINT nanoparticles: The impact of PEG density on protein binding, macrophage association, biodistribution, and pharmacokinetics. *Nano Lett.* **2012**, *12*, 5304–5310. [CrossRef] [PubMed]

43. Gref, R.; Lück, M.; Quellec, P.; Marchand, M.; Dellacherie, E.; Harnisch, S.; Blunk, T.; Müller, R.H. 'Stealth' corona-core nanoparticles surface modified by polyethylene glycol (PEG): Influences of the corona (PEG chain length and surface density) and of the core composition on phagocytic uptake and plasma protein adsorption. *Colloids Surf. B* **2000**, *18*, 301–313. [CrossRef]

44. Fang, C.; Shi, B.; Pei, Y.Y.; Hong, M.H.; Wu, J.; Chen, H.Z. In vivo tumor targeting of tumor necrosis factor-α-loaded stealth nanoparticles: Effect of MePEG molecular weight and particle size. *Eur. J. Pharm. Sci.* **2006**, *27*, 27–36. [CrossRef] [PubMed]
45. Bazile, D.; Prud'homme, C.; Bassoullet, M.T.; Marlard, M.; Spenlehauer, G.; Veillard, M. Stealth Me. PEG-PLA nanoparticles avoid uptake by the mononuclear phagocytes system. *J. Pharm. Sci.* **1995**, *84*, 493–498. [CrossRef] [PubMed]
46. Walkey, C.D.; Olsen, J.B.; Guo, H.; Emili, A.; Chan, W.C. Nanoparticle size and surface chemistry determine serum protein adsorption and macrophage uptake. *J. Am. Chem. Soc.* **2012**, *134*, 2139–2147. [CrossRef] [PubMed]
47. Sheng, Y.; Yuan, Y.; Liu, C.; Tao, X.; Shan, X.; Xu, F. In vitro macrophage uptake and in vivo biodistribution of PLA-PEG nanoparticles loaded with hemoglobin as blood substitutes: Effect of PEG content. *J. Mater. Sci. Mater. Med.* **2009**, *20*, 1881–1891. [CrossRef] [PubMed]
48. Liebscher, J.; Mrówczyński, R.; Scheidt, H.A.; Filip, C.; Hădade, N.D.; Turcu, R.; Bende, A.; Beck, S. Structure of polydopamine: A never-ending story? *Langmuir* **2013**, *29*, 10539–10548. [CrossRef] [PubMed]
49. Liu, T.; Kim, K.C.; Lee, B.; Chen, Z.; Noda, S.; Jang, S.S.; Lee, S.W. Self-polymerized dopamine as an organic cathode for Li- and Na-ion batteries. *Energy Environ. Sci.* **2017**, *10*, 205–215. [CrossRef]
50. Lee, H.; Rho, J.; Messersmith, P.B. Facile conjugation of biomolecules onto surfaces via mussel adhesive protein inspired coatings. *Adv. Mater.* **2009**, *21*, 431–434. [CrossRef] [PubMed]
51. Hinckley, D.A.; Seybold, P.G. A spectroscopic/thermodynamic study of the rhodamine B lactone ⇌ zwitterion equilibrium. *Spectrochim. Acta Part A* **1988**, *44*, 1053–1059. [CrossRef]
52. Mchedlov-Petrossyan, N.O.; Vodolazkaya, N.A.; Doroshenko, A.O. Ionic equilibria of fluorophores in organized solutions: The influence of micellar microenvironment on protolytic and photophysical properties of rhodamine B. *J. Fluoresc.* **2003**, *13*, 235–248. [CrossRef]
53. Butcher, K.J.; Hurst, J. Aromatic amines as nucleophiles in the Bargellini reaction. *Tetrahedron Lett.* **2009**, *50*, 2497–2500. [CrossRef]

Review

Catechol-Based Hydrogel for Chemical Information Processing

Eunkyoung Kim [1,2], Zhengchun Liu [3], Yi Liu [1,2], William E. Bentley [1,2] and Gregory F. Payne [1,2,*]

1   Institute for Biosystems and Biotechnology Research, University of Maryland, 5115 Plant Sciences Building, College Park, MD 20742, USA; ekim@umd.edu (E.K.); yliu123@umd.edu (Y.L.); bentley@umd.edu (W.E.B.)
2   Fischell Department of Bioengineering, University of Maryland, College Park, MD 20742, USA
3   Hunan Key Laboratory for Super Microstructure and Ultrafast Process, School of Physics and Electronics, Central South University, Changsha 410083, China; liuzhengchunseu@126.com (Z.L.)
*   Correspondence: gpayne@umd.edu; Tel.: +301-405-8389; Fax: +301-314-9075

Academic Editors: Marco d'Ischia and Daniel Ruiz-Molina
Received: 31 May 2017; Accepted: 23 June 2017; Published: 3 July 2017

Abstract: Catechols offer diverse properties and are used in biology to perform various functions that range from adhesion (e.g., mussel proteins) to neurotransmission (e.g., dopamine), and mimicking the capabilities of biological catechols have yielded important new materials (e.g., polydopamine). It is well known that catechols are also redox-active and we have observed that biomimetic catechol-modified chitosan films are redox-active and possess interesting molecular electronic properties. In particular, these films can accept, store and donate electrons, and thus offer redox-capacitor capabilities. We are enlisting these capabilities to bridge communication between biology and electronics. Specifically, we are investigating an interactive redox-probing approach to access redox-based chemical information and convert this information into an electrical modality that facilitates analysis by methods from signal processing. In this review, we describe the broad vision and then cite recent examples in which the catechol–chitosan redox-capacitor can assist in accessing and understanding chemical information. Further, this redox-capacitor can be coupled with synthetic biology to enhance the power of chemical information processing. Potentially, the progress with this biomimetic catechol–chitosan film may even help in understanding how biology uses the redox properties of catechols for redox signaling.

Keywords: catechol; chitosan; hydrogel; information processing; redox-capacitor; electrochemistry

1. Introduction

1.1. Background: Redox-Active Catecholic Materials

Biology uses materials to perform diverse and important functions, and our understanding of these materials enables biomimetic efforts. For instance, studies of the adhesive properties of mussel proteins identified catecholic residues as critical for both surface binding (adhesion) and protein crosslinking (cohesion), and this knowledge enabled biomimetic materials with broad applications (e.g., polydopamine) [1–5]. Catecholic residues are also redox-active, which means that catechol residues can be switched between oxidized (quinone) and reduced (hydroquinone) states by exchanging electrons with diffusible mediators [6–8]. There are several studies to show that the redox activities of catecholic residues can be relevant to some biological functions such as redox buffering [9,10], antioxidant protection [11], metal chelation [12], redox signaling [13] and electron transport [6,14,15]. Although some attempts have been made to study the redox activities of catecholic residues [7,16–18], the mechanistic understanding is still lacking because there are few suitable techniques. We are investigating the redox activities of biomimetic catecholic materials and their biological relevance using a recently developed electrochemical reverse engineering method. We believe these redox-active catecholic materials may offer unique technological opportunities for processing chemical information. Thus, our approach may not be truly biomimetic since we are not guided by an understanding of how biology uses melanin's redox properties to perform functions. We have rather observed intriguing redox properties of biomimetic catecholic materials and we are pursuing technological applications that may or may not be related to their biological function.

1.2. Vision: A New Paradigm for Accessing Chemical Information

Over the last 50 years, advances in information processing transformed our lives by changing the way we access, analyze, and transmit information. Each new device seems to be incrementally cheaper, smaller, faster, more powerful, and easier to use than its predecessor. However, this trend does not extend to instruments that acquire and process chemically-related information. For instance, when we need to access critical chemical information in real time, we still rely on dogs to sniff-out this information. Why have the advances in information processing not been extended to the acquisition and processing of chemically-based information? We believe one issue is that the current paradigm of chemical information is too limited, viewing chemical information through the lens of analytical chemistry and characterizing chemical information in terms of chemical composition and concentration [19]. In essence, this paradigm specifies instrument-intensive approaches (e.g., high performance liquid chromatography–mass spectrometry (HPLC–MS)) to access chemical information and incremental advances are generally accompanied by increasing costs and complexity [20–22].

We suggest an alternative paradigm for chemical information processing: one that accesses the power of information processing by searching "chemical space" for global signatures. We envision that this search will be rapid, cheap and convenient, but will lack the granular details of chemical composition and concentration that are targets of conventional analytical chemistry and omics approaches [22]. Thus, we envision a signal processing paradigm that is complementary to (not a replacement for) the current paradigm that focuses on composition and concentration [23–26].

As illustrated in Figure 1a, our signal processing approach is approximately analogous to sonar. Sonar uses a transmitted signal (pressure wave) to propagate through a medium in search of physical information of nearby objects. Interaction with such objects generates a reflected wave that contains information of the objects (e.g., their presence, size, shape and motion).

Figure 1b illustrates our approach to interactively probe for chemical information in a local environment. There are three key features of our interactive electrochemical approach. First is the use of diffusible redox-active mediators (electron shuttles) that serve to transmit redox "signals" that can propagate through the local environment in search of redox-based chemical information. Electron transfer interactions between the mediators and the local environment (i.e., redox reactions) will be detected when the mediator returns to the electrode and engages in electrochemical reactions that serve to transduce the redox information into an electrical output. The second feature illustrated in Figure 1b is use of complex electrical inputs that allows redox-probing to be tailored in search of specific types of information. As will be discussed, the resulting electrical outputs contain information of the mediators' redox interactions [27,28] and can be analyzed using approaches adapted from signal processing [29,30]. The third feature illustrated in Figure 1b is the use of thin hydrogel film coatings that are used to facilitate signal processing. A catechol-based redox-capacitor is one such film that has proven to be especially useful for processing redox-based chemical information [31,32].

In this review, we will focus on the fabrication and properties of the catechol-based redox-capacitor and we will cite several examples illustrating the value of this capacitor for processing redox-based chemical information.

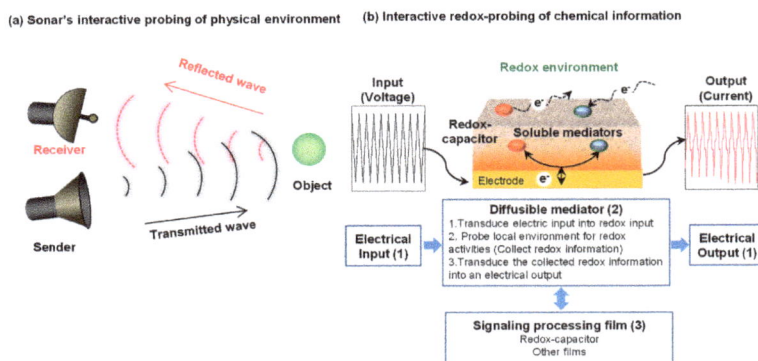

Figure 1. Interactive probing of a local environment. (**a**) Sonar analogy. (**b**) Interactive electrochemical probing consists of: (**1**) complex inputs/outputs to tailor the interactive probing; (**2**) diffusible mediators (electron shuttles) to transmit redox signals that can probe information; and (**3**) signaling processing film to facilitate information processing.

2. Fabrication of Catechol-Based Redox-Capacitor

Figure 2a illustrates that the catechol-based redox-capacitor film is fabricated in two steps from catechols and the aminopolysaccharide chitosan [33,34]. The first step is the electrodeposition of a thin chitosan hydrogel film at an electrode surface. This electrodeposition step enlists chitosan's pH-responsive self-assembling properties [35] and uses cathodic electrolytic reactions to generate the localized high pH that induces chitosan's neutralization and gel formation [36,37]. In the second step, the chitosan-coated electrode is immersed in a catechol-containing solution and the catechol is then oxidized to generate a reactive o-quinone that grafts to chitosan through non-enzymatic reactions. Biologically, enzymes (e.g., tyrosinase) catalyze catechol oxidation [1,38,39] while Figure 2a shows that catechols can also be electrochemically oxidized by biasing the electrode to serve as an anode. Importantly, the chitosan film is a hydrogel that allows diffusion of the catechol reactant and o-quinone product. It is also important that the o-quinone product is reactive and quickly grafts to chitosan's primary amines. These quinone grafting reactions are complex and incompletely characterized and likely involve Michael-type adduct and Schiff base chemistries [22,40,41], as suggested in Figure 2b.

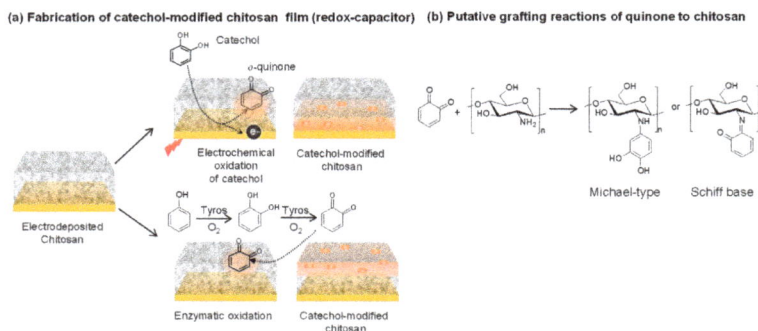

Figure 2. Fabrication of catechol-based redox-capacitor films. (**a**) Catechol is electrochemically oxidized or enzymatically oxidized by tyrosinase (Tyros) and the diffusible oxidation product (o-quinone) grafts to the nucleophilic amines of the aminopolysachharide chitosan. (**b**) Quinone grafting likely involves Michael-type and Schiff base adduct formation.

3. The Catechol–Chitosan Film Can Accept, Store and Donate Electrons

Early studies with melanin suggested that it possesses conducting and semiconductor properties [42,43] and the obvious question was: Are the catechol–chitosan films conducting? Our experimental results indicated that these films are not conducting: electrons do not flow in response to an applied voltage and there does not appear to be direct exchange of electrons with the underlying electrode. This is not surprising given that the films are relatively thick (~1 µm when wet) and the catechols may be too far from the electrode to directly exchange electrons.

The next question was: Are the catechol–chitosan films redox-active? Specifically, can the grafted moieties be switched between oxidized states and reduced states? One challenge to assessing the redox switching capabilities of a non-conducting film is what mechanism can be used to transfer electrons from the electrode to the grafted moieties of the film. As illustrated in Figure 3a, we decided to test whether diffusible mediators (i.e., electron shuttles) could be used to engage the catechol–chitosan film in redox-cycling reactions. We found that one mediator, $Ru(NH_3)_6Cl_3$ (Ru^{3+}), could engage in reductive redox-cycling to transfer electrons from the electrode to the film to convert oxidized moieties (Q) to reduced moieties (QH_2) and thereby charge the film with electrons. A second mediator, ferrocene dimethanol (Fc), could engage in oxidative redox-cycling to transfer electrons from the film to the electrode to convert the film's QH_2 to Q and thereby discharge the film.

Importantly, we observed that we could immerse the film-coated electrode into a solution containing both mediators and sequentially engage it in reductive and oxidative redox-cycling reactions if we imposed cyclic voltage inputs as illustrated in Figure 3b. It is also important to note that the "flow" of electrons is constrained by thermodynamics, as illustrated in Figure 3c. The results from these initial studies demonstrated that the catechol–chitosan films are redox-active and can be switched between reduced or oxidized states by exchanging electrons with mediators.

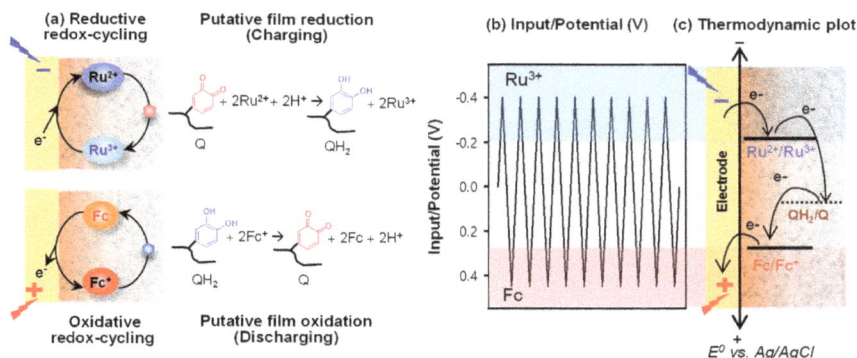

Figure 3. Proposed redox-cycling mechanism of redox-capacitor. (**a**) The reductive redox-cycling reaction of Ru^{3+} can reduce the quinone moieties (charge, QH_2) and the oxidative redox-cycling reaction of ferrocene dimethanol (Fc) can oxidize the catechol moieties (discharge, Q). (**b**) Electrical input potential. (**c**) Thermodynamics requires electrons to be transferred from more reducing (more negative) potentials to more oxidizing (more positive) potentials.

The observation that the catechol–chitosan films are redox-active but non-conducting leads to two interesting concepts. First, the fact that catechol–chitosan films can accept, store and donate electrons essentially means that they are redox-capacitors. As will be discussed, we use these redox-capacitor properties for information processing in aqueous environments. Second, electrons "flow" through the catechol–chitosan films via distinct electron transfer steps (with intermediates) and not as a "sea" of electrons (as in electron currents in wires). This mechanism is consistent with biological electron transfer reactions (e.g., in the respiratory chain) [44,45] in which electron transfer occurs through distinct stable intermediates. Interestingly, quinones, which are the putative redox-active moieties in our capacitor film, are also important redox-active intermediates in the biological electron transfer chains of both respiration (ubiquinone) [14,15] and photosynthesis (plastoquinone) [13].

4. Molecular Electronic Properties of the Catechol–Chitosan Redox-Capacitor

As a result of its redox activities, the catechol–chitosan redox-capacitor offers interesting molecular electronic properties and we highlight four such properties: amplification, partial rectification, gating, and steady response [31,46–48]. The first three properties are illustrated by the results in Figure 4a. In this experiment, the electrode coated with the catechol–chitosan film was immersed in a solution containing both mediators (Fc and Ru^{3+}) and a cyclic voltage input was imposed as illustrated by the left plot in Figure 4a. The middle plot of Figure 4a shows the current output response for these studies. The rightmost plot in Figure 4a shows a conventional cyclic voltammogram (CV) which is an alternative representation of the input–output data in which time is not explicitly shown. One control in Figure 4a is incubation of a catechol–chitosan film in the buffer solution without mediators. The output currents for this control show no discernible peaks, which is expected because the catechol–chitosan films are non-conducting. A second control is an electrode coated with a chitosan film (without catechol modification) and immersed in a solution containing both mediators. Results for this control show small output peak currents for Fc and Ru^{3+}: these mediators can diffuse through the chitosan film and undergo electron exchange with the underlying electrode. When the electrode coated with the catechol–chitosan film was tested in solutions containing Fc and Ru^{3+}, the output peak currents for Fc-oxidation and Ru^{3+}-reduction were considerably amplified. Amplification of the mediator currents is consistent with the redox-cycling mechanisms of Figure 3a.

Interestingly, the amplification observed in Figure 4a occurs primarily in one direction for each mediator. The Fc-oxidation current is greatly amplified while the Fc-reduction current is not amplified. The Ru^{3+}-reduction current is greatly amplified but the Ru^{3+}-oxidation current is not amplified. This partial rectification of the mediator currents is consistent with the thermodynamic plot in Figure 3c, which indicates that Fc can engage in oxidative redox-cycling but not reductive redox-cycling. Similarly, Ru^{3+} can engage in reductive but not oxidative redox-cycling.

To understand the gating property, it is useful to recognize that the currents observed in Figure 4a do not result from direct electron transfer between the film and electrode, but rather are the result of electron transfer between the mediators and electrode. Amplification of these peak currents occurs because the mediators redox-cycle with the film and shuttle electrons between the film and electrode. One requirement for redox-cycling is illustrated by the thermodynamic plot in Figure 3c: a reductive redox-cycling mediator (e.g., Ru^{3+}) must have a redox potential (E^0) that is more reducing than that of the film, and an oxidative redox-cycling mediator (e.g., Fc) must have a E^0 that is more oxidizing than that of the film. A second requirement for redox-cycling is illustrated by the input voltage curve of Figure 3b: reductive redox-cycling can only occur if the imposed voltage is more reducing than the E^0 of the reducing mediator, and oxidative redox-cycling can only occur if the imposed voltage is more oxidizing than the E^0 of the oxidizing mediator (see below for details). In brief, the mediators' E^0 plays a critical role in determining if the film can be charged or discharged with electrons, and mediators with differing values can be used to shift the voltage that must be imposed at the electrode to initiate the redox-cycling reactions (i.e., E^0 of the mediator serves as a gate).

The ability of the catechol–chitosan film to generate steady output responses is illustrated by the experiment in Figure 4b in which an oscillating voltage input was imposed over several hours in the presence of both mediators. The output response (or CV representation) for the electrode coated with the catechol–chitosan film shows that the output currents for both Fc-oxidation and Ru^{3+}-reduction were amplified (compared to a control chitosan film) and these amplifications were nearly constant (i.e., steady) over time. From a chemical standpoint, the steady output current responses of Figure 4b indicate that the catechol–chitosan film can be repeatedly switched between oxidized and reduced states. From a signal processing standpoint, steady oscillating inputs and outputs (i.e., sine waves) are integral to the coding and decoding of information.

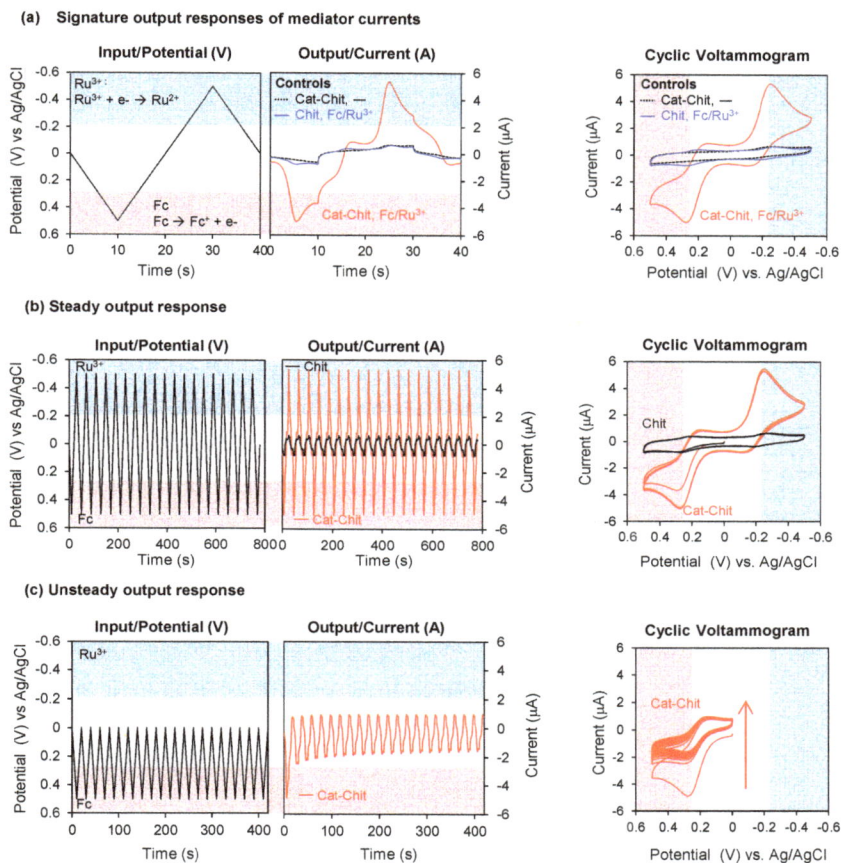

Figure 4. Molecular electronic properties of redox-capacitor. (**a**) Input–output curves and cyclic voltammogram show that the redox-cycling reaction of redox-capacitor with mediators results in output currents that are amplified, partially rectified and gated. (**b**) The output current of mediator is steady over time under unperturbed environment. (**c**) The output current becomes unsteady in the limited range of input potential. Cat-Chit: Catechol-modified chitosan film; Fc: Ferrocene dimethanol. Reproduced from [31] by permission of The Royal Society of Chemistry, 2014.

In contrast to the case of steady input–output, Figure 4c shows an example of unsteady response characteristics. In this unsteady case, the left plot of Figure 4c shows a more limited potential range was imposed (+0.5~0 V) to provide the oxidative voltages required to oxidize Fc but to provide reducing voltages that are insufficient to reduce Ru^{3+}. Figure 4c shows the current output response for this unsteady case has no Ru^{3+}-reduction peaks and the amplified currents of Fc-oxidation decrease progressively over time. Presumably, this decay in Fc-oxidation currents occurs because the catechol–chitosan film is progressively depleted of electrons during the repetitive Fc-redox-cycling reaction, but this film cannot be replenished with electrons because the imposed voltage is never sufficiently reductive for Ru^{3+}-reductive redox-cycling. This unsteady output current response also illustrates the gating function of Ru^{3+}: since the imposed voltage remains too oxidative (relative to the E^0 for Ru^{3+}), then reductive redox-cycling mechanism cannot be engaged to recharge the film with electrons. Under this condition, no Ru^{3+}-reduction currents are observed and the progressive discharging of the film leads to a progressive decay in Fc-oxidation currents.

The results of Figure 4 illustrate four important molecular electronic properties of the catechol–chitosan redox-capacitor that we use for chemical information processing. One additional feature of this redox-capacitor is illustrated in Table 1. Specifically, the catechol–chitosan redox-capacitor has been observed to accept electrons from a broad range of reductants and donate electrons to various oxidants. This broad ability to exchange electrons with oxidants and reductants indicates that this redox-capacitor has a somewhat generic ability to access redox information (i.e., various different types of mediators can be used). From a chemical standpoint, the broad ability of the catechol–chitosan capacitor to exchange electrons means that it possesses redox catalytic properties and is capable of transferring electrons from reductants to oxidants in response to thermodynamic driving forces. However, we should note that not all redox-active chemicals can exchange electrons with the catechol–chitosan redox-capacitor. Thus, while thermodynamic plots may suggest what reactions are possible, some reactions do not occur within relevant timescales because redox-reactions can have significant kinetic barriers.

Table 1. Tested redox chemicals that can redox-interact with catechol–chitosan films.

Reductants to Donate Electrons to Catechols	Oxidants to Accept Electrons from Catechols
nicotinamide adenine dinucleotide phosphate (NADPH) [49], glutathione (GSH), ascorbic acid [49], pyocyanin (PYO) [50], paraquat [51], $Ru(NH_3)_6Cl_3$ (Ru^{3+}) [47]	O_2 [49], acetosyringone (AS) [49], clozapine [52], acetaminophen [53], p-aminophenol (p-AP) [54], K_2IrCl_6 (Ir^{4+}) [55], ferrocene dimethanol (Fc) [47]

5. Examples of Chemical Information Processing Using Catechol-Based Redox-Capacitor

As suggested in Figure 1b, we believe that interactive redox-probing can provide access to chemical information, and we are especially focused on accessing chemical information relevant to redox biology. It is well-known that biology uses redox reactions for energy harvesting [56,57] (e.g., electron transfer in the respiratory chain), biosynthesis (e.g., nicotinamide adenine dinucleotide phosphate (NADPH) serves as a diffusible reductant) [10,58,59] and immune response (i.e., the oxidative burst) [60–62]. Redox is also emerging as an important signaling modality in biology with the use of diffusible extracellular signaling molecules (e.g., H_2O_2) [63–65] and redox-based receptor mechanisms (e.g., cysteine-based sulfur switches) [66–68]. In addition, redox is recognized as important in biological homeostasis with suggestions that oxidative stress is essentially redox dysregulation. Potentially, probing a local biological environment may reveal information of the redox activities and the redox context [69–71].

To illustrate the potential of redox-probing to access complex biological information, we cite recent studies on the development of a blood test for oxidative stress [55]. Specifically, as illustrated in Figure 5, an iridium (Ir) salt was added to serum to serve as an oxidant to detect reducing activities. Subsequent measurements of the amount of Ir reduced could be used to determine an "Ir-reducing capacity": the lower the Ir-reducing capacity, the greater the oxidative stress. We observed that this measure of oxidative stress could correlate to clinical indicators of schizophrenia and thus this serum assay may aid in the diagnosis and assessment of symptom severity. The important point of this example is that an information processing approach was shown to rapidly access chemical information from serum and this information appears to have considerable clinical utility. Traditional approaches to develop serum tests for schizophrenia generally use panels of analytes, have high costs, and have been unsuccessful in the clinic. The Ir-mediated signal processing approach is essentially a reverse engineering approach in which the Ir mediator is used to probe the serum sample for redox-based information of oxidative stress. This approach does not rely on knowledge of underlying biological mechanisms but probes serum for redox information in a somewhat unbiased way in search of global signatures of relevant information.

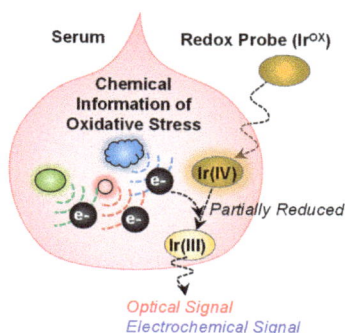

Figure 5. Redox-probing to access chemical information of oxidative stress. The redox mediator (K$_2$IrCl$_6$, Ir^{OX}) can probe chemical information relevant to oxidative stress in blood. The information can be transmitted into optical and electrochemical modalities. Adapted with permission from [55]. Copyright (2017) American Chemical Society.

5.1. Interactive Redox-Probing of Biothiols

Thiols are important moieties in biology because the thiol of glutathione is important for antioxidant protection [72–76] and the thiols of the cysteine residues of proteins are able to serve as redox-responsive crosslinks [74]. Thiols also possess unusual chemical properties and tend to self-assemble onto gold surfaces through gold-thiol interactions rather than transferring their electrons through redox-reactions [77,78]. These chemical properties have made it difficult to detect thiols by electrochemical methods. For instance, the self-assembly of thiols tends to "foul" gold electrodes with insulating monolayer regions (i.e., patches on the gold surface) [79–81].

We examined whether a redox-probing approach could be used to detect the presence of the biothiol glutathione [27]. In initial studies, we immersed a gold electrode in solutions containing both the Fc and Ru^{3+} mediators and observed that the addition of small amounts of glutathione to this solution resulted in an attenuation of the Fc-oxidation currents. This attenuation is consistent with the self-assembly of the biothiol and a blocking of some of the electrode area to limit Fc-oxidation. When the gold electrode was coated with the catechol–chitosan redox-capacitor film and oscillating voltage inputs were imposed, as illustrated in Figure 6a, the Fc and Ru^{3+} currents were both amplified and the glutathione-induced signal attenuation was more easily detected. Presumably, glutathione can diffuse through the catechol–chitosan and self-assemble onto the underlying electrode as illustrated in Figure 6b. Importantly, Figure 6c shows that quantitative analysis of this signal attenuation could be correlated to glutathione concentration and a linear correlation extended over five orders of magnitude in concentration.

This is an unusual example in the sense that we are using an electrochemical approach to detect the presence glutathione but the method is not based on a redox-reaction between glutathione and either mediator, or between glutathione and the redox-capacitor. Detection is rather believed to result because of the unique chemical capability of thiols to self-assemble onto gold and attenuate the electrode's ability to oxidize Fc. To provide confirmatory evidence for this mechanism, we tested whether a strong reducing potential that is known to disassemble thiols from gold could be used to reverse the attenuation as illustrated in Figure 6d. We observed that indeed the use of such a reducing potential reversed the attenuated Fc currents and this sequence of attenuation and reversal could be repeated multiple times.

This work illustrates two important points. First, the catechol–chitosan redox-capacitor generates amplified signals that facilitate detection. Second, complex electrical inputs can be imposed to probe for specific information: we used oscillating inputs to generate steady outputs that facilitated quantification of attenuation and we used step changes to reducing potentials to test for biothiol disassembly and a recovery of the Fc-oxidation current. This latter point illustrates that complex electrical input signals can be designed to test specific chemical hypotheses.

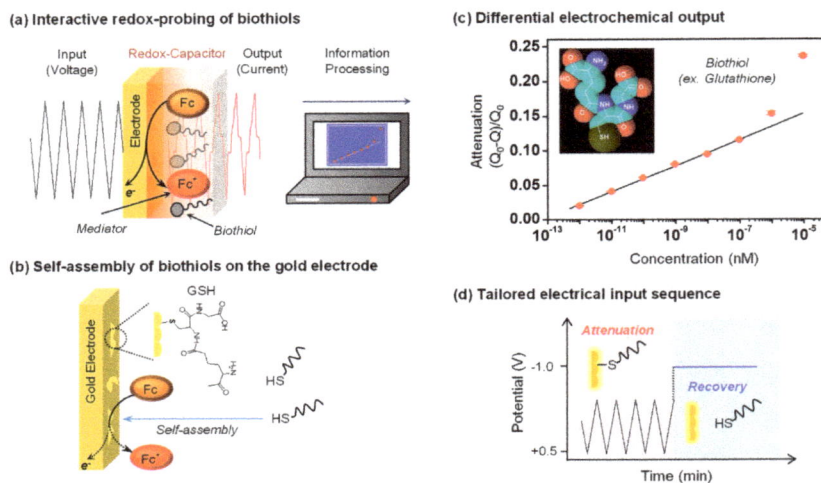

Figure 6. Interactive redox-probing of biothiols. (**a**) The electrochemical information processing approach allows the acquisition of chemical information of biothiols. (**b**) Thiols can self-assemble on gold and potentially attenuate electrochemical signals. (**c**) The quantitative analysis of signal attenuation could be correlated to glutathione (GSH) concentration. Q: Ferrocene dimethanol (Fc)-oxidative charge with biothiol; Q_0: Fc-oxidative charge without biothiol. (**d**) Tailored electrical input sequence to test the hypothesis that GSH self-assembly on gold attenuates Fc-oxidation. Adapted with permission from [27]. Copyright (2016) American Chemical Society.

5.2. Detection of a Redox-Active Bacterial Metabolite

Biology often uses diffusible redox-active metabolites to perform functions: immune cells generate reactive oxygen species to defend against pathogen attack (e.g., the oxidative burst) and H_2O_2 is emerging as a diffusible signaling molecule [82–84]. Attention has also been focused on other redox-active metabolites such as phenazines, which are among the most-studied redox-active bacterial metabolites [85–87]. These metabolites are believed to: (1) allow the producing bacteria to mediate signaling among cells (i.e., quorum sensing) [88]; (2) transfer electrons outside the cell for redox-balancing [89]; and (3) maintain redox homeostasis of multicellular biofilms [90]. One of the phenazines, pyocyanin (PYO), is also a virulence factor for the opportunistic pathogen *Pseudomonas aeruginosa* [85,91]. Importantly, *P. aeruginosa* is emerging as one of the most significant pathogens of nosocomial (hospital acquired) infections, especially for burn patients [92–94]. A rapid detection of this pathogen could be integral to successfully identifying and treating infections in this vulnerable patient population.

Figure 7a shows that *P. aeruginosa* secretes the redox-active metabolite PYO that can diffuse through the redox-capacitor film [50]. The diffused PYO can be electrochemically reduced at the electrode when the cathodic potential is applied and then the reduced PYO can diffuse back into the film and donate its electrons to the redox-capacitor film. Thus, PYO can undergo reductive redox-cycling with the redox capacitor film and yield amplified output currents that could facilitate detection of *P. aeruginosa*.

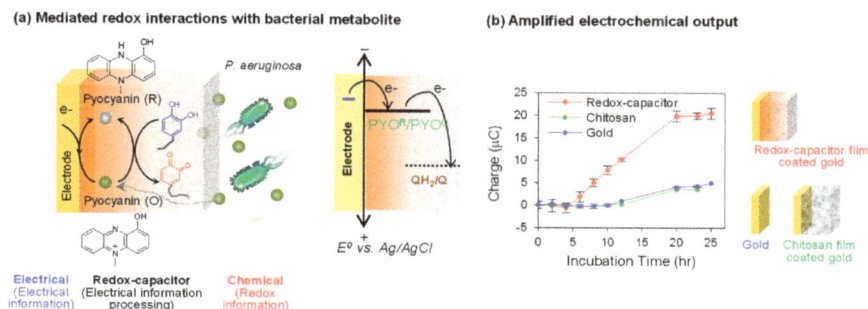

Figure 7. Electrochemical detection of the redox-active bacterial metabolite pyocyanin (PYO). (**a**) Schematic shows that the redox-capacitor can amplify the reduction current of PYO due to the reductive redox-cycling reaction. QH_2/Q: Catechol(QH_2) moieties/quinone (Q) moieties of catechol–chitosan film. (**b**) Compared with bare gold and chitosan-coated electrode, the redox-capacitor can sensitively detect the PYO production by bacteria. Adapted with permission from [50]. Copyright (2013) American Chemical Society.

To illustrate the sensitive detection of PYO production, we immersed an electrode coated with the catechol–chitosan redox-capacitor into a growing bacterial culture and intermittently performed in situ electrochemical measurement (using a chronocoulometric technique). Figure 7b compares that the charge transfer (a measure of the number of electrons transferred across the electrode) for PYO-reduction against two controls, a bare gold electrode and an electrode coated with an unmodified chitosan film. Both controls show very small charge transfer for PYO-reduction compared to the amplified output generated by the electrode coated by the catechol–chitosan redox-capacitor film.

In summary, the results in Figure 7 demonstrate that the redox-active metabolite (PYO) can undergo redox interactions with the electrode and the capacitor film, and these interactions lead to amplified electrical output signals. In essence, these redox interactions serve to convert chemical information of the bacterial generation of PYO into an electrical output. Potentially, this amplified detection might allow the early detection of infections by the opportunistic pathogen, *P. aeruginosa*. In addition, this measurement may be useful for studying the spatiotemporal dynamics of PYO generation in complex systems (e.g., bacterial biofilms).

5.3. A Global Analysis of Redox Context

Figure 1b suggests that redox-probing and signal processing can provide a new paradigm for accessing redox-based chemical information. Our initial effort to measure global redox information is illustrated in Figure 8a in which we immersed an electrode coated with a catechol–chitosan redox-capacitor into a complex bacterial culture and imposed cyclic potential inputs [95]. As noted, these electrical inputs are transduced into redox "transmissions" by the mediators that diffuse through the film into the local environment to probe for redox information (i.e., to assess the redox context). For the example in Figure 8, we added two redox-active biological mediators: the bacterial phenazine pyocyanin that undergoes reductive redox-cycling for film-charging, and the plant phenolic acetosyringone (AS) that can undergo oxidative redox-cycling for film discharging. As an aside, it is useful to note that both molecules are believed to perform signaling functions in biology: PYO for bacterial quorum sensing and AS for a plant innate immune response [96–98]. In this example, the catechol–chitosan redox-capacitor films serve to manipulate the redox signals in ways that facilitate interpretation (e.g., the capacitor amplifies and partially rectifies the currents).

To evaluate the signal processing approach, we exposed this capacitor-coated electrode to different redox contexts based on whether the experimental system did or did not have a living population of *Escherichia coli* (biotic or abiotic) or whether there was or was not oxygen present (aerobic or anaerobic). Figure 8b shows a typical CV output response for this experiment and illustrates the signal analysis approach for analyzing the signal. As illustrated in Figure 8b, the CV signal was divided into three regions which were operationally assigned to the specific chemical processes of PYO-reduction, PYO-oxidation and AS-oxidation (note these assignments are important for signal processing but are approximations of the underlying chemistries) [53]. The currents (I) in these regions were integrated with respect to time (t) to determine the charge transfer ($Q = \int I dt$) in these three regions. These values were then used to generate either a rectification ratio for pyocyanin (RR_{PYO}) or the fraction of electrochemical oxidation occurring in the pyocyanin region (F_{PYO}). Using these analytical values, the individual CVs for the four different redox contexts were compared as illustrated in Figure 8c. The correlation plot of Figure 8c shows the ability to discern these four redox contexts (see original paper [95] for details).

In summary, the results in Figure 8 illustrate that redox-probing and signal processing can provide global signatures capable of discerning difference in redox context. Potentially, this analysis could provide a new approach to extract redox-based chemical information from systems that are not well understood and are difficult to probe by conventional methods (e.g., the microbiome [99]).

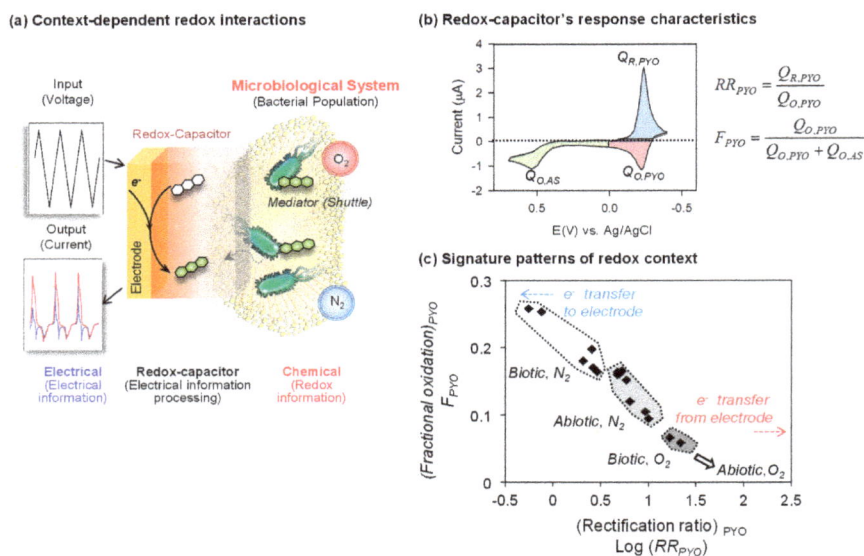

Figure 8. Global analysis of redox context. (**a**) Schematic illustrates that cyclic electrical inputs–outputs are transferred via mediators to the aqueous/biological system, while the catechol–chitosan interface "processes" this information. (**b**) Parameters calculated from cyclic voltammograms that correlate data based on rectification of pyocyanin (PYO) currents and the fraction of total electrochemical oxidation that is attributed to PYO. $Q_{R,PYO}$: PYO-reductive charge; $Q_{O,PYO}$: PYO-oxidative charge; $Q_{O,AS}$: Acetosyrigone (AS)-oxidative charge. (**c**) Two parameters (rectification ratio for pyocyanin (RR_{PYO}) and fraction of electrochemical oxidation occurring in the pyocyanin region (F_{PYO})) show the correlation for the four experimental contexts. Adapted with permission from [95]. Copyright (2013) American Chemical Society.

5.4. Coupling Redox-Probing with Synthetic Biology to Access Biochemical Signals

As observed in previous example, some biologically important chemicals are redox-active (e.g., signaling molecules PYO and AS). In these cases, electrochemical methods can be used for direct detection. However, not all molecules are redox-active and in these cases, electrochemistry cannot be directly employed for detection. An emerging approach is to enlist synthetic biology (synbio) to create engineered cells that can recognize a specific chemical and transduce this recognition event into a redox-based signal. For instance, an important bacterial quorum sensing molecule, autoinducer-2 (AI-2), is not redox-active and a synbio reporter cell has been constructed to recognize AI-2 and convert this chemical information into a redox signal that can be electrochemically detected [23,54]. Figure 9a illustrates that this *E. coli* reporter cell transduces the AI-2 molecular input into the expression of the enzyme β-galactosidase (β-gal) that can convert a redox-inactive substrate (*p*-aminophenly β-D-galactopyranoside (PAPG)) into a redox-active product (*p*-aminophenol (PAP)).

Figure 9a also shows that a dual-film system is used to interface these *E. coli* reporter cells adjacent to an electrode. Specifically, these cells are entrapped within a Ca^{2+}-alginate bio-hydrogel film that is electroaddressed on top of the catechol–chitosan redox-capacitor film [100–102]. The redox-inactive PAPG substrate is purposefully added to the system, and it can diffuse into the dual-film system. When this dual film system is exposed to AI-2, the reporter cells express the β-gal enzyme that converts PAPG into the redox-active PAP product that can diffuse into the redox-capacitor where it can undergo oxidative redox-cycling reactions. Figure 9b shows experimental results demonstrating that the oxidative charge transfer is considerably larger when the dual-film was exposed to AI-2 (compared to the control which the dual-film was not exposed to AI-2).

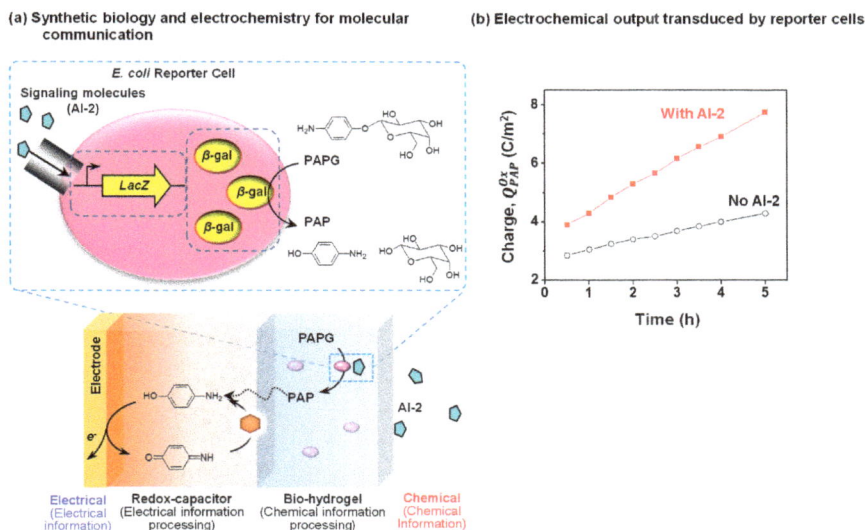

(a) Synthetic biology and electrochemistry for molecular communication

(b) Electrochemical output transduced by reporter cells

Figure 9. Coupling redox-probing with synthetic biology to access biochemical signals. (**a**) Schematic shows the *Escherichia coli* reporter cell created to detect autoinducer-2 (AI-2) and transduce this information from the redox-inactive substrate *p*-aminophenly β-D-galactopyranoside (PAPG) into a redox-active intermediate that is electrochemically detected. The redox-capacitor converts *p*-aminophenol (PAP) into an amplified electrochemical output. (**b**) The electrochemical output (oxidative charge, Q^{Ox}) of the dual-film containing *E. coli* reporter cells shows faster increase in the presence of the signaling molecule AI-2 compared with its absence. Reproduced with permission from [54], published by John Wiley and Sons, 2017.

In summary, the dual-film system serves to process the chemical information of AI-2 into an electrical signal by first using a synbio construct to transduce the AI-2 chemical input into a redox intermediate (PAP), and then using the catechol–chitosan redox-capacitor to convert this redox intermediate into an amplified electrical output. There are two broad features of this work. First, the work demonstrates the coupling of synthetic biology, thin film technology, and electrochemistry to convert chemical information into electrical signals. Second, this coupling is enabled by the use of redox as an intermediate modality that bridges the chemical modality of biology and the electrical modality of devices. Redox can bridge these modalities because it shares features of both the molecular and electrical modalities.

6. Conclusions and Future Perspectives

Over the past half-century, microelectronics and information technology transformed the way we process information, but these advances have had a relatively small impact on accessing and understanding the nature of chemical information. A key limitation to interfacing advanced electronic technology with biology is to identify a suitable means to span the chemical modalities of biology and the electrical modality of modern devices. We suggest redox provides a means to span these modalities and propose a new paradigm of interactively probing a local environment for redox-based chemical information in a manner that is analogous to sonar. In this approach, we impose electrical inputs and purposely add redox mediators that can diffuse into the environment and transduce electrical inputs into redox signals (e.g., redox transmission). These mediators probe for redox information in the environment and this information is converted at the electrode into electrical signals that can be "decoded" by signal processing strategies to extract the chemical information. In this approach, thin hydrogel films coated onto an electrode can perform important information processing operations.

We summarize several examples to show that a catechol-based redox-capacitor film can serve to process the redox information generated from various biological systems. We envision the catechol–based redox-capacitor could be useful for processing redox-based chemical information because: (1) it facilitates electron exchange with a broad range of oxidants and reductants, and thus may provide a means to globally sample redox information; (2) it is easily assembled on the electrode by chitosan electrodeposition and electrochemical grafting of catechols; and (3) it possesses unique molecular electronic properties (amplification, partial rectification, gating, switching, and steady response) that serve to process electrochemical information.

We anticipate that this new paradigm for accessing chemical information using a signal processing approach could provide insights on: (1) the oxidative/reductive stresses being exerted on a biological system (e.g., due to inflammation or tumor therapy); (2) the oxidative/reductive actions being taken by a biological system (e.g., oxidative burst or redox signaling); (3) the redox-mediated or regulation process (e.g., for disulfide bond formation/cleavage); and (4) the redox interactions that occur among various chemical components (e.g., between redox-cycling drugs and antioxidants in our diet). Obviously, more tests are needed to validate this proposed signal processing approach and also to understand the utility of the information gained from this interactive redox-probing.

Acknowledgments: This work has been supported by National Science Foundation (NSF) (DMREF-1435957) and Defense Threat Reduction Agency (DTRA) (HDTRA1-13-1- 0037).

Author Contributions: This review describes results from a multi-investigator collaboration. E.K. and G.F.P. were responsible for the redox-capacitor studies; Z.L. was responsible for the interactive redox-probing of biothiols; Y.L. was responsible for coupling redox-probing with synthetic biology; W.E.B. and G.F.P. were responsible for the conceptualization of many of the advances described in this paper; and G.F.P. and E.K. wrote the manuscript.

Conflicts of Interest: The authors declare no conflict of interest.

References

1. Lee, H.; Dellatore, S.M.; Miller, W.M.; Messersmith, P.B. Mussel-inspired surface chemistry for multifunctional coatings. *Science* **2007**, *318*, 426–430. [CrossRef] [PubMed]

2. Liu, Y.; Ai, K.; Lu, L. Polydopamine and its derivative materials: Synthesis and promising applications in energy, environmental, and biomedical fields. *Chem. Rev.* **2014**, *114*, 5057–5115. [CrossRef] [PubMed]
3. Ryu, J.H.; Hong, S.; Lee, H. Bio-inspired adhesive catechol-conjugated chitosan for biomedical applications: A mini review. *Acta Biomater.* **2015**, *27*, 101–115. [CrossRef] [PubMed]
4. Lee, H.; Lee, B.P.; Messersmith, P.B. A reversible wet/dry adhesive inspired by mussels and geckos. *Nature* **2007**, *448*, 338–341. [CrossRef] [PubMed]
5. Wilker, J.J. The iron-fortified adhesive system of marine mussels. *Angew. Chem. Int. Ed.* **2010**, *49*, 8076–8078. [CrossRef] [PubMed]
6. Klupfel, L.; Piepenbrock, A.; Kappler, A.; Sander, M. Humic substances as fully regenerable electron acceptors in recurrently anoxic environments. *Nat. Geosci.* **2014**, *7*, 195–200. [CrossRef]
7. Scott, D.T.; McKnight, D.M.; Blunt-Harris, E.L.; Kolesar, S.E.; Lovley, D.R. Quinone moieties act as electron acceptors in the reduction of humic substances by humics-reducing microorganisms. *Environ. Sci. Technol.* **1998**, *32*, 2984–2989. [CrossRef]
8. Lovley, D.R.; Coates, J.D.; Blunt-Harris, E.L.; Phillips, E.J.P.; Woodward, J.C. Humic substances as electron acceptors for microbial respiration. *Nature* **1996**, *382*, 445–448. [CrossRef]
9. Jacobson, E.S.; Hong, J.D. Redox buffering by melanin and Fe(II) in *Cryptococcus neoformans*. *J. Bacteriol.* **1997**, *179*, 5340–5346. [CrossRef] [PubMed]
10. Noctor, G. Metabolic signalling in defence and stress: The central roles of soluble redox couples. *Plant Cell Environ.* **2006**, *29*, 409–425. [CrossRef] [PubMed]
11. Sarna, T.; Plonka, P.M. Biophysical studies of melanin. In *Biomedical EPR, Part A: Free Radicals, Metals, Medicine, and Physiology*; Eaton, S.R., Eaton, G.R., Berliner, L.J., Eds.; Springer: Boston, MA, USA, 2005; pp. 125–146.
12. Schweigert, N.; Zehnder, A.J.B.; Eggen, R.I.L. Chemical properties of catechols and their molecular modes of toxic action in cells, from microorganisms to mammals. *Environ. Microbiol.* **2001**, *3*, 81–91. [CrossRef]
13. Foyer, C.H.; Noctor, G. Redox signaling in plants. *Antioxid Redox Signal.* **2013**, *18*, 2087–2090. [CrossRef]
14. Sarewicz, M.; Osyczka, A. Electronic connection between the quinone and cytochrome c redox pools and its role in regulation of mitochondrial electron transport and redox signaling. *Physiol. Rev.* **2014**, *95*, 219–243. [CrossRef] [PubMed]
15. Ernster, L.; Dallner, G. Biochemical, physiological and medical aspects of ubiquinone function. *Biochim. Biophys. Acta* **1995**, *1271*, 195–204. [CrossRef]
16. Maurer, F.; Christl, I.; Hoffmann, M.; Kretzschmar, R. Reduction and reoxidation of humic acid: Influence on speciation of cadmium and silver. *Environ. Sci. Technol.* **2012**, *46*, 8808–8816. [CrossRef] [PubMed]
17. Aeschbacher, M.; Graf, C.; Schwarzenbach, R.P.; Sander, M. Antioxidant properties of humic substances. *Environ. Sci. Technol.* **2012**, *46*, 4916–4925. [CrossRef] [PubMed]
18. Kim, J.H.; Lee, M.; Park, C.B. Polydopamine as a biomimetic electron gate for artificial photosynthesis. *Angew. Chem. Int. Ed.* **2014**, *53*, 6364–6368. [CrossRef]
19. Steinfeld, J.I.; Wormhoudt, J. Explosives detection: A challenge for physical chemistry. *Annu. Rev. Phys. Chem.* **1998**, *49*, 203–232. [CrossRef] [PubMed]
20. Madsen, R.; Lundstedt, T.; Trygg, J. Chemometrics in metabolomics—A review in human disease diagnosis. *Anal. Chim. Acta* **2010**, *659*, 23–33. [CrossRef] [PubMed]
21. Bosque-Sendra, J.M.; Cuadros-Rodríguez, L.; Ruiz-Samblás, C.; de la Mata, A.P. Combining chromatography and chemometrics for the characterization and authentication of fats and oils from triacylglycerol compositional data—A review. *Anal. Chim. Acta* **2012**, *724*, 1–11. [CrossRef]
22. Karoui, R.; Downey, G.; Blecker, C. Mid-infrared spectroscopy coupled with chemometrics: A tool for the analysis of intact food systems and the exploration of their molecular structure—quality relationships—A review. *Chem. Rev.* **2010**, *110*, 6144–6168. [CrossRef] [PubMed]
23. Liu, Y.; Kim, E.; Li, J.; Kang, M.; Bentley, W.E.; Payne, G.F. Electrochemistry for bio-device molecular communication: The potential to characterize, analyze and actuate biological systems. *Nano Commun. Netw.* **2017**, *11*, 76–89. [CrossRef]
24. Li, J.; Liu, Y.; Kim, E.; March, J.C.; Bentley, W.E.; Payne, G.F. Electrochemical reverse engineering: A systems-level tool to probe the redox-based molecular communication of biology. *Free Radic. Biol. Med.* **2017**, *105*, 110–131. [CrossRef] [PubMed]

25. Röck, F.; Barsan, N.; Weimar, U. Electronic nose: Current status and future trends. *Chem. Rev.* **2008**, *108*, 705–725. [CrossRef] [PubMed]

26. Li, Z.; Suslick, K.S. Portable optoelectronic nose for monitoring meat freshness. *ACS Sens.* **2016**, *1*, 1330–1335. [CrossRef]

27. Liu, Z.; Liu, Y.; Kim, E.; Bentley, W.E.; Payne, G.F. Electrochemical probing through a redox capacitor to acquire chemical information on biothiols. *Anal. Chem.* **2016**, *88*, 7213–7221. [CrossRef] [PubMed]

28. Kim, E.; Liu, Y.; Ben-Yoav, H.; Winkler, T.E.; Yan, K.; Shi, X.; Shen, J.; Kelly, D.L.; Ghodssi, R.; Bentley, W.E.; et al. Fusing sensor paradigms to acquire chemical information: An integrative role for smart biopolymeric hydrogels. *Adv. Healthc. Mater.* **2016**, *5*, 2595–2616. [CrossRef] [PubMed]

29. Giner-Sanz, J.J.; Ortega, E.M.; Pérez-Herranz, V. Total harmonic distortion based method for linearity assessment in electrochemical systems in the context of EIS. *Electrochim. Acta* **2015**, *186*, 598–612. [CrossRef]

30. Lisdat, F.; Schäfer, D. The use of electrochemical impedance spectroscopy for biosensing. *Anal. Bioanal. Chem.* **2008**, *391*, 1555. [CrossRef] [PubMed]

31. Kim, E.; Leverage, W.T.; Liu, Y.; White, I.M.; Bentley, W.E.; Payne, G.F. Redox-capacitor to connect electrochemistry to redox-biology. *Analyst* **2014**, *139*, 32–43. [CrossRef] [PubMed]

32. Gray, K.M.; Kim, E.; Wu, L.-Q.; Liu, Y.; Bentley, W.E.; Payne, G.F. Biomimetic fabrication of information-rich phenolic-chitosan films. *Soft Matter* **2011**, *7*, 9601–9615. [CrossRef]

33. Wu, L.Q.; McDermott, M.K.; Zhu, C.; Ghodssi, R.; Payne, G.F. Mimicking biological phenol reaction cascades to confer mechanical function. *Adv. Funct. Mater.* **2006**, *16*, 1967–1974. [CrossRef]

34. Wu, L.Q.; Ghodssi, R.; Elabd, Y.A.; Payne, G.F. Biomimetic pattern transfer. *Adv. Funct. Mater.* **2005**, *15*, 189–195. [CrossRef]

35. Morrow, B.H.; Payne, G.F.; Shen, J. pH-Responsive self-assembly of polysaccharide through a rugged energy landscape. *J. Am. Chem. Soc.* **2015**, *137*, 13024–13030. [CrossRef] [PubMed]

36. Wu, L.-Q.; Gadre, A.P.; Yi, H.; Kastantin, M.J.; Rubloff, G.W.; Bentley, W.E.; Payne, G.F.; Ghodssi, R. Voltage-dependent assembly of the polysaccharide chitosan onto an electrode surface. *Langmuir* **2002**, *18*, 8620–8625. [CrossRef]

37. Fernandes, R.; Wu, L.-Q.; Chen, T.; Yi, H.; Rubloff, G.W.; Ghodssi, R.; Bentley, W.E.; Payne, G.F. Electrochemically induced deposition of a polysaccharide hydrogel onto a patterned surface. *Langmuir* **2003**, *19*, 4058–4062. [CrossRef]

38. Sugumaran, M.; Hennigan, B.; Obrien, J. Tyrosinase catalyzed protein polymerization as an in vitro model for quinone tanning of insect cuticle. *Arch. Insect Biochem. Physiol.* **1987**, *6*, 9–25. [CrossRef]

39. Kramer, K.J.; Kanost, M.R.; Hopkins, T.L.; Jiang, H.; Zhu, Y.C.; Xu, R.; Kerwin, J.L.; Turecek, F. Oxidative conjugation of catechols with proteins in insect skeletal systems. *Tetrahedron* **2001**, *57*, 385–392. [CrossRef]

40. Aberg, C.M.; Chen, T.; Olumide, A.; Raghavan, S.R.; Payne, G.F. Enzymatic grafting of peptides from casein hydrolysate to chitosan. Potential for value-added byproducts from food-processing wastes. *J. Agric. Food Chem.* **2004**, *52*, 788–793. [CrossRef] [PubMed]

41. Kerwin, J.L.; Whitney, D.L.; Sheikh, A. Mass spectrometric profiling of glucosamine, glucosamine polymers and their catecholamine adducts: Model reactions and cuticular hydrolysates of *Toxorhynchites amboinensis* (Culicidae) pupae. *Insect Biochem. Mol. Biol.* **1999**, *29*, 599–607. [CrossRef]

42. McGinness, J.; Corry, P.; Proctor, P. Amorphous semiconductor switching in melanins. *Science* **1974**, *183*, 853–855. [CrossRef] [PubMed]

43. Meredith, P.; Sarna, T. The physical and chemical properties of eumelanin. *Pigment Cell Res.* **2006**, *19*, 572–594. [CrossRef] [PubMed]

44. Newman, D.K.; Kolter, R. A role for excreted quinones in extracellular electron transfer. *Nature* **2000**, *405*, 94–97. [CrossRef] [PubMed]

45. Hernandez, M.E.; Newman, D.K. Extracellular electron transfer. *Cell. Mol. Life Sci.* **2001**, *58*, 1562–1571. [CrossRef] [PubMed]

46. Kim, E.; Liu, Y.; Bentley, W.E.; Payne, G.F. Redox capacitor to establish bio-device redox-connectivity. *Adv. Funct. Mater.* **2012**, *22*, 1409–1416. [CrossRef]

47. Kim, E.; Liu, Y.; Shi, X.-W.; Yang, X.; Bentley, W.E.; Payne, G.F. Biomimetic approach to confer redox activity to thin chitosan films. *Adv. Funct. Mater.* **2010**, *20*, 2683–2694. [CrossRef]

48. Kim, E.; Xiong, Y.; Cheng, Y.; Wu, H.-C.; Liu, Y.; Morrow, H.B.; Ben-Yoav, H.; Ghodssi, R.; Rubloff, W.G.; Shen, J.; et al. Chitosan to connect biology to electronics: Fabricating the bio-device interface and communicating across this interface. *Polymers* **2015**, *7*, 1–46. [CrossRef]
49. Kim, E.; Liu, Y.; Baker, C.J.; Owens, R.; Xiao, S.; Bentley, W.E.; Payne, G.F. Redox-cycling and H_2O_2 generation by fabricated catecholic films in the absence of enzymes. *Biomacromolecules* **2011**, *12*, 880–888. [CrossRef] [PubMed]
50. Kim, E.; Gordonov, T.; Bentley, W.E.; Payne, G.F. Amplified and in situ detection of redox-active metabolite using a biobased redox capacitor. *Anal. Chem.* **2013**, *85*, 2102–2108. [CrossRef] [PubMed]
51. Kim, E.; Leverage, W.T.; Liu, Y.; Panzella, L.; Alfieri, M.L.; Napolitano, A.; Bentley, W.E.; Payne, G.F. Paraquat–melanin redox-cycling: Evidence from electrochemical reverse engineering. *ACS Chem. Neurosci.* **2016**, *7*, 1057–1067. [CrossRef] [PubMed]
52. Ben-Yoav, H.; Winkler, T.E.; Kim, E.; Chocron, S.E.; Kelly, D.L.; Payne, G.F.; Ghodssi, R. Redox cycling-based amplifying electrochemical sensor for in situ clozapine antipsychotic treatment monitoring. *Electrochim. Acta* **2014**, *130*, 497–503. [CrossRef]
53. Liu, Y.; Kim, E.; White, I.M.; Bentley, W.E.; Payne, G.F. Information processing through a bio-based redox capacitor: Signatures for redox-cycling. *Bioelectrochemistry* **2014**, *98*, 94–102. [CrossRef] [PubMed]
54. Liu, Y.; Tsao, C.-Y.; Kim, E.; Tschirhart, T.; Terrell, J.L.; Bentley, W.E.; Payne, G.F. Using a redox modality to connect synthetic biology to electronics: Hydrogel-based chemo-electro signal transduction for molecular communication. *Adv. Healthc. Mater.* **2017**, *6*, 1600908. [CrossRef] [PubMed]
55. Kim, E.; Winkler, T.E.; Kitchen, C.; Kang, M.; Banis, G.; Bentley, W.E.; Kelly, D.L.; Ghodssi, R.; Payne, G.F. Redox probing for chemical information of oxidative stress. *Anal. Chem.* **2017**, *89*, 1583–1592. [CrossRef] [PubMed]
56. Logan, B.E. Exoelectrogenic bacteria that power microbial fuel cells. *Nat. Rev. Microb.* **2009**, *7*, 375–381. [CrossRef] [PubMed]
57. Scherer, S. Do photosynthetic and respiratory electron transport chains share redox proteins? *Trends Biochem. Sci.* **1990**, *15*, 458–462. [CrossRef]
58. Deplancke, B.; Gaskins, H.R. Redox control of the transsulfuration and glutathione biosynthesis pathways. *Curr. Opin. Clin. Nutr. Metab. Care* **2002**, *5*, 85–92. [CrossRef] [PubMed]
59. Circu, M.L.; Aw, T.Y. Reactive oxygen species, cellular redox systems, and apoptosis. *Free Radic. Biol. Med.* **2010**, *48*, 749–762. [CrossRef] [PubMed]
60. Levine, A.; Tenhaken, R.; Dixon, R.; Lamb, C. H_2O_2 from the oxidative burst orchestrates the plant hypersensitive disease resistance response. *Cell* **1994**, *79*, 583–593. [CrossRef]
61. Nürnberger, T.; Scheel, D. Signal transmission in the plant immune response. *Trends Plant Sci.* **2001**, *6*, 372–379. [CrossRef]
62. Nathan, C. Neutrophils and immunity: Challenges and opportunities. *Nat. Rev. Immunol.* **2006**, *6*, 173–182. [CrossRef] [PubMed]
63. Schieber, M.; Chandel, N.S. ROS function in redox signaling and oxidative stress. *Curr. Biol.* **2014**, *24*, R453–R462. [CrossRef] [PubMed]
64. Torres, M.A.; Jones, J.D.G.; Dangl, J.L. Reactive oxygen species signaling in response to pathogens. *Plant Physiol.* **2006**, *141*, 373–378. [CrossRef] [PubMed]
65. Boudsocq, M.; Willmann, M.R.; McCormack, M.; Lee, H.; Shan, L.; He, P.; Bush, J.; Cheng, S.-H.; Sheen, J. Differential innate immune signalling via Ca^{2+} sensor protein kinases. *Nature* **2010**, *464*, 418–422. [CrossRef] [PubMed]
66. Klomsiri, C.; Karplus, P.A.; Poole, L.B. Cysteine-based redox switches in enzymes. *Antioxid. Redox Signal.* **2010**, *14*, 1065–1077. [CrossRef] [PubMed]
67. Brandes, N.; Schmitt, S.; Jakob, U. Thiol-based redox switches in eukaryotic proteins. *Antioxid. Redox Signal.* **2008**, *11*, 997–1014. [CrossRef] [PubMed]
68. Groitl, B.; Jakob, U. Thiol-based redox switches. *Biochim. Biophys. Acta* **2014**, *1844*, 1335–1343. [CrossRef] [PubMed]
69. Jones, D.P. Redefining oxidative stress. *Antioxid. Redox Signal.* **2006**, *8*, 1865–1879. [CrossRef] [PubMed]
70. Schafer, F.Q.; Buettner, G.R. Redox environment of the cell as viewed through the redox state of the glutathione disulfide/glutathione couple. *Free Radic. Biol. Med.* **2001**, *30*, 1191–1212. [CrossRef]

71. Levonen, A.-L.; Hill, B.G.; Kansanen, E.; Zhang, J.; Darley-Usmar, V.M. Redox regulation of antioxidants, autophagy, and the response to stress: Implications for electrophile therapeutics. *Free Radic. Biol. Med.* **2014**, *71*, 196–207. [CrossRef] [PubMed]

72. Lin, M.T.; Beal, M.F. Mitochondrial dysfunction and oxidative stress in neurodegenerative diseases. *Nature* **2006**, *443*, 787–795. [CrossRef] [PubMed]

73. Halliwell, B. Oxidative stress and neurodegeneration: Where are we now? *J. Neurochem.* **2006**, *97*, 1634–1658. [CrossRef] [PubMed]

74. Stocker, R.; Keaney, J.F. Role of oxidative modifications in atherosclerosis. *Physiol. Rev.* **2004**, *84*, 1381. [CrossRef] [PubMed]

75. Valko, M.; Leibfritz, D.; Moncol, J.; Cronin, M.T.D.; Mazur, M.; Telser, J. Free radicals and antioxidants in normal physiological functions and human disease. *Int. J. Biochem. Cell Biol.* **2007**, *39*, 44–84. [CrossRef] [PubMed]

76. Dringen, R. Metabolism and functions of glutathione in brain. *Prog. Neurobiol.* **2000**, *62*, 649–671. [CrossRef]

77. Chen, Z.; Zheng, H.; Lu, C.; Zu, Y. Oxidation of L-cysteine at a fluorosurfactant-modified gold electrode: Lower overpotential and higher selectivity. *Langmuir* **2007**, *23*, 10816–10822. [CrossRef] [PubMed]

78. Tang, H.; Chen, J.; Nie, L.; Yao, S.; Kuang, Y. Electrochemical oxidation of glutathione at well-aligned carbon nanotube array electrode. *Electrochim. Acta* **2006**, *51*, 3046–3051. [CrossRef]

79. Leopold, M.C.; Bowden, E.F. Influence of gold substrate topography on the voltammetry of cytochrome c adsorbed on carboxylic acid terminated self-assembled monolayers. *Langmuir* **2002**, *18*, 2239–2245. [CrossRef]

80. Boubour, E.; Lennox, R.B. Insulating properties of self-assembled monolayers monitored by impedance spectroscopy. *Langmuir* **2000**, *16*, 4222–4228. [CrossRef]

81. Janek, R.P.; Fawcett, W.R.; Ulman, A. Impedance spectroscopy of self-assembled monolayers on Au(111): Sodium ferrocyanide charge transfer at modified electrodes. *Langmuir* **1998**, *14*, 3011–3018. [CrossRef]

82. Apel, K.; Hirt, H. Reactive oxygen species: Metabolism, oxidative stress, and signal transduction. *Annu. Rev. Plant Biol.* **2004**, *55*, 373–399. [CrossRef] [PubMed]

83. Abraham, S.N.; St. John, A.L. Mast cell-orchestrated immunity to pathogens. *Nat. Rev. Immunol.* **2010**, *10*, 440–452. [CrossRef] [PubMed]

84. Iriti, M.; Faoro, F. Review of innate and specific immunity in plants and animals. *Mycopathologia* **2007**, *164*, 57–64. [CrossRef] [PubMed]

85. Price-Whelan, A.; Dietrich, L.E.P.; Newman, D.K. Rethinking 'secondary' metabolism: Physiological roles for phenazine antibiotics. *Nat. Chem. Biol.* **2006**, *2*, 71–78. [CrossRef] [PubMed]

86. Okegbe, C.; Sakhtah, H.; Sekedat, M.D.; Price-Whelan, A.; Dietrich, L.E.P. Redox eustress: Roles for redox-active metabolites in bacterial signaling and behavior. *Antioxid. Redox Signal.* **2011**, *16*, 658–667. [CrossRef] [PubMed]

87. Jacob, C.; Jamier, V.; Ba, L.A. Redox active secondary metabolites. *Curr. Opin. Chem. Biol.* **2011**, *15*, 149–155. [CrossRef] [PubMed]

88. Dietrich, L.E.P.; Price-Whelan, A.; Petersen, A.; Whiteley, M.; Newman, D.K. The phenazine pyocyanin is a terminal signalling factor in the quorum sensing network of *Pseudomonas aeruginosa. Mol. Microbiol.* **2006**, *61*, 1308–1321. [CrossRef] [PubMed]

89. Price-Whelan, A.; Dietrich, L.E.P.; Newman, D.K. Pyocyanin alters redox homeostasis and carbon flux through central metabolic pathways in *Pseudomonas aeruginosa* PA14. *J. Bacteriol.* **2007**, *189*, 6372–6381. [CrossRef] [PubMed]

90. Dietrich, L.E.P.; Okegbe, C.; Price-Whelan, A.; Sakhtah, H.; Hunter, R.C.; Newman, D.K. Bacterial community morphogenesis is intimately linked to the intracellular redox state. *J. Bacteriol.* **2013**, *195*, 1371–1380. [CrossRef] [PubMed]

91. Dietrich, L.E.P.; Teal, T.K.; Price-Whelan, A.; Newman, D.K. Redox-active antibiotics control gene expression and community behavior in divergent bacteria. *Science* **2008**, *321*, 1203. [CrossRef] [PubMed]

92. Branski, L.K.; Al-Mousawi, A.; Rivero, H.; Jeschke, M.G.; Sanford, A.P.; Herndon, D.N. Emerging infections in burns. *Surg. Infect.* **2009**, *10*, 389–397. [CrossRef] [PubMed]

93. Pruitt, J.B.A.; McManus, A.T.; Kim, S.H.; Goodwin, C.W. Burn wound infections: Current status. *World J. Surg.* **1998**, *22*, 135–145. [PubMed]

94. Tredget, E.E.; Shankowsky, H.A.; Rennie, R.; Burrell, R.E.; Logsetty, S. *Pseudomonas* infections in the thermally injured patient. *Burns* **2004**, *30*, 3–26. [CrossRef] [PubMed]

95. Kim, E.; Gordonov, T.; Liu, Y.; Bentley, W.E.; Payne, G.F. Reverse engineering to suggest biologically relevant redox activities of phenolic materials. *ACS Chem. Biol.* **2013**, *8*, 716–724. [CrossRef] [PubMed]

96. Baker, C.J.; Mock, N.M.; Smith, J.M.; Aver'yanov, A.A. The dynamics of apoplast phenolics in tobacco leaves following inoculation with bacteria. *Front. Plant Sci.* **2015**, *6*, 649. [CrossRef] [PubMed]

97. Baker, C.J.; Whitaker, B.D.; Roberts, D.P.; Mock, N.M.; Rice, C.P.; Deahl, K.L.; Aver'yanov, A.A. Induction of redox sensitive extracellular phenolics during plant–bacterial interactions. *Physiol. Mol. Plant Pathol.* **2005**, *66*, 90–98. [CrossRef]

98. Jacyn Baker, C.; Roberts, D.P.; Mock, N.M.; Whitaker, B.D.; Deahl, K.L.; Aver'yanov, A.A. Apoplastic redox metabolism: Synergistic phenolic oxidation and a novel oxidative burst. *Physiol. Mol. Plant Pathol.* **2005**, *67*, 296–303. [CrossRef]

99. Espey, M.G. Role of oxygen gradients in shaping redox relationships between the human intestine and its microbiota. *Free Radic. Biol. Med.* **2013**, *55*, 130–140. [CrossRef] [PubMed]

100. Shi, X.-W.; Tsao, C.-Y.; Yang, X.; Liu, Y.; Dykstra, P.; Rubloff, G.W.; Ghodssi, R.; Bentley, W.E.; Payne, G.F. Electroaddressing of cell populations by co-deposition with calcium alginate hydrogels. *Adv. Funct. Mater.* **2009**, *19*, 2074–2080. [CrossRef]

101. Cheng, Y.; Luo, X.; Tsao, C.-Y.; Wu, H.-C.; Betz, J.; Payne, G.F.; Bentley, W.E.; Rubloff, G.W. Biocompatible multi-address 3D cell assembly in microfluidic devices using spatially programmable gel formation. *Lab On A Chip* **2011**, *11*, 2316–2318. [CrossRef] [PubMed]

102. Liu, Y.; Kim, E.; Ghodssi, R.; Rubloff, G.W.; Culver, J.N.; Bentley, W.E.; Payne, G.F. Biofabrication to build the biology–device interface. *Biofabrication* **2010**, *2*, 022002.

MDPI

St. Alban-Anlage 66

4052 Basel, Switzerland

Tel. +41 61 683 77 34

Fax +41 61 302 89 18

http://www.mdpi.com

Biomimetics Editorial Office

E-mail: biomimetics@mdpi.com

http://www.mdpi.com/journal/biomimetics

www.ingramcontent.com/pod-product-compliance
Lightning Source LLC
Chambersburg PA
CBHW051847210326
41597CB00033B/5801